10th Aceh International Workshop and Expo on Sustainable Tsunami Disaster Recovery (AIWEST-DR 2016)

IOP Conference Series: Earth and Environmental Science
Volume 56

Bada Aceh, Indonesia
22 – 24 November 2016

Editors:

Ella Meilianda
Aiko Sakurai
Elizabeth Maly
Rina Suryani Oktar

ISBN: 978-1-5108-3712-6
ISSN: 1755-1307

Printed from e-media with permission by:

Curran Associates, Inc.
57 Morehouse Lane
Red Hook, NY 12571

Some format issues inherent in the e-media version may also appear in this print version.

Printed by Curran Associates, Inc. (2017)

For permission requests, please contact the Institute of Physics
at the address below.

Institute of Physics
Dirac House, Temple Back
Bristol BS1 6BE UK

Phone: 44 1 17 929 7481
Fax: 44 1 17 920 0979

techtracking@iop.org

Additional copies of this publication are available from:

Curran Associates, Inc.
57 Morehouse Lane
Red Hook, NY 12571 USA
Phone: 845-758-0400
Fax: 845-758-2633
Email: curran@proceedings.com
Web: www.proceedings.com

Table of contents

Volume 56

10th Aceh International Workshop and Expo on Sustainable Tsunami Disaster Recovery (AIWEST-DR 2016)

22–24 November 2016, Banda Aceh, Indonesia

Accepted papers received: 14 February 2017
Published online: 1 March 2017

Preface

Science and Technology in Disaster Management

IOP Conf. Series: Earth and Environmental Science **56** (2017) 011001 doi:10.1088/1755-1315/56/1/011001

PREFACE

The 10th Aceh International Workshop and Expo on Sustainable Tsunami Disaster Recovery, AIWEST-DR 2016, was organizsed from 22 to 24 November 2016. The now bi-annual AIWEST-DR is renewed with the theme "Entering the Second Decade of the 2004 Indian Ocean Tsunami: Are We Better Prepared?". Tsunami and Disaster Mitigation Research Center (TDMRC) of Syiah Kuala University was profoundly enjoying the collaboration with the International Research Institute of Disaster Science (IRIDeS) of Tohoku University in Japan as the organizing committee.

AIWEST-DR conference is meant for academics, professionals, researchers and students to meet bi-annually to advance and share knowledge and progress of research works related to sustainable disaster recovery. This proceeding of AIWEST-DR 2016 contain papers written by academics, professionals, researchers and students from at least five countries including Japan, Indonesia, Philipines, USA, Germany and Thailand. Various research works related to the conference theme were presented during the conference, which categorized into the following specific topics:

A. Science and Technology in Disaster Management
B. Disaster Risk Reduction Efforts in Disaster Affected Area

Topic A (Best Practices in Disaster Management) discusses about various approaches to have better insight on disaster phenomena, particularly tsunami events, such as by using numerical simulations, spatial pattern analysis, behaviour of reinforced concrete structures from giant earthquakes, and also to observe disaster behaviour through geophysical properties as well as application on mobile technology to increase tsunami awareness.

Topic B (Disaster Risk Reduction Efforts in Disaster Affected Area) encapsulates education as an approach towards community resilience, either from medical perspective, introduction of disaster management into curricula, as well as assessment in school disaster preparedness at the tsunami-prone cities. Furthermore, this topic also focuses on Disaster Risk Reduction policies and regulations, Inter-Organizational network during disasters, Indigeneous knowledge management to enhance community resilience, the nexus of science-practice for survey and the concept of green city after experiencing tsunami and other disasters. This section also deals with observations of the post-tsunami recovery efforts persistance and effects, housing reconstruction process, post-disaster as windows of opportunities for "building back better", disaster budgeting and conceptual framework for evaluating tsunami resilience.

The organization of AIWEST-DR 2016 would not have been possible without the support of many institutes and individuals. First and foremost, we are very grateful to the financial sponsors of AIWEST-DR 2016: *the Ministry of Research, Technology and Higher Education of Republic of Indonesia* through the provision of TDMRC as the Indonesian Center of Excellence in Tsunami Disaster Mitigation; and *IRIDeS of Tohoku University*.

The conference was organized by people from the Tsunami and Disaster Mitigation Research Center (TDMRC) of Syiah Kuala University and IRIDeS of Tohoku University. We also thank to our colleagues and extended institutional networks who were kindly supporting and guiding us through a steering commitee. Most importantly, we were humbly thankful for all the people and students in the organizing committee and beyond who had been tirelessly preparing for the success of the conference and to keep working afterwards on preparing this proceeding.

Last but not least, we thank all the authors for their excellent contributions to these proceeding and all the participants for their contributions to this successful conference, either in the form of oral presentation, posters, expo material, or in the form of their attention to the presentations and contributions to the discussions.

We hope that everyone enjoys reading the proceeding and that the papers inspire you in your research or other professional activities.

Ella Meilianda, Tsunami and Disaster Mitigation Research Center (TDMRC) of Syiah Kuala University, Indonesia
Aiko Sakurai, International Research Institute of Disaster Science (IRIDeS) of Tohoku University, Japan
Elizabeth Maly, International Research Institute of Disaster Science (IRIDeS) of Tohoku University, Japan
Rina Suryani Oktari Tsunami and Disaster Mitigation Research Center (TDMRC) of Syiah Kuala University, Indonesia

Peer review statement

All papers published in this volume of *IOP Conference Series: Earth and Environmental Science* have been peer reviewed through processes administered by the proceedings Editors. Reviews were conducted by expert referees to the professional and scientific standards expected of a proceedings journal published by IOP Publishing.

IOP Conf. Series: Earth and Environmental Science **56** (2017) 012001 doi:10.1088/1755-1315/56/1/012001

Increasing tsunami risk awareness via mobile application

N Leelawat[1], A Suppasri[1], P Latcharote[1], F Imamura[1], Y Abe[1] and K Sugiyasu[1]

[1] International Research Institute of Disaster Science, Tohoku University, 468-1 Aramaki-Aza Aoba, Aoba-ku, Sendai 980-0845 Japan

Abstract. In the information and communication technology era, smartphones have become a necessity. With the capacity and availability of smart technologies, a number of benefits are possible. As a result, designing a mobile application to increase tsunami awareness has been proposed, and a prototype has been designed and developed. The application uses data from the 2011 Great East Japan Tsunami. Based on the current location determined by a GPS function matched with the nearest point extracted from the detailed mesh data of that earlier disaster, the application generates the inundation depth at the user's location. Thus, not only local people but also tourists visiting the affected areas can understand the risks involved. Application testing has been conducted in an evacuation experiment involving both Japanese and foreign students. The proposed application can be used as a supplementary information tool in tsunami evacuation drills. It also supports the idea of smart tourism: when people realize their risks, they possess risk awareness and hence can reduce their risks. This application can also be considered a contribution to disaster knowledge and technology, as well as to the lessons learned from the practical outcome.

1. Introduction

Through data coverage and the capability of mobile technology, it is possible to utilize the everyday technology, such as mobile applications, in a disaster risk reduction context. Many mobile applications have been developed to warn and provide information during disasters [1, 2, 3]. Some applications related to tsunami disasters have been developed; for example, a mobile application for estimating the building damage from a tsunami [2]. That application has various uses: ordinary people use it to help their decision-making when a tsunami warning is announced; drill participants also use it to support their decision-making; disaster risk analysts and researchers use it to help write their reports; town planners or local disaster policy makers use it in designing disaster preparedness plans; and engineers and architects use it to consider the height, function, and material when constructing buildings [2].

The purpose of this study is to create a practical Information and Communication Technology (ICT) application for tsunami disaster risk reduction. This application is expected to overcome the domain problem of long-term tsunami risk awareness level.

The following section provides background; the third presents the research design and methodology; the fourth describes the mobile application development; the fifth describes the experiment and its results; and the last discusses the findings and conclusions.

2. Background

As van der Akker [4] suggested, some typical steps, including a literature review and expert consultation, are necessary for a preliminary investigation. Our expert consultation included the former head and current officer of the Crisis Management Division at Kesennuma City, Miyagi Prefecture, Japan. All participants experienced the 2011 Great East Japan Tsunami. Although many people and agencies possessed a high level of risk awareness due to the danger and damage of the 2011 tsunami, the concerned local organizations began nonetheless to consider how to maintain such an awareness level for future generations. Thus, some activities such as disaster education and disaster evacuation drills were necessary.

Disaster experience can help people prepare for the future events [5]. Disaster education and information campaigns are necessary to alert people to an awareness of disaster risks [6]. Despite the high risk awareness of a tsunami following the 2011 Great East Japan Tsunami, with the passing of time, maintaining a high level of tsunami risk awareness is a problem, especially for new generations and new visitors to the affected areas. What we can do, especially for this new generation, is to provide preparedness and practical exercises. Disaster drills are important activities that allow participants to practice how to proceed after receiving a warning alarm. Nevertheless, we cannot directly evoke the image of a disaster for people who have not experienced it. In addition, such people may not understand the importance of conducting the practice drills. They are probably difficult to proceed during a normal situation because they cannot imagine their surroundings being flooded in a tsunami. We might take advantage of Augmented Reality (AR) to allow participants to see, hear, and feel the situation of a tsunami during a practice drill. In any event, as implementing development costs may be high, this might be considered as the domain problem of this study.

Kendra and Wachtendorf [6] identified and described the concept of three terms: "community", "innovation", and "community innovation". Community innovation could be "a feature of pre-disaster mitigation and preparedness and of post-disaster response and recovery" [6, p. 317]. This study proposes a new, practical application that is simple but efficient. Thus, some practical measures must be taken to increase and maintain tsunami risk awareness via a special tool used in disaster evacuation.

Experts will tell you that a variety of requirements are needed to increase tsunami awareness. We examined feasibility by keeping in mind the available data and technology. As a result, we designed a tool that will show the inundation depth at users' current location so that they can imagine the height of a tsunami exactly where they are. When used in a tsunami drill exercise, this tool is expected to increase users' realization, which in turn will increase their tsunami risk awareness.

3. Research Design

3.1. Methodology

Following the design science approach [7], this study develops an instantiation as the design artefact. Research used software development to create a prototype. Based on the availability of inundation depth data in the affected area of the 2011 Great East Japan Tsunami, we designed, and developed a user-friendly disaster mobile application prototype using Xcode, an integrated development environment, with Swift language.

By checking the nearest detailed 100-m mesh data at the current location detected by the GPS function, the application retrieved inundation depth data of the closest location point from its Core Data (i.e., database).

After we designed the user interface of the application, we discussed it with experienced experts in the disaster and emergency management department of Kesennuma City to obtain feedback before

IOP Conf. Series: Earth and Environmental Science **56** (2017) 012001 doi:10.1088/1755-1315/56/1/012001

proceeding to the implementation phase. After the implementation, we developed the application prototype. Then, we conducted the evacuation experiment.

3.2. Study areas
For this prototype, only data from Kesennuma City and Matsushima Town in Miyagi Prefecture were included. The study area selection was chosen according to the following criteria. First, Kesennuma City experienced various levels of tsunami height in 2011. Second, Matsushima Town, although it did not experience such a high tsunami, is a tourist area frequented by many Japanese and foreigners.

4. Mobile Application Development

This sub-section describes our mobile application analysis and design in terms of their scope, feature, scenario, process flow, database, and user interface.

4.1. Scope
This application is designed to allow users to monitor previous inundation depths at their current location.

4.2. Feature
This application checks the current location via the GPS function, then shows the map and inundation depth data.

4.3. Scenario
This application provides tsunami inundation depth information for all tsunami drill participants.

4.4. Database
The database of this application contains the inundation depth of each position (latitude and longitude). The data were extracted from 100-meter mesh data using ArcGIS.

4.5. User interface
To make our application simple and user-friendly, it was designed to contain only three views: (1) *Home* (*see* Figure 1); (2) *Tsunami check* (*see* Figure 2); and (3) Application information.

Figure 1. User interface-Home.

Figure 2. User interface-Tsunami Check.

5. Evacuation Experiment

5.1. Experiment design

To test this application, we arranged an evacuation experiment and selected Matsushima Town as our testing area. Matsushima Town is a popular travel destination for both Japanese and foreign visitors, as it has one of the three most scenic views in Japan. It was also one of the areas affected by the 2011 Great East Japan Tsunami, which damaged the town's infrastructure, in particular the central tourist area.

The experiment was prepared in July 2016 and conducted in August 2016. Table 1 shows the demographics of experiment participants, which included 2 Japanese and 3 British high school students. We included foreigners because we also wanted to investigate how they would handle an evacuation drill. The day before the experiment, all participants were given a short lecture covering basic knowledge of natural disasters and disaster mitigation.

The starting point was set at Fukuura Island, in Matsushima Bay, connected to the inland area by a 252-meter bridge. Although damaged by the 2011 tsunami, the bridge has been repaired. The experiment design setting is described in Table 2. Participants were given 10 minutes to evacuate to a place they considered safe. GPS loggers were used to track their evacuation routes, together with video cameras recording their evacuation behavior. In addition, following the suggestion that the pilot test might involve interviews or focus groups [8], we conducted interviews. Feedback and opinions were recorded after the evacuation to preserve participants' feelings and experiences.

Table 1. Demographic profile of participants.

No.	Nationality	Grade	Gender
Sample 01	Japanese	12	Male
Sample 02	Japanese	11	Female
Sample 03	British	11	Female
Sample 04	British	11	Female
Sample 05	British	11	Male

Table 2. Experiment design setting.

No.	Experiment	Participants	Area	Internal device	External device
Exp. 01	Tsunami evacuation individually (10 minutes)	Sample 01, Sample 02	Matsushima Town	-	GPS logger[a], Video recorder[b] (Sample 01)
Exp. 02	Tsunami evacuation individually (10 minutes)	Sample 03, Sample 04, Sample 05	Matsushima Town	-	GPS logger[a], Video recorder[c]
Exp. 03	Tsunami evacuation as a group (10 minutes)	Group 01 (Sample 01, Sample 02)	Matsushima Town	Mobile application	GPS logger[a], Video recorder[b]
Exp. 04	Tsunami evacuation as a group (10 minutes)	Group 02 (Sample 01, Sample 02, Sample 03)	Matsushima Town	Mobile application	GPS logger[a], Video recorder[c]

[a] Recorded all samples.
[b] Recorded only Sample 01.
[c] Recorded only Sample 05.

5.2. Results

The participants had never visited this area before testing, and they were from different schools. Thus, they were similar to travelers. The only difference was that Japanese participants could read tsunami evacuation-related signs written in Japanese. Most of the signs in that area were written in both Japanese and English. Limited information was given to participants, including a scenario of a forthcoming tsunami. The overall results are shown in Table 3.

Table 3. Experiment results.

No.	Sample/Group	Sample behavior	Result
Exp. 01	Sample 01	Evacuated to high ground in the island area	Completed (Safe)
Exp. 01	Sample 02	Evacuated to high ground in the island area	Completed (Safe)
Exp. 02	Sample 03	Followed Sample 05	Completed (Safe)
Exp. 02	Sample 04	Followed Sample 05	Completed (Safe)
Exp. 02	Sample 05	Evacuated to inland area using coastal road heading to evacuation area	Completed (Safe)
Exp. 03	Group 01	Evacuated to inland area climbing up to high ground	Completed (Safe)
Exp. 04	Group 02	Evacuated to inland area climbing up to high ground	Completed (Safe)

During Exp. 01, both Sample 01 and Sample 02 checked the map before making decisions. We found from Exp. 01 that Samples 01 and 02 evacuated separately from the starting point. However, both decided to find a route to the island. The route taken by Sample 01 in Exp. 01 is shown in Figure 3. Sample 01 decided to evacuate first to the middle of the island by climbing the hill because it was higher ground than the starting point. With limited information, Sample 01, via trial and error, encountered many dead ends. However, stopping at every sign and information board, he gathered as much information as he could.

During an interview, he mentioned that he wanted to climb to the highest area possible to observe the situation. As a result, he finally managed to reach the highest area on the island, a location higher than the inundation depth.

Figure 3. Evacuated route of Sample 01 in Exp. 01: Evacuation to the island.

In Exp. 02, after checking the map as Sample 01 and Sample 02 did in Exp. 01, Samples 03, 04, and 05 decided to evacuate as a group. Sample 05 became a group leader and was followed by Samples 03 and 04. He and the group decided to cross the bridge to remain far away from the coast line. The evacuated route taken by Sample 05 in Exp. 02 is shown in Figure 4. After crossing the bridge, Sample 05 and the others checked the evacuation shelter sign, then began to go in the direction indicated. They went along the coastal road and then headed to the temple officially designated as the evacuation area. Samples 03 and 04 followed Sample 05 to that destination.

Interviews indicate that they decided to evacuate as a group because they were not familiar with the area or the language (Japanese). Sample 05, the fastest in the group, became the leader and could run to read the signs faster than the others. Although they could not read Japanese, they had no problem understanding the signs, since there were also pictures and some important keywords in English.

As a result, in 10 minutes, Samples 05 and 03 reached the temple designated as the evacuation area. Sample 04 reached the area near the temple. Considering the inundation depth, all were safe from the tsunami.

Figure 4. Evacuated route of Sample 05 in Exp. 02: Evacuated to inland area.[a]

[a] The satellite signal was delayed during the first period when Sample 05 crossed the bridge.

After that, we conducted Exp. 03 and 04. The difference from previous experiments was that each group was provided with tablets on which the proposed application prototype had been installed (*see* Figure 5).

In the Exp. 03, Samples 01 and 02 were assigned to evacuate together to safe place within 10 minutes. They could use the application provided to make their decision. Figure 6 shows their evacuated route. This time, Samples 01 and 02 decided to go to the inland area. After crossing the bridge, they immediately took a route to the high ground area. Using their application, they checked the inundation depth after reaching high ground and until they reached a safe area.

Figure 5. Apple iPad tablets with the application prototype.

Figure 6. Evacuated route of Sample 01 in Exp. 03: Evacuated to inland area.

Exp. 04 was the last to be conducted. Figure 7 shows the evacuation route of Samples 03, 04, and 05. Although they crossed the bridge to the inland area, they did not use the same route as in Exp. 02, but tried to find a way to climb up the hill as soon as they reached land. They kept checking the application to monitor the inundation depth until they reached the top of the hill, where the application indicated an inundation depth of 0.0 meters. At that location, they mentioned that, in a real situation, they would have climbed to the top of the pavilion located on the hill.

One interesting point in Exp. 04 was that specific roles were allocated within the group. Sample 05 was the leader of the group, Sample 03 was assigned to carry the tablet to monitor the application, and Sample 04 helped Sample 05 decide on the route.

Figure 7. Evacuated route of Sample 02 in Exp. 04: Evacuated to inland area.

When the experiment was over, all GPS tracks were collected and mapped into the satellite map on *Google Earth*. Everyone gathered to see the others' evacuated routes. We allowed all samples to explain their decision-making processes.

6. Discussion and Conclusions

This study attempted to solve the problem of reduced awareness of tsunami risk from a long-term perspective. Our research team designed a practical application using the available technology and data. We developed a mobile application prototype to inform users using previous tsunami risk data based on the 2011 Great East Japan Tsunami. In accordance with the action priorities in the Sendai Framework for Disaster Risk Reduction 2015-2030 [9], this mobile application prototype can support first priority action by allowing users to understand the risks involved when faced with a tsunami. This application is suitable for inclusion in tsunami disaster drills. Drill participants can use this mobile application to estimate tsunami inundation at their location and then check inundation levels again once they change location or evacuate elsewhere. Despite its simplicity, the application provides fundamental information for a disaster evacuation drill.

In addition to local users, visitors to tsunami-affected areas will also learn how many meters the inundation depth was in the previous tsunami. Thus, the application enhances tourist safety, which, in turn, supports the concept of smart tourism. As a result, there are at least three scenarios for using this application: (1) drill participants use it as a supplementary tool to check their risk during drill exercises; (2) the local population uses it to understand their risk once they remain in tsunami-affected areas; and (3) tourists or visitors use it to realize their tsunami risk when they enter previous-tsunami-affected areas. In creating the idea from an available resource, this study contributes to a practice that provides risk awareness through peoples' smartphones. This application is expected to become one of the ICT methods used to enhance awareness of tsunami disasters, along with smart tourism and active evacuation drill concepts.

There are many interesting findings from our experiment. Evacuation practice without inundation depth information (i.e., Exp. 01 and 02) sometimes limited decision-making. We found that before providing inundation depth information, students paid attention to arriving at the assigned evacuation area by following the signs but without considering alternate locations. In a real life situation, it might not be safe to use a coastline route (as chosen by Samples 03 to 05 in Exp. 02). On the other hand, when such information was known, they were likely to find high ground that was somehow nearer. As was observed in Exp. 03 and 04, students immediately chose to climb to high ground after crossing the bridge.

Although some samples could not use the Japanese language, they encountered no language barrier in understanding the signs showing routes to evacuation areas, as these were mainly pictures. To provide the necessary evacuation information, the use of a universal language (i.e., pictures) remains one of the best practices.

Examining the actions performed in Exp. 02, we found that leadership is also an important factor. The evacuation action in a real life situation is mostly performed as a group process, probably with family members, friends, colleagues, or others [5]. When practicing a leadership role during a drill, students can understand how they should lead, follow, or respond in a group.

In Exp. 03 and 04, we observed that there was communication among samples, as one sample in each group was assigned to keep monitoring the inundation depth using the application. With such information, all samples agreed that the disaster felt more realistic.

Nevertheless, based on the results of our evacuation experiment, we discovered two limitations. First, location conditions, such as road traffic, limited evacuation action. Second, the tested prototype required access to an Internet signal.

A future study is planned in which further experiments will be conducted, as well as product evaluation via usability testing following Biel et al. [10]. Moreover, comments and feedback from the experiment just described will be taken into account in creating the final version of the mobile application before launching it on the market.

References

[1] Leelawat N, Pee LG and Iijima J 2013 *Proc. 12th Int. Conf. on Mobile Business (Berlin)* (Atlanta, GA: AIS) paper 15
[2] Leelawat N, Suppasri A, Park J, Charvet I, Latcharote P, Imamura F, Iijima J, Abe Y 2016 *Int. Conf. on Urban Risks Proc. (Lisbon)* (Lisbon: European Centre on Urban Risks) pp 267-274
[3] Sung S J 2011 *Proc. 8th Int. Telecommunications Society Asia-Pacific Regional Conf.* (Taipei) (EconStor)
[4] van den Akker J 1999 *Design approaches and tools in education and training*, eds J van den Akker, R M Branch, K Gustafson, N Nieveen and T Plomp (Dordrecht: Springer Netherlands) chapter 1 pp 1-14
[5] Sorensen J H and Sorensen B V 2007 *Handbook of Disaster Research*, eds H Rodríguez, E L Quarantelli and R R Dynes (New York, NY: Springer Science+Business Media) chapter 11 pp 183-199

[6] Kendra J M and Wachtendorf T 2007 *Handbook of Disaster Research*, eds H Rodríguez, E L Quarantelli and R R Dynes (New York, NY: Springer Science+Business Media) chapter 18 pp 316-334
[7] Johannesson P and Perjons R 2014 *An introduction to design science* (Springer International Publishing Switzerland)
[8] van Teijlingen E R and Hundley V 2001 *social research UPDATE*, ed N Gilbert (Guildford: Department of Sociology, University of Surrey) issue 35
[9] United Nations 2015 *Sendai Framework for Disaster Risk Reduction 2015-2030* (Geneva: United Nations International Strategy for Disaster Reduction)
[10] Biel B, Grill T and Gruhn V 2010 *J. Syst. Software* **83** 2031-44

Acknowledgments
This study was partly supported by the JSPS Grant-in-Aid for Young Scientists (B) "Applying developed fragility functions for the Global Tsunami Model (GTM)" (No. 16K16371; PI: Assoc. Prof. Anawat Suppasri); the Inter-Graduate School Doctoral Degree Program on Science for Global Safety (G-Safety) of Tohoku University; and Tokio Marine & Nichido Fire Insurance Co., Ltd. through the International Research Institute of Disaster Science (IRIDeS) at Tohoku University. The authors would like to thank the Center for Spatial Information Science, University of Tokyo, for data; Technical Assistant Kumagai Saiichi (IRIDeS, Tohoku University) for his kind support; British and Japanese students for having actively participated in our evacuation experiment; teachers at Japanese and British schools for their kind support of our activities; and the Crisis Management Division of Kesennuma City for their kind collaboration.

Analysis of tsunami disaster map by Geographic Information System (GIS): Aceh Singkil-Indonesia

A Farhan[1,3] and H Akhyar[2,3,*]

[1]Department of Physic Education, FKIP Syiah Kuala University, Darussalam, Banda Aceh 23111, Indonesia
[2]Department of Mechanical Engineering, Faculty of Engineering, Syiah Kuala University, Jl. Syech Abdurrauf No. 7, Darussalam, Banda Aceh 23111, Indonesia
[3] Tsunami and Disaster Mitigation Research Center (TDMRC), Syiah Kuala University, Jl. Prof. Dr. Ibrahim Hasan, Gampong Pie, Banda Aceh, 23233, Indonesia

*E-mail: akhyar@unsyiah.ac.id

Abstract. Tsunami risk map is used by stakeholder as a base to decide evacuation plan and evaluates from disaster. Aceh Singkil district of Aceh- Indonesia's disaster maps have been developed and analyzed by using GIS tool. Overlay methods through algorithms are used to produce hazard map, vulnerability, capacity and finally created disaster risk map. Spatial maps are used topographic maps, administrative map, SRTM. The parameters are social, economic, physical environmental vulnerability, a level of exposed people, parameters of houses, public building, critical facilities, productive land, population density, sex ratio, poor ratio, disability ratio, age group ratio, the protected forest, natural forest, and mangrove forest. The results show high-risk tsunami disaster at nine villages; moderate levels are seventeen villages, and other villages are shown in the low level of tsunami risk disaster.

1. Introduction

Aceh Province-Indonesia is one of an area that has a high level of tsunami hazard, according to history great tsunami affected on December 26, 2004. A tsunami map should be done with multi-criteria approach corresponds with the assessment site. Geographic information systems (GIS) are a tool that would be used for analyzing and visualize the level of the tsunami risk. The Indonesian government through BNPB (Indonesian National Disaster Bureau) has regulations to standardize the maps through Perka No.2 in 2012. The regulations discussed the concept of how to develop hazard, vulnerability, capacity, risk or disaster maps in Indonesia [1].

Identify impact tsunami disaster in Tohoku 2011 and reconstruction to build tsunami-resilient community. Tsunami inundation zone recognized and measurement by satellite remote-sensing processing and field measurement [2]. Tsunami inundation numerical models have been an interpretation by using open source GIS and databases to assessment for two coastal areas of Oman. Simple software is used to estimate tsunami run-up to carry out tsunami inundation numerical models by a deterministic and probabilistic approach. Worst-case scenarios are applied a tsunami risk assessment by a deterministic and probabilistic approach [3]. A method for tsunami inundation and impact assessment is used by integrating GIS applications, and multi-criteria investigation through Analytical Hierarchy Process (AHP) has been studied in Ofunato City, Iwate Prefecture, Japan. The parameters were used Digital Elevation Model (DEM), satellite imagery for remote-sensing analysis and field data collected. The GIS analysis consisted coastline distance created from a coastline vector map. The remote-sensing analysis has consisted of slope and elevation created from Aster GDEM version 2, vegetation density generated from ALOS ANVIR-2 image. AHP processed by weighting the parameter using a pair-wise and comparing with five iterations of a normalized matrix. Five classes of tsunami inundation map and the weighted overlay have been used to produce the tsunami vulnerability map. The final tsunami map has good agreement with the observation and historical data. It is useful for disaster mitigation, evacuation, strategy and management of preliminary disaster information [4].

Smart phone application is introduced as D-Aid app has been developed to visualize the information of victim amount, necessities, and dangerous after the disaster. This application allowed communities to order the complexity of differentiation, integration, and collaboration data [5]. Developing a system of spatial information to support tsunami evacuation action planning by using geo-information technology is discussed. GIS applications are presented through a GIS model builder as a tool in this system, and it used to minimize damages by Tsunami and developed run-up plans [6]. Tsunami thematic maps with GIS layer for the area Indonesia derived from analyzing several key parameters of the tsunami risk for the southern region of Sumatra, Java and Bali have been made by Strunz, et al, 2011 [7]. Detection of ground motion that resulting to earthquake locally by using remote sensing and GIS tools has been investigated [8]. It conducted by evaluating satellite imagery, topography, Digital Elevation Model (DEM) for extracting parameters geomorphometric, SRTM (Shuttle Radar Topography Mission) and ASTER-data.

The purpose of this work is to analyze the tsunami disaster at Seaboard area in Aceh Singkil district Aceh province Indonesia using the GIS tools. The tsunami risk-disaster map is needed by the government and the public people to understand about their region on the status of the tsunami-vulnerable or risk disaster. The government of Aceh Singkil through BPBD (regional disaster management bureau) would decide evacuation plan and evaluated the physical and infrastructure development in a safe area, especially for the risk zone according to the results in tsunami risk map. Government and stakeholders of Aceh Singkil required this tsunami disaster map that will be used as a reference to mitigating greater damage and more victims. Furthermore, tsunami map is used as a reference for region development planning area that can minimize the victim, physical and economic destruction. Many investigations and analysis about disaster maps have been done, but in this paper discussed about creating the tsunami maps refer to the Perka BNPB No.2 at 2012 (Indonesia laws from National Disaster Bureau) as the basis for the analysis of these maps.

2. Methods

Aceh Singkil is one of a district in Aceh Province of Indonesia and located on the Western Coast (Fig.1). Disaster risk map is a function of three interrelated aspects, such as hazard, vulnerability, and capacity. Data required to produce the tsunami risk map consists of administrative maps of the study area, digital Rupa Bumi Indonesia Map (Topographic map) with scale 1:50.000 from BIG (geospatial information bureau), SRTM (The Shuttle Radar Topography Mission) Aceh Singkil, shapefile of tsunami area that affected on December 24, 2005. The GPS, digital camera, computer with GIS software is used for visualization and reports. Data collection during implementation is obtained from the field survey with supported by FGD (Focus disaster group, involving local government and stakeholder). Secondary data related tsunami histories, seismic, climate, demographic are collected by the FGD.

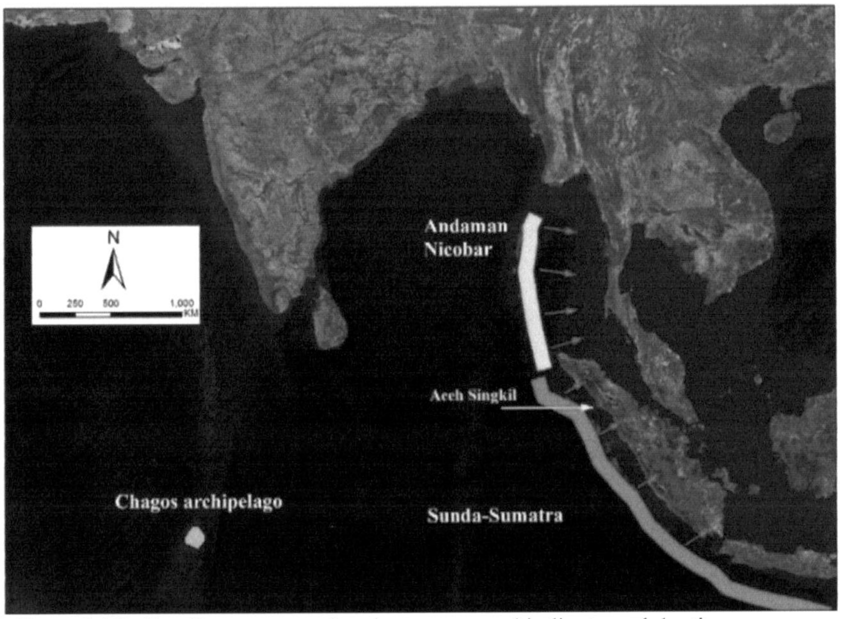

Figure 1. Potential Indian Ocean tsunamigenic sources, red indicates subduction zones, green partly strike-slip faults, and orange volcanic centers [3].

The tsunami hazard map is obtained by analyzing the SRTM elevation data and overlay with tsunami shapefile. Vulnerability map developed from some parameters, which are social, economic, physical environmental vulnerability, and level of exposed people. Capacity maps obtained from the field assessment result. Disaster-risk map was derived by analyzing the hazard maps, vulnerability, and capacity with the disaster algorithms. All levels of hazard, vulnerability, capacity and disaster risk-map can be seen in three classes, the scoring starting from zero to one index (0-0.3333 indicate for low, 0.3334-0.6666 for medium, and 0.6667-0.9999 for high levels of hazard).

2.1. Social vulnerability

Social vulnerability is obtained from analysis Eq.1 with parameters is population density, sex ratio, poor-ratio, disability ratio and age group ratio [1]. The data obtained from the center of Indonesia's statistics (BPS) and cross-checked with existing data at the sub-district institutions [9].

$$\begin{aligned} Social\ Vulnerability &= \left(0.6 * \frac{\log\left(\frac{Population\ Density}{0.01}\right)}{\log\left(\frac{100}{0.01}\right)}\right) + (0.1 * Sex\ Ratio) \\ &+ (0.1 * Disability\ Ratio) + (0.1 * Poor\ Ratio) + (0.1 * Poor\ Ratio) \end{aligned} \quad (1)$$

Table 1. Parameter for analysis social vulnerability [1].

Parameter	Percentage (%)	Class			Score
		Low	Medium	High	
Population density	60	<500 people/km^2	500-1000 people/km^2	>1000 people/km^2	Class/ Max. index of classes
Sex ratio (10%) Poor ratio (10%) Disability ratio (10%) Group ages ratio (10%)	40	< 20 %	20-40 %	>40%	

2.2. Economic vulnerability

The economic vulnerability data is analyzed Eq.2 by scoring productive land with PDRB (gross regional domestic product) [1,9].

$$Economic\ Vulnerability = (0.6 * Productive\ Land\ Score) + (0.4 * PDRB\ Score) \quad (2)$$

Table 2. Parameter for analysis economic vulnerability [1].

Parameter	Percentage (%)	Class			Score
		Low (IRD)	Medium (IRD)	High (IRD)	
Productive land	60	< 50 M	50-200 M	> 200 M	Class/ Max. index of classes
PDRB (gross domestic income)	60	< 50 M	50-200 M	> 200 M	

2.3. Physical vulnerability

The physical vulnerability is analyzed by using Eq.3, which is parameters of houses, public building and critical facilities in that area [1,9].

$$Physic\ vulnerability = (0.4 * house\ score) + (0.3 * public\ facility\ score) + (0.3 * critical\ facility\ score) \quad (3)$$

Table 3. Parameter for analysis physical vulnerability [1].

Parameter	Percentage (%)	Class			Score
		Low (IRD)	Medium (IRD)	High (IRD)	
House	40	< 400 M	400-800 M	> 800 M	Class/ Max. index of classes
Public facilities	30	< 500 M	500 M-1 B	> 1 B	
Critical facilities	30	< 500 M	500 M-1 B	> 1 B	

2.4. Environment vulnerability

Environmental vulnerability map could be analyzed by using Eq.4. The parameters used are the protected forest, natural forest, and mangrove forest [1,9].

$$Environment\ vulnerability = (0.3 * main\ forest\ score) + (0.3 * buffering\ forest\ score) + (0.4 * mangrove\ score) \quad (4)$$

Table 4. Parameter for analysis environment vulnerability [1].

Parameter	Percentage (%)	Class			Score
		Low	Medium	High	
Main forest	30	< 20 ha	20-50 ha	> 50 ha	Class/ Max. index of classes
Buffering forest	30	< 25 ha	25-75 ha	> 75 ha	
Mangrove	40	< 10 ha	10-30 ha	> 30 ha	

Finally, every vulnerability maps overlay using tsunami vulnerability algorithms, as shown in Eq.5 [1].

$$\begin{aligned} &Tsunami\ Vulnerability \\ &\quad = (0.4 * Social\ Score) + (0.25 * Economic\ Score) + (0.25 * Physical\ Score) \\ &\quad + (0.1 * Environment\ Score) \qquad (5) \end{aligned}$$

Data from field assessment used to produce the capacity map. Parameters are reviewed from the availability of local rules and society for disaster management, availability of early warning systems, the study of disaster risk, availability of disaster education, and a factor of essential risk reduction, development of preparedness funding, logistical and social networks.

The disaster risk map obtained from the overlay of those maps such as hazard, vulnerability and capacity using Eq.6 [1].

$$Risk = \sqrt[3]{Hazard * Vulnerability * (1 - Capasity)} \qquad (6)$$

3. Results

Aceh Singkil district consists of 11 sub-district and 120 villages. As mention before that disaster risk can be seen as the interaction of hazard (H), vulnerability (V) and capacity (C). Disaster risk map is a function of hazard map and vulnerability map then divided by capacity map.

3.1. Hazard Map

Analysis hazard map derived from the weighted overlay spatial method that obtained the hazard map for Aceh Singkil district as presented in Figure 2-5 and Table 5 shows data attribute of the hazard map.

Table 5. Affected villages by tsunami hazard analysis of Aceh Singki district.

No.	Villages	Sub-district	Level Hazard
1	Kampung Baru	Singkil Utara	Medium
2	Teluk Rumbia	Singkil	Medium
3	Suka Makmur	Singkil	High
4	Selok Aceh	Singkil	Medium
5	Rantau Gedang	Singkil	Medium
6	Pulau Sarok	Singkil	High
7	Pasar Singkil	Singkil	High
8	Kuta Simboling	Singkil	High
9	Kilangan	Singkil	High
10	Kampung Ujung	Singkil	High
11	Ujung Sialit	Pulau Banyak Barat	High
12	Suka Makmur	Pulau Banyak Barat	High
13	Haloban	Pulau Banyak Barat	High
14	Asan Tola	Pulau Banyak Barat	High
15	Teluk Nibung	Pulau Banyak	High
16	Pulau Balai	Pulau Banyak	High
17	Pulau Baguk	Pulau Banyak	High
18	Kuala Baru Sungai	Kuala Baru	High
19	Kuala Baru Laut	Kuala Baru	High

20	Kayu menang	Kuala Baru	High

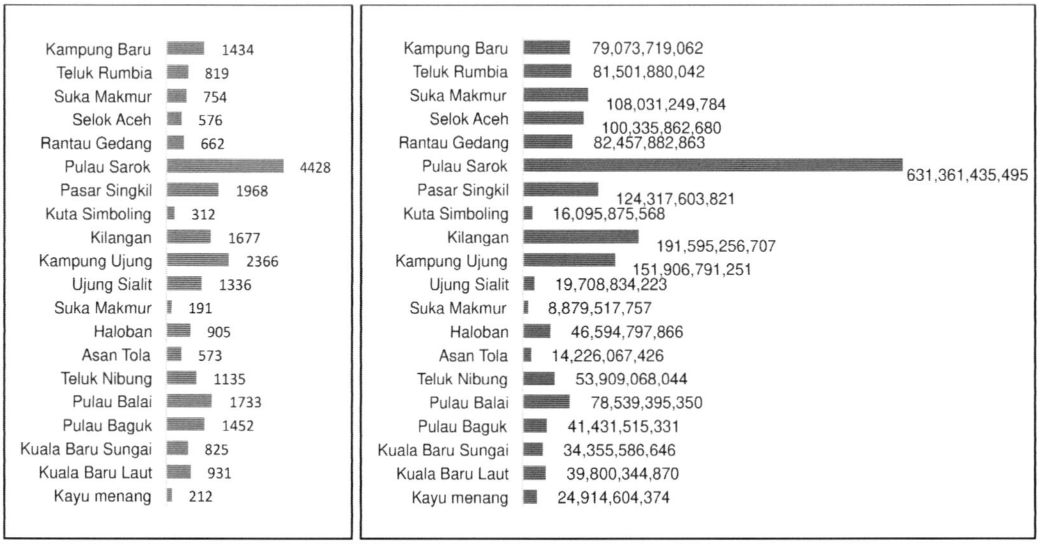

Figure 2. (a) The number exposures people, and (b) the number loss of property in Indonesia rupiah (IRD) by the tsunami in Aceh Singkil District.

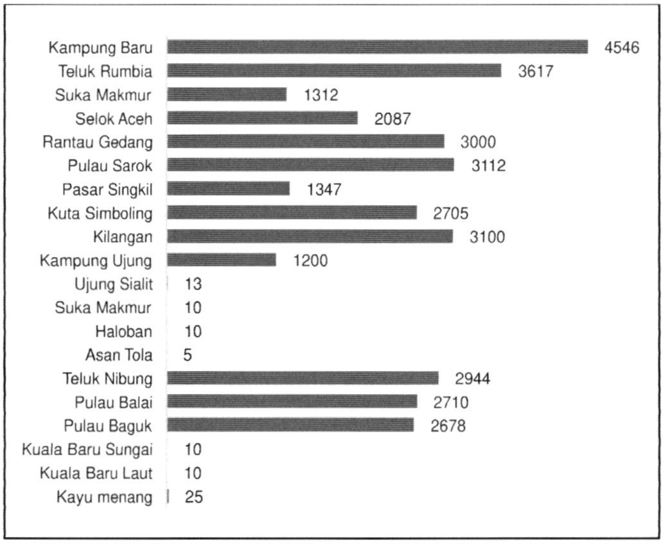

Figure 4. The number environmental damages (in hectares) by tsunami in Aceh Singkil District.

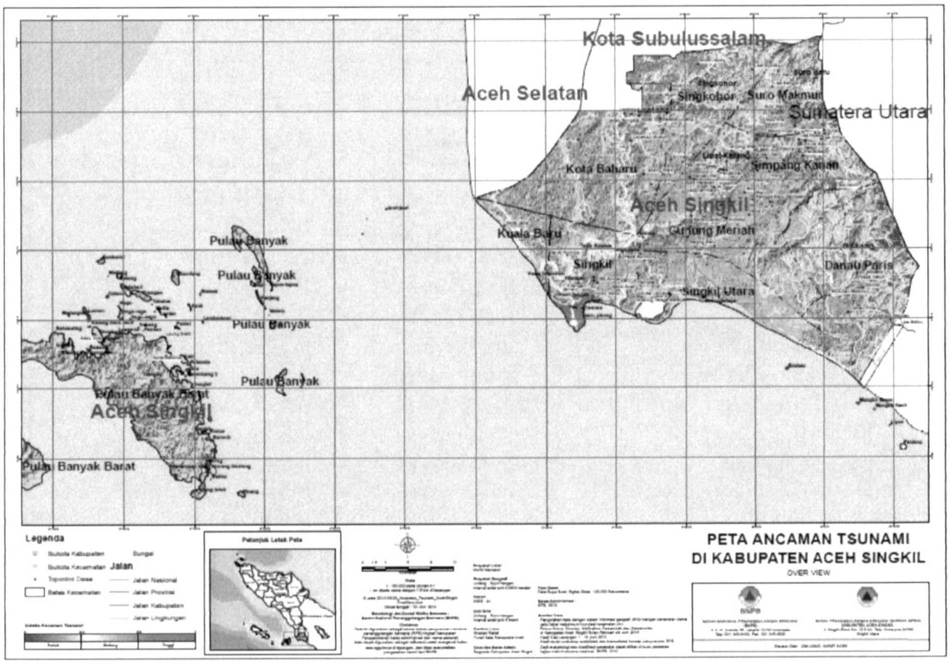

Figure 5. Tsunami hazard map of Aceh Singkil district.

3.2. Vulnerability Map

Vulnerability map obtained from the analysis of Eq.5, which is the function of social, economic, physical environmental vulnerability maps.

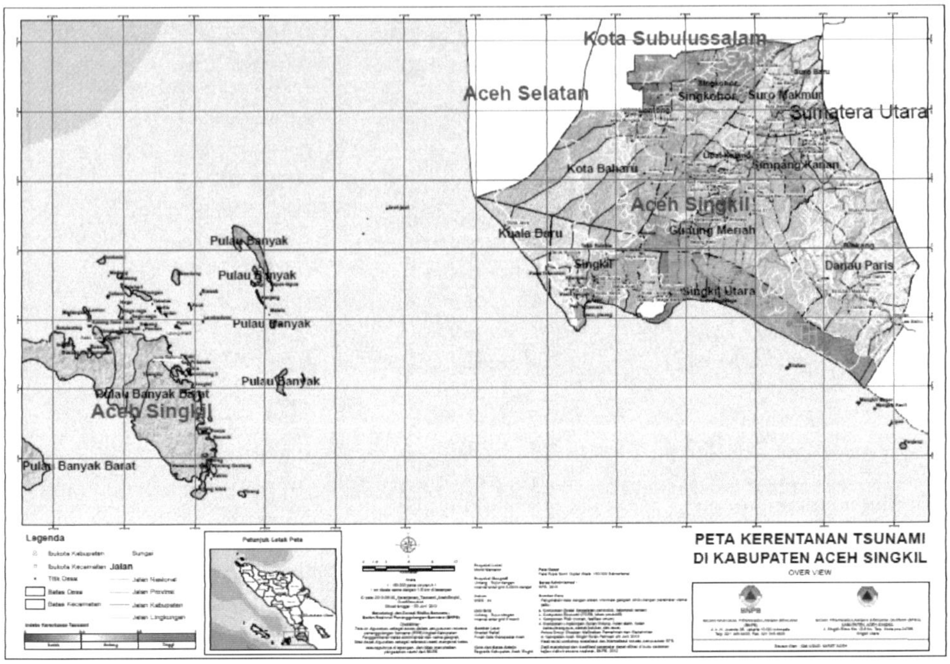

Figure 5. Tsunami vulnerability map of Aceh Singkil district.

Table 6. Tsunami vulnerability analysis of Aceh Singki district.

No.	Village	District	Score of Vulnerability
1	Biskang	Danau Paris	0.535495701
2	Danau Pinang	Danau Paris	0.461265483
3	Lae Balno	Danau Paris	0.586344095
4	Napa Galuh	Danau Paris	0.546612169
5	Sikoran	Danau Paris	0.541781929
6	Sikoran	Danau Paris	0.541360338
7	Sintuban Makmur	Danau Paris	0.477901334
8	Situbuh-tubuh	Danau Paris	0.48832011
9	Blok 15	Gunung Meriah	0.570432587
10	Blok 31	Gunung Meriah	0.565524608
11	Blok VI Baru	Gunung Meriah	0.566602279
12	Blok-18	Gunung Meriah	0.4776248
13	Bukit Harapan	Gunung Meriah	0.600208703
14	Cingkam	Gunung Meriah	0.582186825
15	Gunung Lagan	Gunung Meriah	0.616451945
16	Labuhan Kera	Gunung Meriah	0.866676384
17	Lae Butar	Gunung Meriah	0.568397106
18	Pandan Sari	Gunung Meriah	0.52978637
19	Panjahitan	Gunung Meriah	0.844867753
20	Parangusan	Gunung Meriah	0.580782128
21	Pertampakan	Gunung Meriah	0.553414103
22	Rimo	Gunung Meriah	0.581953473
23	Sangga Beru Selulusan	Gunung Meriah	0.54326786
24	Sebatang	Gunung Meriah	0.578144281
25	Seping Baru	Gunung Meriah	0.594345651
26	Sianjo Anjo Mariah	Gunung Meriah	0.545775538
27	Sidorejo	Gunung Meriah	0.565676981
28	Suka Makmur	Gunung Meriah	0.565840334
29	Tanah Bara	Gunung Meriah	0.569982995
30	Tanah Merah	Gunung Meriah	0.610487654
31	Tanjung Betik	Gunung Meriah	0.735914427
32	Tulaan	Gunung Meriah	0.570796653
33	Tunas Harapan	Gunung Meriah	0.559873303
34	Butar	Kota Baharu	0.582758115
35	Danau Bungara	Kota Baharu	0.577525485
36	Ladang Bisik	Kota Baharu	0.575509359
37	Lapahan Buaya	Kota Baharu	0.53258096
38	Lentong	Kota Baharu	0.58642335
39	Muara Pea	Kota Baharu	0.578505403
40	Mukti Lincir	Kota Baharu	0.593910628
41	Samardua	Kota Baharu	0.591318392
42	Selakar Udang	Kota Baharu	0.498659883
43	Sumber Mukti	Kota Baharu	0.680789344
44	Kayu menang	Kuala Baru	0.545247721
45	Kuala Baru Laut	Kuala Baru	0.485089986
46	Kuala Baru Sungai	Kuala Baru	0.500595827
47	Suka Jaya	Kuala Baru	0.492308018
48	Pulau Baguk	Pulau Banyak	0.592257278
49	Pulau Balai	Pulau Banyak	0.591116185
50	Teluk Nibung	Pulau Banyak	0.590309904

No.	Village	District	Score of Vulnerability
51	Asan Tola	Pulau Banyak Barat	0.456958104
52	Haloban	Pulau Banyak Barat	0.508909934
53	Suka Makmur	Pulau Banyak Barat	0.454336341
54	Ujung Sialit	Pulau Banyak Barat	0.431392243
55	Cibubukan	Simpang Kanan	0.504398997
56	Guha	Simpang Kanan	0.501111879
57	Kain Golong	Simpang Kanan	0.526618017
58	Kuta Batu	Simpang Kanan	0.516278185
59	Kuta Kerangan	Simpang Kanan	0.558259101
60	Kuta Tinggi	Simpang Kanan	0.507004074
61	Lae Gambir	Simpang Kanan	0.498543318
62	Lae Gecih	Simpang Kanan	0.486677509
63	Lae Nipe	Simpang Kanan	0.449025086
64	Lae Riman	Simpang Kanan	0.501566912
65	Lipat Kajang Atas	Simpang Kanan	0.505769515
66	Lipat Kajang Bawah	Simpang Kanan	0.503993469
67	Pakiraman	Simpang Kanan	0.479452431
68	Pandan Sari	Simpang Kanan	0.490480513
69	Pangi	Simpang Kanan	0.481259507
70	Pertabas	Simpang Kanan	0.478078867
71	Serasah	Simpang Kanan	0.500619462
72	Siantas	Simpang Kanan	0.621277369
73	Sidodadi	Simpang Kanan	0.49079077
74	Sidorejo	Simpang Kanan	0.534967596
75	Silatong	Simpang Kanan	0.517212023
76	Tanjung Mas	Simpang Kanan	0.490916299
77	Tugan	Simpang Kanan	0.510911153
78	Tuh-Tuhan	Simpang Kanan	0.470977567
79	Ujung Limus	Simpang Kanan	0.509004261
80	Kampung Ujung	Singkil	0.606505959
81	Kilangan	Singkil	0.546755182
82	Kuta Simboling	Singkil	0.553028432
83	Pasar Singkil	Singkil	0.56073654
84	Paya Bumbung	Singkil	0.575881339
85	Pemuka	Singkil	0.624308996
86	Pulau Sarok	Singkil	0.585589132
87	Rantau Gedang	Singkil	0.551363639
88	Selok Aceh	Singkil	0.566720326
89	Siti Ambia	Singkil	0.631526966
90	Suka Damai	Singkil	0.588602634
91	Suka Makmur	Singkil	0.543270359
92	Takal Pasir	Singkil	0.4232003
93	Teluk Ambon	Singkil	0.581406981
94	Teluk Rumbia	Singkil	0.553646977
95	Ujung Bawang	Singkil	0.59655226
96	Gosong Telaga Barat	Singkil Utara	0.565797866
97	Gosong Telaga Selatan	Singkil Utara	0.653946101
98	Gosong Telaga Timur	Singkil Utara	0.621074894
99	Gosong Telaga Utara	Singkil Utara	0.634045763
100	Kampung Baru	Singkil Utara	0.663399385
101	Ketapang Indah	Singkil Utara	0.64978343

No.	Village	District	Score of Vulnerability
102	Telaga Bakti	Singkil Utara	0.646259117
103	Lae Pinang	Singkohor	0.560919767
104	Lae Sipola	Singkohor	0.624787686
105	Mukti Jaya	Singkohor	0.595385918
106	Pea Jambu	Singkohor	0.425526898
107	Singkohor	Singkohor	0.661189349
108	Sri Kayu	Singkohor	0.598723174
109	Alur Linci	Suro Makmur	0.464752781
110	Bulu Ara	Suro Makmur	0.499503722
111	Bulu Sema	Suro Makmur	0.614149945
112	Keras	Suro Makmur	0.514232565
113	Ketangkuhan	Suro Makmur	0.517698735
114	Lae Bangun	Suro Makmur	0.544607071
115	Mandumpang	Suro Makmur	0.519536902
116	Pangkalan Sulampi	Suro Makmur	0.49291591
117	Siompin	Suro Makmur	0.541958043
118	Sirimo Mungkur	Suro Makmur	0.475702412
119	Suro Baru	Suro Makmur	0.508986688

3.3. Capacity Map

The capacity index is calculated based on the indicator from the Hyogo Framework for Actions (HFA). It is obtained by carrying out FGD for some actors in disaster management in an area. Based on the evaluation of achievement indicators of resilience area that would divide it into five levels (Table 7), there are:

Level 1 is the area had small achievements in disaster risk reduction by implementing some advanced actions in plans or policies.

Level 2 is the area has implemented several actions in disaster risk-reduction achievements with still sporadic from lack of institutional commitment and systematic policy.

Level 3 is the government's commitment and related to some community disaster risk-reduction in an area has been reached and supported by a systematic policy, but the achievements obtained by the commitment, and the policy have not been comprehensive still not sufficiently to reduce the risk of disaster.

Level 4 is supported by full commitment and overall policy on disaster risk reduction in the area has success to gain the achievements. However, it recognized there are still limitations in commitment, financial resources or operational capacity in the implementation of disaster risk-reduction efforts within the region.

Level 5 is the overall achievement has been the success with the full commitment and capacity to all community and government levels.

Table 7. Capacity index components [1].

No.	Indicator	Index			Data source
		Low	Medium	High	
1	Local institutional rules and disaster management	Disaster resilient of level 1 and 2	Disaster resilient of level 3	Disaster resilient of level 4 and 5	Forum Discussion Group (National government, National-International non-government, Stakeholder, University)
2	Early Warning and Study of Disaster Risk Reduction				
3	Education of disaster				
4	Basic factor of risk reduction				
5	Preparedness development on the all sector				

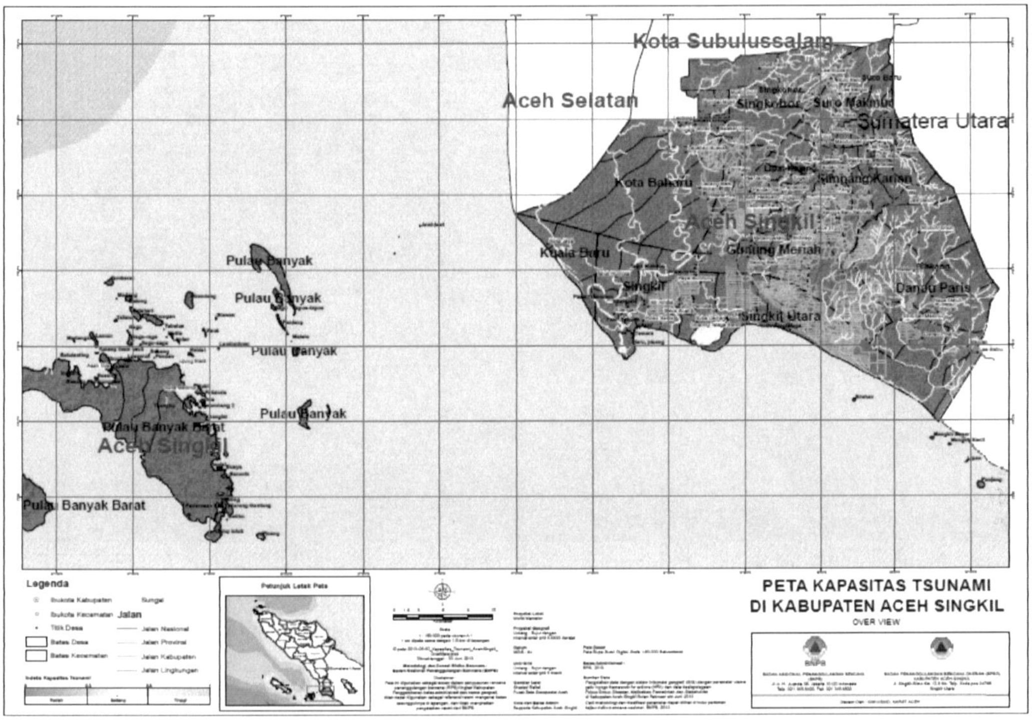

Figure 6. Tsunami capacity map of Aceh Singkil district.

3.4. Tsunami risk disaster map

Disaster risk map based on hazard, vulnerability and capacity map could be seen as a function of interaction from those maps. They are used to obtain the level of an area disaster risk by calculating the potential of people exposures, loss-property, and environment damage area. Hazard map analyzed by SRTM and tsunami affected maps, vulnerability map analyzed by the overlay of social, economic, physic and environmental damage and capacity map analyzed by evaluation of achievement indicators. The result from the analysis by the weighted overlay obtained tsunami level risk map for Aceh Singkil district as shown in Fig.7 and Table 8.

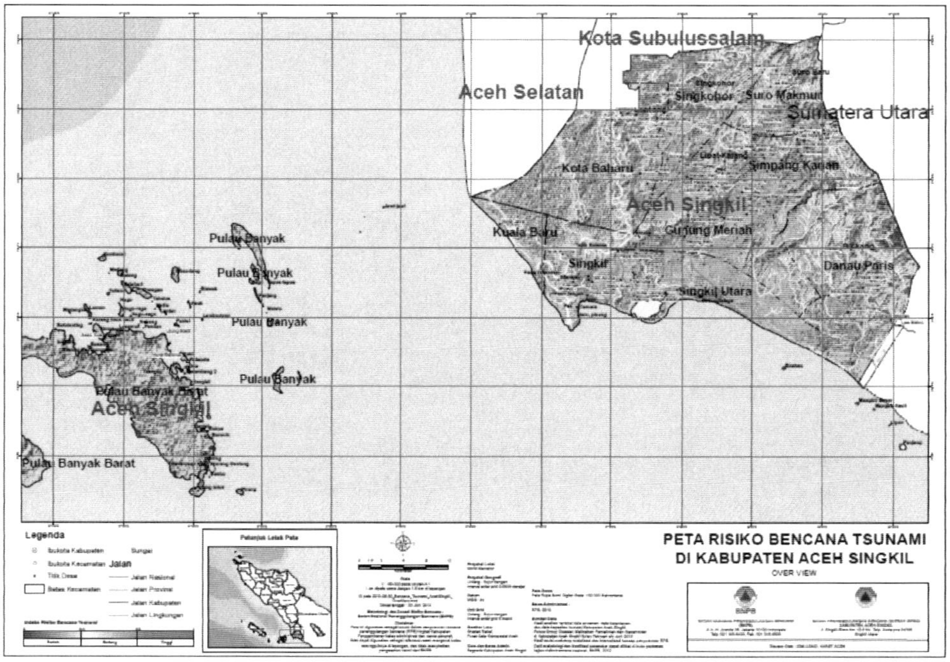

Figure 7. Tsunami risk disaster map of Aceh Singkil district.

Table 8. Tsunami risk disaster analysis of Aceh Singki district.

No.	Villages	Sub-district	Level of Risk
1	Kayu menang	Kuala Baru	Medium
2	Suka Jaya	Kuala Baru	Medium
3	Kuala Baru Sungai	Kuala Baru	Medium
4	Pulau Baguk	Pulau Banyak	Medium
5	Teluk Nibung	Pulau Banyak	High
6	Ujung Sialit	Pulau Banyak Barat	Medium
7	Suka Makmur	Pulau Banyak Barat	High
8	Haloban	Pulau Banyak Barat	Medium
9	Asan Tola	Pulau Banyak Barat	Medium
10	Teluk Rumbia	Singkil	Medium
11	Pulau Sarok	Singkil	Medium
12	Pasar Singkil	Singkil	Medium
13	Kilangan	Singkil	Medium
14	Paya Bumbung	Singkil	Medium
15	Siti Ambia	Singkil	Medium
16	Pemuka	Singkil	Medium
17	Rantau Gedang	Singkil	High
18	Kampung Ujung	Singkil	High
19	Ujung Bawang	Singkil	High
20	Kampung Baru	Singkil Utara	High
21	Ketapang Indah	Singkil Utara	Medium
22	Telaga Bakti	Singkil Utara	Medium
23	Gosong Telaga Barat	Singkil Utara	Medium
24	Gosong Telaga Timur	Singkil Utara	High
25	Gosong Telaga Utara	Singkil Utara	High
26	Gosong Telaga Selatan	Singkil Utara	High

4. Conclusion

Study of tsunami hazard, vulnerability, capacity, and disaster maps for Aceh Singkil district with three levels risk (high, medium and low) is analyzed in this paper base on Perka No.2. The results show sixteen villages have high and four villages have the medium level of hazard. The tsunami vulnerability analysis of Aceh Singkil district can be concluded that the largest number exposure people (4,428 people) and loss property (IDR 631,361,435,495) at Pulau Sarok, and the highest number of environmental damages showed at Kampung Baru village (4,546 ha).

High tsunami risk-disaster shows at Teluk Nibung, Suka Makmur, Rantau Gedang, Kampung Ujung, Ujung Bawang, Kampung Baru, Gosong Telaga Timur, Gosong Telaga Utara, and Gosong Telaga Selatan villages. Some villages in moderate level are Kayu menang, Suka Jaya, Kuala Baru Sungai, Pulau Baguk, Ujung Sialit, Haloban, Asan Tola, Teluk Rumbia, Pulau Sarok, Pasar Singkil, Kilangan, Paya Bumbung, Siti Ambia, Pemuka, Ketapang Indah, Telaga Bakti, and Gosong Telaga Barat. The low level of tsunami risk shows for other remaining villages in Aceh Singkil District.

Acknowledgments
This research was made possible by a Grant from the International Organization for Migration (IOM)-Aceh, and the financial support is greatly appreciated. We would like to thank Mr. Abdillah Imron Nasution and FGD members of Aceh Singkil for support during this research.

References

[1] Peraturan kepala badan nasional penanggulangan bencana nomor 02 tahun 2012 tentang pedoman umum pengkajian risiko bencana, 2012: 1-62 (in Indonesia).
[2] S. Koshimura, S. Hayashi, H. Gokon, The impact of the 2011 Tohoku earthquake tsunami disaster and implications to the reconstruction, Soils and Foundations, 2014; 54(4):560–572.
[3] J. Browning, N. Thomas, An assessment of the tsunami risk in Muscat and Salalah, Oman, based on estimations of probable maximum loss, International Journal of Disaster Risk Reduction, 2016, 16: 75–87.
[4] A.B. Sambah, F. Miura, Integration of Spatial Analysis for Tsunami Inundation and Impact Assessment, Journal of Geographic Information System, 2014, 6, 11-22.
[5] L.C. Schunke, L.P.L. de Oliveira, M. Cardoso and M.B. Villamil, D-Aid – An App to Map Disasters and Manage Relief Teams and Resources, Procedia Computer Science, 2015, 51: 2898–2902.
[6] A.N. Liyanage and H. Lee, A Schematic Approach for GIS Application for Tsunami Disaster Management, ASEA/DRBC, Springer-Verlag Berlin Heidelberg, 2012, CCIS 340: 395–400.
[7] Strunz, G., Post, J., Zosseder, K., Wegscheider, S., Muck, M., Riedlinger, T., Mehl, H., Dech, S., Birkmann, J., Gebert, N., Harjono, H., Anwar, H.Z., Sumaryono, Khomarudin, R. M., and Muhari, A. Tsunami risk assessment in Indonesia, 2011, Nat. Hazards Earth Syst. Sci., 11, 67–82.
[8] Theilen-Willige, B., and Burnett, F.B., Detection of local site conditions influencing earthquake shock and secondary effects in the valparaiso area in central-chile using remote sensing and gis methods, Journal of Tsunami Society International, 2011, Vol.30 (3), pp.1-32.
[9] Singkil Dalam Angka 2015, Katalog BPS: 1102001.11020, Badan Pusat Statistik (BPS) Kabupaten Aceh Singkil (in Indonesia).

Science-practice nexus for landslide surveying: technical training for local government units in Indonesia

A L Alam[1], S Hespiantoro[1], D Dyar[2], D Balzer[3], D Kuhn[3], J Torizin[3], M Fuchs[3], S Kastl[2,3] and J Anhorn[2,3]

[1] PPSDM GEOMINERBA – Centre for Development of Human Resources in Geology, Mineral Resources and Coal, Jl. Jendral Sudirman No. 623, Bandung, 40211, Indonesia
[2] GEORISK Project, Jl. Cisitu Lama No. 37, Bandung, 40135, Indonesia
[3] BGR – German Federal Institute for Geosciences and Natural Resources, Stilleweg 2, Hannover, 30655, Germany

Abstract. The Indonesian archipelago is prone to various geological hazards on an almost day to day basis. In order to mitigate disaster risk and reduce losses, the government uses its unique setup of ministerial training institutions. The Centre for Development of Human Resources in Geology, Mineral and Coal offers different level of technical training to local governments in order to provide them with the necessary means to understand geological hazards, mitigate risks, and hence close the gap between local and national governments. One key factor has been the continuous incorporation of new scientific knowledge into their training curricula. The paper presents benefits and challenges of this science-practice nexus using the standardised landslide survey as one example where mobile technology has been introduced to the training just recently.

1. Introduction

The Indonesian archipelago faces multiple geological hazards and the government needs a comprehensive mechanism to mitigate disaster risks [1,2]. According to the Indonesian disaster statistics of the Centre for Research on the Epidemiology of Disasters (CRED), landslides are the third most frequent disaster, with floods on top, followed by forest fires. Adding up to more than 1.500 events in the year 2016 only (until August) [3]. The hazard specific Global Assessment Report on Disaster Risk Reduction (GAR) on landslide risk in Indonesia [4], states that according to Christanto et al. [5] during the period 1981-2007 the average number of events per year was 49, with the annual distribution closely related to the rainfall variations.

The Hyogo Framework of Action (HFA) with the third Priority for Action on knowledge, innovation, and education includes the following two key elements, which are of particular interest to this paper. First, "Develop training and learning programmes in disaster risk reduction targeted at specific sectors (development planners, emergency managers, local government officials, etc.)" [6]. Second, "Promote the use, application and affordability of recent information, communication and space-based technologies and related services, as well as earth observations, to support disaster risk reduction, particularly for training and for the sharing and dissemination of information among different categories of users" [6].

Based on the biannual HFA progress reports there has been considerable progress in Indonesia on these matters since the formulation of the HFA agenda [7–10]. For example, disaster risk reduction has been included into the national education curriculum and trainings for local governments are provided regularly. Still, the Indonesian National Disaster Management Authority (BNPB) reports in 2015, that "One of the remaining challenges in this regard [core indicator 2 of priority for action 3] include the need to enhance coordination among relevant agencies from the national down to the local levels. The government needs to advocate further the integration of DRR [Disaster Risk Reduction] and recovery concepts into school education and DM [Disaster Management] training and exercises, particularly at the district/city governments as the actual service providers" [7]. In the "new" Sendai Framework for Disaster Risk Reduction 2015-2030, the issue of government employees as service providers to the wider community has again been raised as an important task [11]. It states now in Priority 1 "Understanding disaster risk": there is a great need to "Build the knowledge of government officials at all levels, civil society, communities and volunteers, as well as the private sector, through sharing experiences, lessons learned, good practices and training and education on disaster risk reduction, including the use of existing training and education mechanisms and peer learning;" [11].

This reflects the ongoing necessity to strengthen the interlinkages, communication channels, and coordination efforts among and between national and local stakeholders including government and non-governmental education institutions in the field of (professional) training to bridge the gap between advanced scientific knowledge on the one hand and implementation on the other hand. In the course of this paper, we refer to this as the science-practice nexus of technical training provision.

Since 2004, the German cooperation project "Mitigation of Georisks", implemented by the German Federal Institute for Geosciences and Natural Resources (BGR) and subordinated agencies of the Ministry of Energy and Mineral Resources of Republic of Indonesia, strengthens public institutions of disaster management and sub-national authorities in Indonesia to improve their services concerning the mitigation of geological risks [12]. In cooperation with the Ministries Human Resources Development Agency (BPSDM) the project focuses since 2015 on the improvement of selected technical trainings [13]. One major result of the project is the development of the Landslide Survey Data Sheet (LSDS) based on international standards [14] together with the Geological Agency of Indonesia (GAI) and its implementation in current training curricula at the Centre for Development of Human Resources in Geology, Mineral and Coal (PPSDM GEOMINERBA). Using international standards for landslide data collection though the LSDS ensures comparability of data on national and international level and fosters harmonization processes on data acquisition, susceptibility mapping and hazard evaluation. The landslide survey data is stored in the Landslide Inventory Database of Indonesia (LIDIA), also developed during the cooperation project [15] and now maintained by the GAI Centre for Volcanology and Mitigation of Geological Hazards. The parameters stored in the database can be used to further analyse landslide susceptibility on various scales.

In order to successfully conduct landslide surveys and apply the LSDS in remote areas, technical trainings for civil servants across the country are indispensable to support the technical capacity at local level. This further serves the national data availability in the field of geological hazards and strengthens the capacity to do landslide susceptibility analysis at GAI. To support instantaneous transfer of survey data from local surveyors to GAI an Android application has been developed in the course of the development cooperation project [16]. The mobile application replaces the paper-based LSDS and is used in the Landslide Survey training of PPSDM GEOMINERBA. Landside survey techniques using mobile applications allow to reach out to remote areas for data collection while preserving high data quality. The necessary technical training using such methods needs to enable local government units to survey independently and maintain minimum standards.

2. Purpose

The paper presents benefits and challenges of existing interlinkages between PPSDM GEOMINERBA and GAI for landslide surveying and database utilization through appropriate training setup and teaching techniques. It highlights the importance of technical trainings for local governmental bodies to support disaster mitigation strategies. Transferring scientific and technical knowledge into formal training processes in the field of geological disaster mitigation has been at the centre of the GEORISK project. In the course of this process, various scientific institutions are involved in reviewing and updating the existing training materials. Simultaneously, implementing personnel (= users) are encouraged to provide their valuable insights from the implementation point of view. In this sense, the science-practice nexus is strengthened through sharing of experiences and working as a team on improving training curricula.

We illustrate this based on the landslide survey trainings offered at PPSDM GEMINERBA compromising of a sequence of two trainings ("Landslide Surveying" and "Survey Data Processing and Interpretation") [17,18]. Therefore the paper first provides insights to the governmental setup of training centres, second the structure of two landslide related trainings is explained using the LSDS and LIDIA as a best-practice example. Third, the rational of introducing mobile technology to further enhance the learning experience and closing in on the science-practice nexus is explained.

3. Institutional Framework of Human Resource Development in Indonesia

In Indonesia, each Ministry has an Agency for Human Resources Development. Its functions encompass all tasks of planning, implementation, monitoring and evaluation of the human resource development activities in their respective working areas. In the case of the Ministry for Energy and Mineral Resources, these working areas are comprised of Geology, Mineral, and Coal; Power, New Energy, Renewable Energy, and Energy Conservation; Oil and Gas. The specific tasks in the geology sector include inter alia provision of training in geological risk mitigation, georisk-sensitive spatial planning and volcanological monitoring.

The training centres are mandated to train Local Government Unit (LGU) personnel. They are eligible to request professional training, which is then performed by the training centres professionals. Often these trainings even take place on the LGUs premises.

The GAI under the same ministry as the Agency for Human Resource Development is mandated among other things to perform research on geological disaster mitigation, mapping of existing hazards, and monitoring of potential geological threats. This includes the assessment of all landslides occurring across the country, providing hazard maps for earthquakes, landslides, and volcanoes, as well as monitoring of volcanic activities. In order to facilitate the surveying and monitoring task of GAI, LGU staff are doing the basic geological survey, If more detailed analysis is necessary, national staff is dispatched to the scene by GAI. Local government employees are therefore often requested to do the initial geological landslide survey. For the purpose of appropriate technical training of these local government employees, GAI also supports PPSDM GEOMINERBA in developing and revising of training curricula and material, as well as during the trainings itself.

While LGU staff needs ready to use methods, the advanced user wants as many details as possible based on the latest scientific achievements. Linking the data user (GAI) with those providing the technical training (PPSDM GEOMINERBA) to the surveyor (LGU) has been of great benefit to the continuous improvement of a practical training setup.

4. Professional Trainings for Local Government Units

The two trainings offered by PPSDM GEOMINERBA in the field of landslide survey and analysis aim to train LGU staff in order to support the government agencies, especially GAI. Learning objectives and training outcomes are defined using a competence-based approach: the Landslide Survey Training uses existing international standards to develop a consistent method for the assessment of landslides in Indonesia. Participants are enabled to perform landslide data acquisition and landslide reporting to the GAI. The second training advances one step further and uses these data, enabling participants to further explore landslide survey data, analyse them according to national regulations, and provide technical recommendations. The following section explains further the layout and content of said trainings and how international surveying methods are incorporated.

The Landslide Survey Training of PPSDM GEOMINERBA is based on a comprehensive and modular landslide assessment workflow that was developed by BGR as a standard operating procedure [19]. In cooperation with the GAI, this standard workflow was adapted to the specific situation and requirements of Indonesia and was implemented by GAI to elaborate and maintain the national landslide inventory of Indonesia. This approach comprises three sub-modules that build upon each other, starting with the "Field Survey Techniques" module that encompasses the knowledge and skills to record and map landslides in the field by using a specific field survey form, called "Landslide Survey Data Sheet (LSDS)" [20]. The second module, the "Digital Landslide Inventory" module, comprises a corresponding database application that is known as the "Landslide Inventory Database of Indonesia (LIDIA)" [15]. This national inventory database application offers the seamless integration of the recorded analogue field data into the digital workflow. The third module, the "Landslide Susceptibility Assessment" [21] module, facilitates the GAI to assess the spatial hazard and risk potential of landslides and supports sustainable land use planning and implementation of mitigation strategies.

In this perspective, the first training course of PPSDM GEOMINERBA enables the participants to record and map landslides in the field, using the LSDS, and transfer the analogue information into the digital LIDIA application.

The LSDS is based on the suggested methods and terminology of the UNESCO Working Party on World Landslide Inventory [14,22,23], serves as a checklist for the documentation and investigation of landslides in the field and aims to achieve a standardized institutional nomenclature. It is organized in three information levels. The first information level is compulsory and contains the minimum required information about the mass movement, namely location, general geological and environmental setting and the date of occurrence. The second information level comprises descriptive information on landslide typology, causes and triggers, and the third information level on damages, losses and casualties [20,24]. The recorded information therefore meets all requirements for a reliable inventory dataset but requires trained personnel with a basic understanding of landslides.

Therefore, the training incorporates a thorough introduction of the theoretical background of landslide mapping and the necessary information to be recorded in the field. This is supplemented by practical exercises in the field, during which the participants apply the imparted knowledge and map different landslides in the surroundings of the training locality. In that way they gather first experiences and are able to discuss results and challenges with the teaching staff. Mixing exercise elements at the landslide location with theoretical classroom sessions provide the basis for a mutual learning environment. In a second step, the participants are introduced to the digital database application LIDIA that is used to store the field data. After that, the participants conduct practical exercises with the landslide database, as they transfer the acquired field data of their "personal" landslide surveys into the LIDIA. Again, using the application in practice deepens the understanding and handling of the database.

The second training course of PPSDM GEOMINERBA on Survey Data Processing and Interpretation provides a more sophisticated approach to analyse data from the LIDIA that currently holds more than 5.900 landslide datasets. State of the art techniques for slope stability analysis and statistical analysis methods, as well as the integration of data to Geographic Information Systems (GIS) are part of the training. The BGR section on Engineering Geological Hazard Assessment also provided recommendations on this training curriculum.

5. Mobile Applications to Enhance Learning Experience

Mobile applications have entered our lives in no time. Smartphones and tablets are becoming more and more central elements of our day-to-day (working) activities, offering a huge variety of connections to information, data, and people. This connectivity has transformed more and more into a bi-directional communication thread, allowing users not just to search for and collect information, but also to share (new) information.

In the field of Disaster Risk Management (DRM) for example the timely information people in affected areas can provide is crucial and has modified the way we deal with disasters, especially if it comes to post-disaster measures. The International Association for Information Systems for Crisis Response and Management (ISCRAM) is a lively community, that shares the primary mission to foster "a community dedicated to promoting research and development, exchange of knowledge and deployment of information systems for crisis management" [25]. Their annual conference is a good example for the continuous effort by researchers and practitioners in developing the highest usable outcome from the Information and Communication Technology (ICT) for DRM.

Recent technological applications allow users to actively engage in the way information and data is collected, for example through providing flood-level data [26], damage state information, or requesting help in the aftermath of a disastrous event. A list of such applications offering two-way-information sharing can be found on the fulcrumapp.com site [27]. The future prospect is that the massive amount of distributed "human sensors" might allow for intelligent data mining leading for example to more accurate flood predictions, better estimations of losses, and improved rescue services.

In order to successfully conduct landslide surveys and apply the LSDS in remote areas of Indonesia, technical trainings for civil servants across the country are indispensable to support the technical capacity at local level. One challenge has always been the huge spatial extent of landslide occurrence and timely surveying of these incidents across the peninsula. To support instantaneous transfer of survey data from local surveyors to GAI an Android application has been developed [11]. The mobile application replaces the paper-based LSDS (booklet and paper-format) and is nowadays used in the Landslide Survey training of PPSDM GEOMINERBA (**Figure 1**). The mobile application allows for a reliable and comprehensive data collection and reduces the risk of human errors during transfer of the data to GAI.

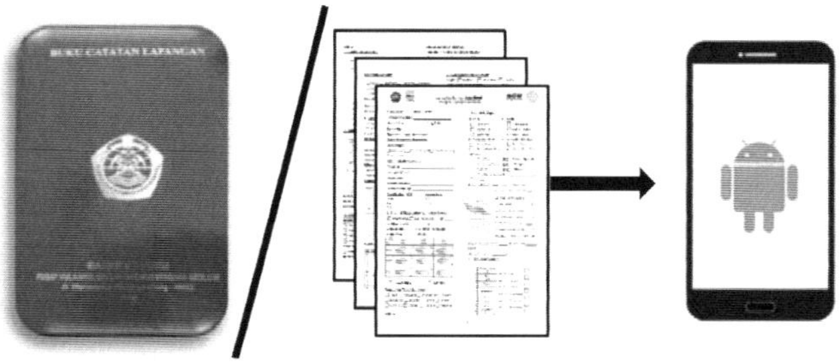

Figure 1. Analogue to Digital Transformation by applying the Android Application.

Most of the current technological developments in the field of mobile technology aims to make work more effective and efficient. The mobile LSDS applications is based on the Android system, was developed with the Java programming language and designed using EXtensible Mark-up Language (XML). The compilation of the program uses Android Studio version 2.1.2. The application is approximately 6 megabyte large and is distributed during the trainings using the exchange android application package (.apk). **Figure 2** provides three screen samples of the final product.

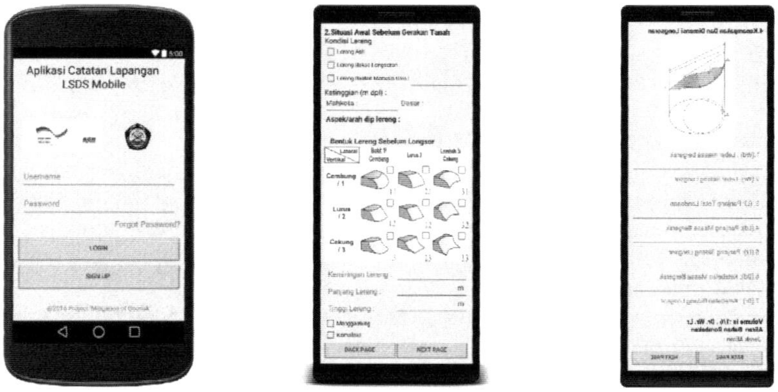

Figure 2. Mock-Up of some pages/screens of the LSDS Android Application.

In general the implementation of the overall Android application is divided into three different menu sections: "Survey Data Form", "Upload Photo or Sketch" and "Send Data". See Table **1** for a detailed explanation of the menu sections.

Table 1. LSDS Android Application Menu.

Application Menu	Description
Survey Data Form	To display the survey sheet that needs to be filled out by the surveyor (12 "pages/screens").
Upload Photo or Sketch	Function to upload a photos and a sketch of the field conditions. Users can also use stored photos from their hand phone's gallery. They are attached to the survey data.
Send Data	Button to send all survey data and attachments via email to the administration of GAI for quality check.

The application needs access to GPS functionalities in order to determine the exact coordinates which are directly stored to the data field. This is considered a major benefit compared to the manual format, where many errors occurred if people had to collect GPS points using additional (handheld-) devices and making notes up to the last digit. In general, the application has been designed to differentiate between mandatory and optional fields for data entry. Hence, submitting incomplete datasets is impossible, as the application forces its users -using colour highlighting- to fill out at least all mandatory data entry fields. These had already been defined in the digital LIDIA, but as enforcement through the paper-based formats is difficult, another main source of errors could be eliminated. Other benefits of the full-digital version are for example in automated field calculations: The survey for instance requires the calculation of the landslide volume utilizing the main dimensions of the landslide area, which is now automated. Additionally, various fields are per se single-option fields. Despite a sophisticated layout of the previous paper-based LSDS, common errors occur if users provide multiple answers instead of single-options. Within the Android App, radio buttons (single option) and check-boxes (multiple choice) have been used in order to account for this and minimize errors.

6. Results

The German Technical Cooperation project "Mitigation of Georisks", supports PPSDM GEOMINERBA to develop trainings for governmental employees in the field of landslide survey using appropriate techniques [5]. During that process 1) eLearning was introduced to allow participants to interact with training experts in geology remotely; 2) an Android application for landslide surveying was developed, which allows to automatically transfer data from the field to the national landslide database LIDIA [3].

This case study of positive collaboration between PPSDM GEOMINERBA and the GAI shows how applied methodologies can support the generation of reliable data in a decentralized way. The use of modern technology allows bridging the gap between data quality and coverage of landslide events given limited resources and their vast spatial distribution in Indonesia.

The ability of LGUs to access these data and do their own analysis introduces a great moment of enhanced (data) ownership. Surveying landslides within their area of responsibility -which allows them to also assess local risks- improves the uplink from local to national level agencies.

A logically structured training setup that makes use of modern eLearning techniques shrinks the distance between so-far distant training and education experts at the national training centre and the remote data collectors. Nevertheless, the practical part of the training including field exercises, group sessions and reporting is an important element, which allows participants to intensively discuss among each other, build relationships and exchange on problems faced and lessons learned during the assessment of landslide events. This methodological mixture between field training and classroom sessions improves the learning experience.

One remaining challenge has been the constant further education of trainers as the training schedule does not allow for time consuming side activities. Therefore, one important additional activity within the German-Indonesian development cooperation has been the provision of Training of Trainers (ToT) to advance the level of expertise within the training institution itself, in order to enable them to provide trainings on the highest possible level.

7. Conclusion

Transferring of scientific and technical knowledge to local government units in order to mitigate geological risks is a challenging task and needs close collaboration among various stakeholders. The Indonesian setup of subordinated ministerial training centres has a huge potential providing a unique interlinked learning environment. It offers top-down and bottom-up ways of learning and teaching by integrating local experiences of landslide risk into the field training, but also the incorporation of the latest scientific innovations.

Technical training for landslide surveying in Indonesia improves local capacities and national efforts to mitigate landslide disaster risk. The need for technical support to perform landslide survey at local level is evident, thus local government units need to receive capacity development through the national training centres using suitable methodologies. Dissemination and upscaling of existing trainings offered in the geological sector is mandatory to further understand geological hazards and mitigate risks in Indonesia.

The use of mobile technology to further enhance data quality and user-friendliness offers great potential to further close the gap between professional training staff and local experts implementing the landslide survey. Nevertheless to close the gap within the science-practice nexus requires commitment and a strong culture of collaborative learning, which is always a challenging task.

References

[1] INFORM 2016 *Country Risk Profile INDONESIA* (Inter-Agency Standing Committee)

[2] Sulaeman 2011 Bencana Geologi Di Indonesia Tahun 2011 *Bul. Vulkanol. Dan Bencana Geol.* **6**

[3] BNPB 2016 Data Dan Informasi Bencana Indonesia *Indones. Disaster Data Inf.*

[4] Cepeda J, Smebye H C, Vangelsten B, Nadim F N and Muslim D 2010 *Landslide Risk in Indonesia* (UNISDR)

[5] Christanto N, Hadmoko D S, Westen C J, Lavigne F, Sartohadi J and Setiawan M A 2009 Characteristic and Behavior of Rainfall Induced Landslides in Java Island, Indonesia : an Overview EGU General Assembly Conference Abstracts vol 11 p 4069

[6] United Nations 2005 *Report of the World Conference on Disaster Reduction Kobe, Hyogo, Japan, 18-22 January 2005* (Kobe, Hyogo: United Nations)

[7] BNPB 2015 *Indonesia: National progress report on the implementation of the Hyogo Framework for Action (2013-2015)* (BNPB)

[8] BNPB 2013 *Indonesia: National progress report on the implementation of the Hyogo Framework for Action (2011-2013)* (BNPB)

[9] BNPB 2011 *Indonesia: National progress report on the implementation of the Hyogo Framework for Action (2009-2011)* (BNPB)

[10] BNPB 2009 *Indonesia: National progress report on the implementation of the Hyogo Framework for Action (2007-2009)* (BNPB)

[11] United Nations 2015 *Sendai Framework for Disaster Risk Reduction 2015-2030*

[12] IFRC 2014 *International Disaster Response Law (IDRL) in Indonesia: An analysis of the impact and implementation of Indonesia's legal framework for international disaster assistance* (Geneva: International Federation of Red Cross and Red Crescent Societies)

[13] GEORISK Project 2015 Mitigation of Georisks: Phase III-2015/2016

[14] WP/WLI (International Geotechnical Societies' UNESCO Working Party on World Landslide Inventory) 1993 A suggested method for describing the activity of a landslide *Bull. Int. Assoc. Eng. Geol. - Bull. Assoc. Int. Géologie Ing.* **47** 53

[15] Balzer D and Kuhn D 2013 *Getting Started With the Landslide Inventory Database Indonesia (LIDIA)* vol Sub-Module Digital Landslide Inventory, ed Bundesanstalt für Gewissenschaften und Rohstoffe (Hannover)

[16] Dyar Dwiqi 2016 Manual for LSDS mobile APP ed GEORISK Project

[17] PUSDIKLAT Geologi 2012 Kurikulum Diklat Teknik Pemeriksaan Gerakan Tanah

[18] PUSDIKLAT Geologi 2016 Kurikulum Diklat Survei Gerakan Tanah

[19] Balzer D 2013 *BGR-Standard Module "Landslides"* vol Sub-Module Field Survey Techniques, ed Bundesanstalt für Gewissenschaften und Rohstoffe (Hannover)

[20] Kuhn D 2013 *Guidelines for Applying the Landslide Survey Data Sheet (LSDS)* vol Sub-Module Field Survey Techniques, ed Bundesanstalt für Gewissenschaften und Rohstoffe (Hannover)

[21] Torizin J 2013 *Bivariate Statistical Methods for Landslide Suscetibility Analysis using ArcGIS* vol Sub-Module Landslide Susceptibility Assessment, ed Bundesanstalt für Gewissenschaften und Rohstoffe (Hannover)

[22] WP/WLI (International Geotechnical Societies' UNESCO Working Party on World Landslide Inventory) 1990 A suggested method for reporting a landslide *Bull. Int. Assoc. Eng. Geol. - Bull. Assoc. Int. Géologie Ing.* **41** 5

[23] WP/WLI (International Geotechnical Societies' UNESCO Working Party on World Landslide Inventory) 1993 *Multilingual Landslide Glossary* (Richmond)

[24] Kuhn D 2013 *Landslide Survey Data Sheet (LSDS)* vol Sub-Module Field Survey Techniques, ed Bundesanstalt für Gewissenschaften und Rohstoffe (Hannover)

[25] ISCRAM 2016 ISCRAM | International Association for Information Systems for Crisis Response and Management

[26] Pakpahan T, Rindengan Y D and Najoan X 2013 Aplikasi Mobile Peta Rawan Bencana Kota Manado Berbasis Android *J. Tek. Elektro Dan Komput.* **2** 77–82

[27] Fulcrum Mobile SOlutions, LLC 2016 Apps for Disaster Response, Disaster Relief, and Support *Disaster Response Apps*

Acknowledgments

The German cooperation project "Mitigation of Georisks", is implemented by the German Federal Institute for Geosciences and Natural Resources (BGR) and subordinated agencies of the Ministry of Energy and Mineral Resources of Republic of Indonesia (KESDM). The project is funded by the German Federal Ministry for Economic Cooperation and Development (BMZ).

IOP Conf. Series: Earth and Environmental Science **56** (2017) 012004 doi:10.1088/1755-1315/56/1/012004

Green city Banda Aceh: city planning approach and environmental aspects

A A Arif[1]

[1]Architecture and Planning Department, Syiah Kuala University, Jl. Syech Abdurrauf, Banda Aceh, 23111, Indonesia

email: azhar_aarif@yahoo.com

Abstract. Banda Aceh as the capital of Aceh Province is the region with the tsunami disaster that occurred on December 26, 2004 the most severe of which over 60% of the city area were destroyed mainly coastal region and settlements. One product plan for rehabilitation and reconstruction of Banda Aceh is made of Banda Aceh as Green City. To realize the Green City Banda Aceh, urban development process should be conducted in a planned and integrated way with attention to spatial and environmental aspects to ensure an efficient urban management and to create a healthy, beautiful and comfortable environment. There is a weakness of the process in urban planning and development that occurred at present where cities tend to minimize the development of green open space and land conversion into a commercial district, residential areas, industrial areas, transport networks and infrastructure and facilities for other cities. Another tendency that occurs is urban environment only developed economically but not ecologically, whereas ecological balance is as important as the development of the economic value of urban areas. Such conditions have caused unbalance of urban ecosystems including increased air temperature, air pollution, declining water table, flooding, salt water intrusion and increased content of heavy metals in the soil. From an ecological perspective, unfavorable microclimate, high-temperature increase due to the lack of trees as a sieve / filter against heavy rain, can cause flooding. These conditions result in inconvienient, arid and less beautiful urban areas. The author identifies the elements contained in the Green City Banda Aceh and how the efforts and approaches must be made toward Green City Banda Aceh.

1. Introduction

As an archipelago with a coastline of about 81,000 km, Indonesia coastal areas have a strategic significance because it is a region of interaction and transition between terrestrial and marine ecosystems that have unique properties and characteristics. The coastal region is very significant for human life on earth. As a transitional area of land and sea that has a unique ecosystem, especially in the field of environment in the context of sustainable development (sustainable development).

One of the natural resources that directly or indirectly holds an important role in human life is the land where the land is a complex environment and have the characteristics such as; climate, relief, soils, hydrology, vegetation and all living creatures that play a role in its use. Land suitability evaluation is the process of land use planning that is essential to compare the requirements demanded by different types of land use, with the nature or quality of land. The use of land is not up to his potential will lead to degradation of land and unsustainable. To avoid this, it is necessary to evaluate the characteristics of the land to support the planning area of the city better.

The tsunami that occurred on December 26, 2004 resulted in damage and great loss for the city of Banda Aceh. Largely very heavy damage occurred in coastal areas resulting in loss of life and economic centers of marine fishermen and changing vegetation patterns and open green spaces in coastal areas. Learning from the tsunami, effectively limiting factor that minimizes runoff tsunami is their protective natural shoreline vegetation *(buffer zone)* by considering the character of coastal geomorphology.

Therefore, the central government and the Aceh provincial government has agreed to set the Spatial Plan of Banda Aceh which in essence is the establishment of a buffer zone *(Buffer Zone)* in the form of green belt width of 300-400 m from the shoreline as well as social forestry and fishing settlement with a width of 600-700 m re-planting mangrove trees (mangrove). The mangrove forest is the habitat for many wildlife species, as the economic function of mangrove forests serve as a source of timber, pulpwood, firewood, char materials, fishing equipment. *Buffer Zone* serves to protect the shoreline of the changes are not desirable as: erosion beach, or sedimentation dialur or harbor cruise, naturally effective coastal protection.

From the above it can be concluded that the coastal areas of Banda Aceh is an area that has the vulnerability and also the strategic potential in terms of aspects of the layout, which is an area that is geographically very important, but there has been little effort to organize the land uses in integrated / unified.

Writing this study aims to evaluate the characteristics of the land *Buffer Zone* Banda Aceh appropriate or not in accordance with the current land use, evaluating the green open land Banda Aceh coastal areas which have been planted or that there is damage or not. Spatially and gives an overview of the areas experiencing the rehabilitation of mangrove forests in the city of Banda Aceh.

2. Research Methods

Location of this research carried out in the coastal region of Banda Aceh which functioned as the *Buffer Zone Area* Kota Banda Aceh. The research was conducted between October 2010 untill December 2010, involving a graduate student Degree in Chemical Engineering, students of Department of Architecture Faculty of Engineering and the Faculty of Agriculture, University of Syiah Kuala. Secondary data were used in this study, which include the administrative map of Banda Aceh, map of Municipality Spatial Planning of Banda Aceh (RTRW) in 2009-2029, and the existing landuse map of Banda Aceh produced in 2010. The primary data were taken by conducting fieldwork utilizing GPS and camera to capture city landscape perspective. Data collected during the survey were analysed using ArcGIS 9.3. This research uses descriptive method with survey to observe of physical condition and making the point coordinates in the field, as well as observations by *Buffer Zone* physical environmental conditions at the sites.

2.1 National legislation

National legislation which is used in the management of coastal areas and urban areas are as follows:

- UU no. 5/1990 on the Conservation of Natural Resources and Ecosystems
- UU no. 26/2007 on Spatial Planning (UUPR)
- The Decree of the Environment No. 45/11/1996 on Sustainable Coastal Program

2.3 Implementation Research

IOP Conf. Series: Earth and Environmental Science **56** (2017) 012004 doi:10.1088/1755-1315/56/1/012004

In general, the research carried out in four phases: preparation, data collection, map overlay techniques and ground checks, sampling technique of point coordinates in the field.

3. Results and Discussion (Banda Aceh Green City*)*

By studying product development scenarios green open space area of Banda Aceh produced by the Government of Banda Aceh in cooperation with JICA institutions, development strategy Banda Aceh *Green City* can be described as follows:

- Maintaining open space and vegetation through conservation, rehabilitation and revitalization, to preserve the environment *(environment).*
- Changing attitudes and behavior for more care and respect for nature and the environment.
- Provide facilities and urban infrastructure that supports environmentally friendly city.
- Adjusting the density of buildings so as to guarantee a water catchment area and green environment.

Development Plan for Banda Aceh *Green City* produced is as follows:

- Conserve / protect the coastal zone by synergizing land use in the area around the beach, which supports the economic, social, cultural and nature conservation / environment.
- Develop a pool of water / retention pond in the city (there) which is part of the drainage system of the city, which also functioned as a green area of the city, recreation area, the area of socio - economic community and the aesthetics of the city.
- The development of city greening in the main structure of the city (roads, rivers) and the open spaces of the city, which became elements of disaster mitigation (track rescue *(escape route)* and gathering areas for evacuation *(escape area),* and protect the city from the effects of climate change micro and create an aesthetic city.
- The implementation of building regulations regarding building density (Coefficient Building Basics / KDB, Coefficient of Land Building / KLB, Building Equivalent Line / GSB and Line Border Rivers / GSS)
- The provision of waste management facilities and waste (by the City Sanitation Department / DKP).
- At the level of spatial planning, as far as possible avoid the allotment of land for industrial heavy / pollutive in the city of Banda Aceh

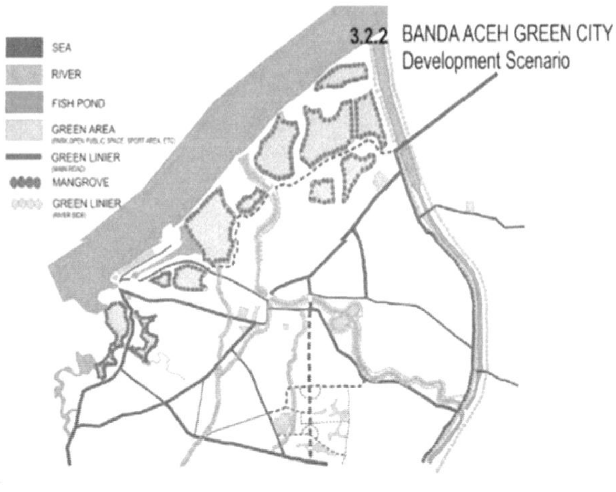

Figure 2. Banda Aceh *Green City* development scenario, JICA, 2006

Regions Research *Green City* in the Coastal city of Banda Aceh can be seen in the map below:

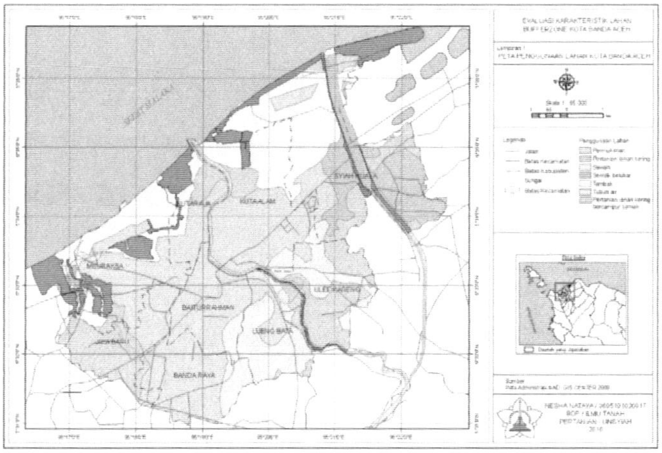

Figure 3. Map of the study area green coastal area of Banda Aceh

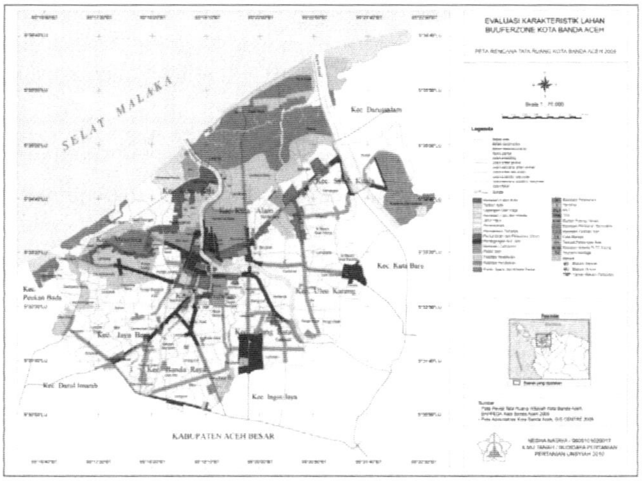

Figure 4. RTRW Kota Banda Aceh

Description of Research Areas of green open space of Banda Aceh in 2006-2008 (during the rehabilitation and reconstruction of Banda Aceh)

Dike River Flood Canal

River Flood Canal levee are in the District of Syiah Kuala exactly starting from Alue Naga village to village Rukoh and Lamnyong. Currently along these canals are flooded vegetation distribution in the form of pine trees. However still linearly with the number still very limited so it can not characterize a green area of the city. The problem today is along the canal flooding has placed temporary shelter for displaced victims of the tsunami that, without supervision will have an impact for the sustainability of the region.

Tibang village, village ponds and dikes Fisherman Seafood

This area is an area of the former pond area and an open area that is relatively hot, because the concept of development in this area is the creation of an urban forest in the area of fishing villages that can help improve the condition of the air due to climatic influences. So, fishing settlement later a settlement located in the forest area of the city. Besides, today in some parts of the village Tibang has started planting mangrove especially around ponds. And in this area there are also marshes (wet land)

Krueng Aceh

The existence of Krueng Aceh that divides the city of Banda Aceh a strong impression on the identity of the city, but the potential is still underutilized and distribution of vegetation along streams and rivers still in groups where some places quite lush but on the other hand even with no vegetation at all.

3.1 River dikes and areas of green open space in the Krueng Aceh

Prototype river dikes to prevent flooding and guides Riverbanks with the adoption of green open space (RTH) and Pedestrian paths.

IOP Conf. Series: Earth and Environmental Science **56** (2017) 012004 doi:10.1088/1755-1315/56/1/012004

Coastal Region

As a result of the tsunami disaster there is little vegetation in coastal areas, only a fraction are mangroves. However mangrove replanting efforts have been started.

3.2 Scenario Concept Approach Green City Banda Aceh

The concept of urban development of green or *Green City* Banda Aceh is to create an urban environment that is strong enough to be able to repair the ecosystem itself, able to adapt to the development of the life of the new city as an environmentally friendly city, and be able to defend the city from other natural forces such as floods, waves and tide and can minimize excessive heat and wind. As the concept of the development of city greening listed in *the Green City* plan scenarios in Banda Aceh, the direction of development of city greening is as follows:

- No major city halls such as rivers and roads that form the linear pattern (Green Linear).
- In the open space of the city (Green Area) as part of the concept of disaster mitigation (escape road and communal areas for evacuation).
- In the conservation area which is the main buffer zone that coastal areas of mangrove forest.

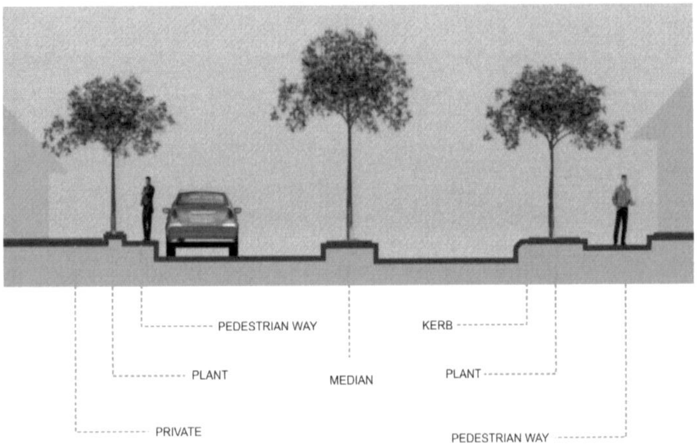

Figure 5. The placement type of vegetation on the main road

3.2 The concept of Crop Type

For plant species concept is based on the appearance and function, the categories are divided as follows:

Category A (on the side of the road and the edge of the RTH)

- General characteristics
  High-category plant canopy cover crops are large or medium
- Appearance
  For shade, firmly rooted, long-lived, can be as a steering road, trunk strong and big and can reduce noise
- Plants; Angsana (*Pterocarpus indicus*), Kidamar (*Agathis domara*), Tamarind (*Tamarindus indicus*), Kihujan (*Samanae saman*), mahogany (*Swieteria mahogany*), Nyamplung (*Calophyllum inophyllum*), Bungur (*Langerstroma spesiosa*)

Category B (on a private area)

- General characteristics
  Modern plants, including the categories of shade trees, canopy size medium
- Appearance
  As a shade tree and the trunk is not too large, need small space as a place to grow, are annual tree.
- Plants; Felicium (Felicium decipiens), Damar (Agathis alba).

Category C (As the hedge)

- General characteristics
  Trees small / medium / shrub has a major attraction in the form of flowers, branches and shape plants

IOP Conf. Series: Earth and Environmental Science 56 (2017) 012004 doi:10.1088/1755-1315/56/1/012004

- Appearance
  Flower color and form high, memorable splendor can be used as an ornamental plant and barrier
- Type of plant; Kasia, Dadap Merah, Flower Butterfly, Peacock Flower, Soka.

Ground cover, flowers or interesting leaf shape

- Appearance
  Alternative land cover other than grass
- Type of plant; Taiwan Beauty, Bird of Paradise.

3.3 Category E (on the edge of the area, the median curb street as well as a steering)

- General characteristics
  Trees with special characters
- Appearance
  As a director, gave the impression and has a distinctive shape and attractive.

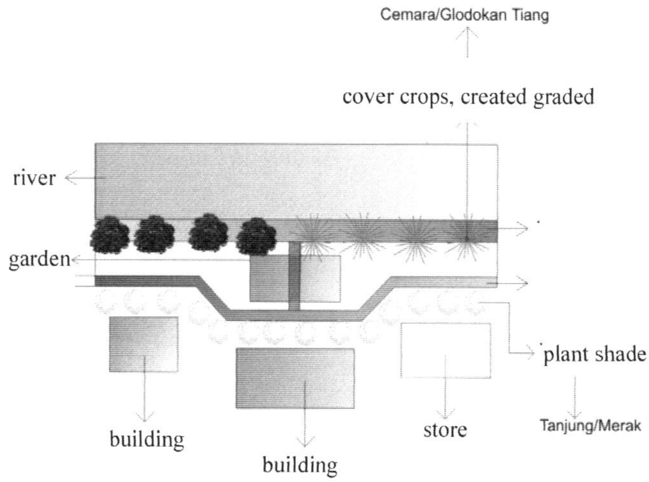

Figure 6. The concept of structuring vegetation village area Lamnyong

3.4 Countryside Lamnyong around the flood channel

For the region around the village Lamnyong River Embankment Canal, plant placement plan is in two locations placements. The first of course is the river basin area, the placement on the river embankment is at the outer side of the dike to create sufficient visibility.

4. Conclusion

Placement of *vegetation on the city's main hall,* which is along the main street with the bearing arrangement of vegetation / plants is between pedestrian and private areas or areas awoke. Placement of vegetation or crops on the curbs and pedestrian paths. Placement of vegetation on the median of the road, on the road with sufficient width and a road with two-way flow of circulation.

The choice of location placement on top of the vegetation will be affected by the width of the existing roads, district GSB also the flow of circulation that occur on roads in the planned planting. Placement of vegetation in the area of the open city directed by planting trees for planting stratified lot of vegetation in this area is expected to regulate the microclimate of the area. Besides the preferred shape of a tree with a wide canopy that can be used as a shelter.

Bibliography

Sarwono.1999. Kesesuaian Lahan dan Perencanaan Tata Guna Tanah. IPB, Bogor

Nesha nataya, 2010, Evaluasi Karakteristik Lahan *Bufferzone* Kota Banda Aceh, Program Studi Ilmu Tanah Fakultas Pertanian Universitas Syiah Kuala,

Laporan perencanaan Green City Banda Aceh, JICA, 2006 RTRW kota Banda Aceh, 2009-2029

Trancik, Roger. 1986, *Finding Lost Space: Theories of Urban Design,* New York: Nostrand Reinhold

Coulomb stress analysis of West Halmahera earthquake mw=7.2 to mount Soputan and Gamalama volcanic activities

G H D Sinaga[1*], M Zarlis[2], M Sitepu[1], R A Prasetyo[3] and A Simanullang[3]

[1]Post Graduate Program, Physics Department, University of Sumatera Utara, Bioteknologi No.1 USU, Medan, North Sumatera, Indonesia
[2]Ilkom-TI Faculty, University of Sumatera Utara, University No.9A USU, Medan, North Sumatera, Indonesia
[3]BMKG Balai Besar Wilayah I, Jl. Ngumban Surbakti No.15 Medan, North Sumatera, Indonesia

*email: goldberdhdsinaga@gmail.com

Abstract. West Halmahera is the convergency of three plates, namely the Philippines plate, the Eurasian plate, and the Pasific plate. The location of the West Halmahera is located in the thress plates, so the Western Halmahera potentially earthquake-prone areas. Some events increased activity of Mount Soputan and Mount Gamalama preceded by a massive earthquake. This research was conducted in the BMKG Region I Medan. This research uses *Coulomb Stress* Model. *Coulomb Stress* Model was used to show increasing and decreasing stress consequence from earthquake in the area of West Halmahera. Data such as the earthquake magnitude, earthquake depth, and Focal Mechanism required as input models. The data obtained from BMKG, Global CMT, and PVMBG. The result of data analyzed show an increase in the *coulomb stress* distribution at Mount Soputan 0.023 bar and 0.007 bar in mountain Gamalama. This stress followed by increased volcanic activity of the mount Soputan and mount Gamalama with freatic eruption type.

1. Introduction

Molucca Sea is located in the western Pacific Ocean, located near Molucca province, Indonesia. This limits the Sea Celebes Sea, the Banda Sea in the south, Halmahera, Seram, Buru, and Celebes. Tectonically, Halmahera Island situated between three plates, namely Philippines plate, the plate, the Eurasian plate and the Pacific plate [4]. Australian plate is located at the south and is bounded on the south by a fault system Sorong, which is a complex transpresif zone that extends to the east over 1500 km, from Papua New Guinea along the northern boundary of the Island of Papua until towards the west about 800 km to Celebes. Eurasia has a fracture eastern boundary in the southern Philippines and continue to fault West Halmahera [16]. Eurasian Plate in the region of Southeast Asia and the Philippines is a complex area and involves a lot of small plates that move semi-independent. West Halmahera island's arm mostly covered by younger volcanic products, so that the development tectonic can not be recognized properly. From earthquake research note that there are two seismic zones were intersected below banioff West Molucca Sea section shows the collision of the arc island. One of Benioff zones being tilted westward beneath the Sangihe arc and Celebes Sea, and the other sloping ramps to the east under Halmahera. Because it allegedly plate Sangihe (Eurasia) subducting eastward and produce mélange in Talaud Island and Halmahera magmatic arc in the west, while to the west subducting plate Halmahera produce mélange in. Talaud Island and magmatic arc in Sangihe Island. Thus becoming the Molucca Sea active earthquake area Mw=6.6 earthquake that occurred August 26 th, 2012, based in northeast Manado. On December 29 th, 2013 an earthquake of Mw=5.7. And in the Molucca Sea region are also frequent aftershocks with a magnitude Mw=5 until now. Unique phenomenon also occurred around the Molucca Sea, the continuity of which there are two volcanoes that form an arc contradictory. Two continuity of this volcano is located on the northern arm of Celebes and Halmahera Islands. The appearance of two continuity of this volcano in line with seismotectonic studies that showed two subduction in the area. [2].

Based on PVMBG website, Soputan is one of the active volcano type stratovolcano located in the district. Tombatu, Minahasa North Celebes. With the location coordinates of 01°06'30 " north latitude and 124°43' west longitude. Soputan has an altitude of 1783.7 meters above sea level. Geomorphology Soputan and surrounding mountains can be grouped into three morphological units which include units of volcanic body morphology, morphological units hills and plains morphology. The seismic data that happens, there are similarities seismic activity before the eruption It shows the characteristics of the eruption of the Soputan is always preceded by earthquakes tremors were followed by a swarm of volcanic earthquakes avalanches and earthquakes. The occurrence of tremors indicate that magma is at a stage movement towards the surface. Occurrence of Earthquake Tremor repeated several times to reflect the occurrence of supply of magma to the surface, which in turn leads to accumulation of pressure at the surface.

Gamalama also one type A strato volcano with an altitude of 1715 m sea surface located in the Molucca Islands, Ternate (Ternate Island), North Molucca province. With the coordinates of 0°48' north latitude and 127° 19'30" east longitude. Gamalama is one of the active volcanoes located on Halmahera Island arcs, the northeast Molucca. The area is estimated as the area meeting some of the plates of which the Pacific Plate, Eurasian and Philippines and other small plates. Ternate Island formed by mount Gamalama take place on the path of subduction (subduction zone) sloping to the east with a small angle. Morphology Gamalama general is flat on the coast, but becomes steeper toward the top.

2. Methods

The data used in parameter earthquake is West Halmahera earthquake mainshock with Mw = 7.2 dated November 15 th , 2016 and depth of 60 km with longitude/latitude of epicenter are 126.49° E /1.95° N (BMKG). Earthquake focal mechanism data with strike/dip/rake = 42°/35°/110° and the type of fault was reverse (Global CMT). And seismicity and focal mechanism data supporting earthquake from November 2004 until December 2014 with a minimum scale of an earthquake of 5.4 magnitude on the location of the North-South Molucca Sea, Halmahera, and the Minahasa peninsula.

Volcanic activity data was taken from PVMBG website. The volcano eruption type based on phreatic and phreatomagmatic in December 2014. This research use coulomb 3.4 software and analyzed coulomb stress method where it is based on static stress change. Static stress changes, the earthquake which resulted in changes in static stress triggering volcano-tectonic earthquakes 8 km under mount Soputan [3] and mount Gamalama thus increasing fault/fracture at the top of the mountain Soputan and mount Gamalama. If the rainfall is high enough, the rain water will be able to enter through the fault/fracture and meet magma, it will eruption. When an earthquake occurs, it changes the state of stress on nearby faults. In order to estimate the state of stress, the Coulomb failure stress is calculated using elastic dislocations on rectangular planes in a homogeneous and isotropic half-space following [13].

Various criteria have been used to characterize the conditions under which failure occurs in rocks. One of the more widely used is the Coulomb failure criterion, which requires that both the shear and normal stress on an incipient fault plane satisfy conditions analogous to those of friction on a pre-existing surface. In the laboratory, confined rocks approximately obey the Coulomb failure conditions, which also appear to explain many field observations [8]. Our approach is similar to those taken by [18], [14], [7], [15], [11], [5], [9], and [19]. In the Coulomb criterion, failure occurs on a plane when the Coulomb stress σ_f exceeds a specific value:

$$\sigma_f = \tau_\beta - \mu(\sigma_\beta - p) \quad (1)$$

Where $\tau\beta$ is the shear stress on the failure plane, $\sigma\beta$ is the normal stress, p is the pore fluid pressure and μ the coefficient of friction. The sign of τ_β must therefore be chosen appropriately. In a system where the x- and y-axes and fault displacements are horizontal, and fault planes are vertical (containing the z direction), stress on a plane at an angle ψ from the x-axis (Fig. 1) is given by

$$\sigma_{11}=\sigma_{xx}\cos^2\psi + 2\sigma_{xy}\sin\psi\cos\psi + \sigma_{yy}\sin^2\psi$$

$$\sigma_{33}=\sigma_{xx}\sin^2\psi - 2\sigma_{xy}\sin\psi\cos\psi + \sigma_{yy}\cos^2\psi$$

$$\tau_{13} = \frac{1}{2}(\sigma_{yy}-\sigma_{xx})\sin^2\psi + \tau_{xy}\cos^2\psi \quad (2)$$

We can now write the change of Coulomb stress for right-lateral σ_f^R and left lateral σ_f^L motion on planes orientated at ψ with respect to the x-axis in the following way.

$$\boldsymbol{\sigma_f^R} = \boldsymbol{\tau_{13}^R} + \mu'\,\sigma_{33}$$

$$\boldsymbol{\sigma_f^L} = \boldsymbol{\tau_{13}^L} + \mu'\,\sigma_{33} \quad (3)$$

The effective friction coefficient is assumed to be 0.4 in all calculations. A μ of 0.4 minimises the calculation error caused by the uncertainty in μ to ± 25% [10]. Failure is facilitated on specified or optimally oriented faults when the Coulomb failure stress, $\sigma\phi$, rises. Unless specified, the optimal fault orientation is defined by the given regional stress field [1].

The accuracy of the Coulomb stress changes due to an earthquake depends mainly on the accuracy of the source parameters of that earthquake (i.e. the location and geometry of the fault rupture, and the amount and sense of slip distribution). The more accurate the source parameters the more reliable results, and thus interpretations can be made. A reliable estimate of the slip distribution and fault geometry is therefore very important for stress transfer calculations. Small differences in slip distribution and fault geometry can lead to significant perturbations in the Coulomb failure stress. Further details of the technique can be found in [10].

3.Result

3.1 Coulomb Stress Change West halmahera November th15 2014

From the data of earthquakes in West Halmahera Mw = 7.2 dated November 15 th , 2015 obtained from BMKG stations and data from the Global CMT Focal Mechamism processed in software Coulomb 3.4. The results of coulomb stress change of the earthquake in four field lobe consists of two lobe fields positive and two lobe field negative. Lobe positive value marked in red is an area of stress increase of 0.01 bar to 0.05 bar in the southwest and northeast. So that the affected area coulomb stress distribution of positive changes produce aftershocks that caused the stress increases due to the earthquake of Mw = 7.2. While the negative lobe marked with blue color which is an area the stress drop of 0.01 bar to 0.05 bar as shown in Figure 1.

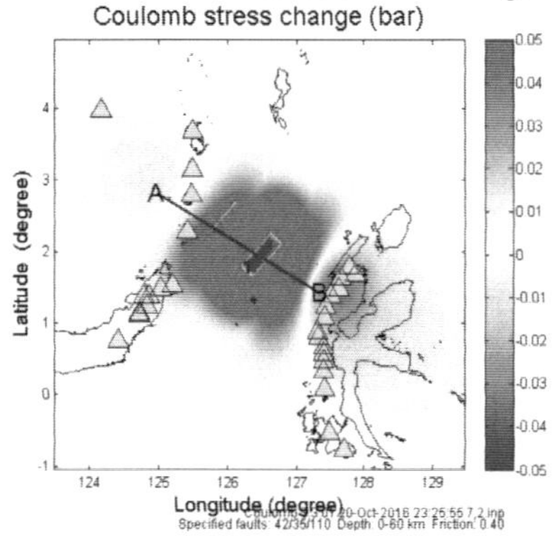

Figure 1. Distribution coulomb stress changes earthquake of 7.2 Mw West Halmahera on November 15th, 2014 at a depth of 60 km

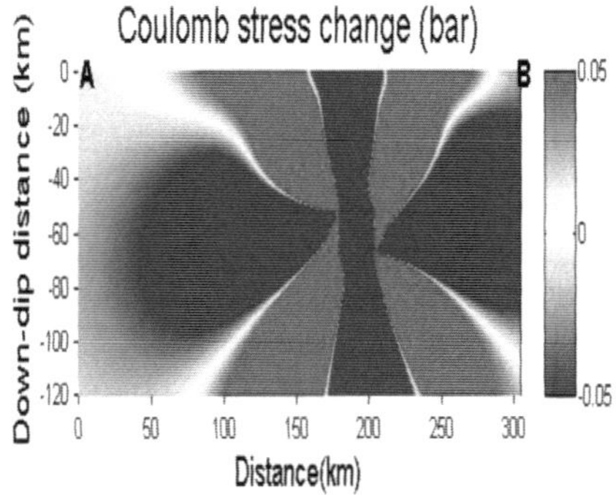

Figure 2. Cross section A-B Coulomb stress changes at a depth of 120 km

IOP Conf. Series: Earth and Environmental Science **56** (2017) 012005 doi:10.1088/1755-1315/56/1/012005

Coulomb stress Changes is positive spread in the depths of 0-120 km. From the results of cross section we can see that the coulomb stress this positive work at a distance of 25-300 km. Regional coulomb increase this stress has a value between 0.01 to 0.05 bar as shown in Figure 1 For areas with decreased coulomb stress is spread in depth at a distance of 0-120 km. From the results of cross section we can see that the positive coulomb stress work at a distance of 0-300 km. The range of values negative coulomb stress is between 0.01 to 0.05 bar as in Figure 2.

3.2 Effects of Changes in Coulomb stress Earthquake West Halmahera December 2004-December 2014 Concerning Volcanic Activity Soputan and Mount Gamalama

From the sum of 112 earthquakes that occurred in the Molucca Sea have red lobe is positive that an increase coulomb stress area of 0.01 bar to 0.05 bar and have a direction east, northeast, south, southwest, and west. While the negative lobe is negative which is a regional decrease in coulomb stress 0.01-0.05 bar to bar.

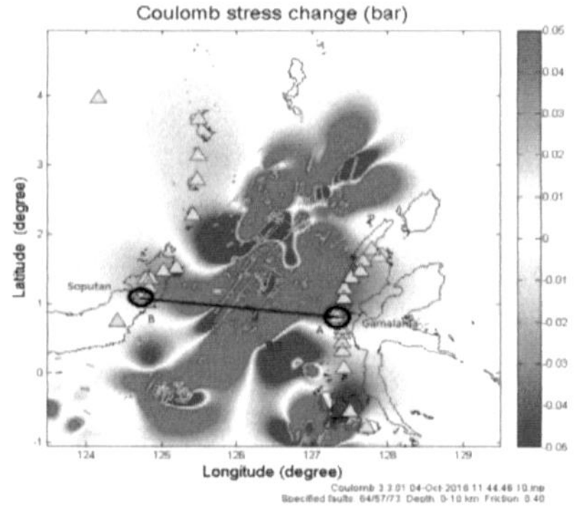

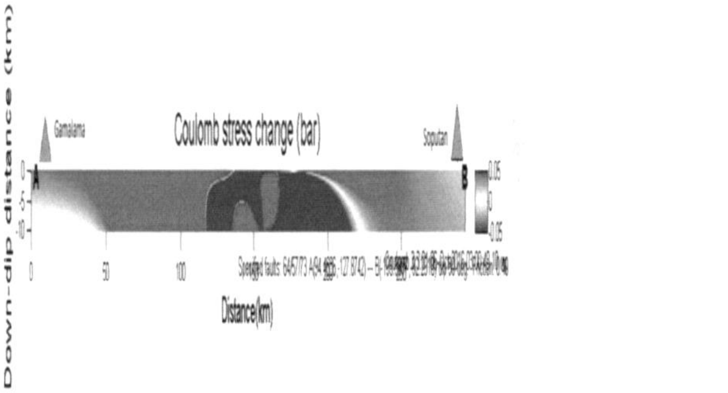

Figure 3. Distribution coulomb stress change West Halmahera 2004-2014 at depth 10 km.

Figure 4. Cross section coulomb stress change West Halmahera Earthquake to depth 10 km

Coulomb stress changes are positive spread in depth from 0 to 10 km. From the results of cross section we can see that the coulomb stress this positive work at distance of 0 to 300 km. Increased stress this area has a value of between 0.01 to 0.05 bar as shown in Figure 3. For areas with decreased coulomb stress at distance of 150 to 225 km. The range of values coulomb negative stress is between 0.01 to 0.05 bar as shown in Figure 4.

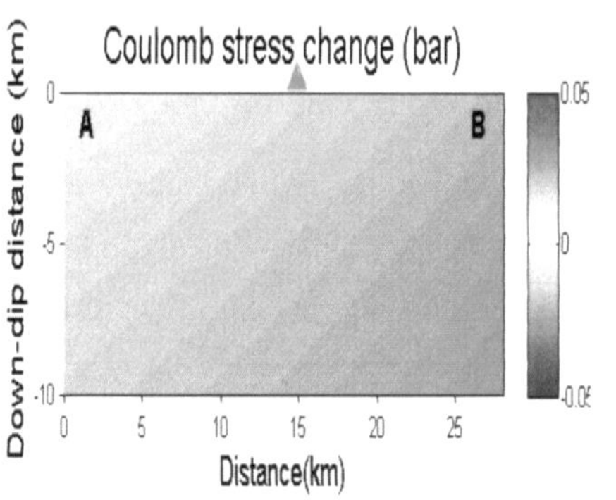

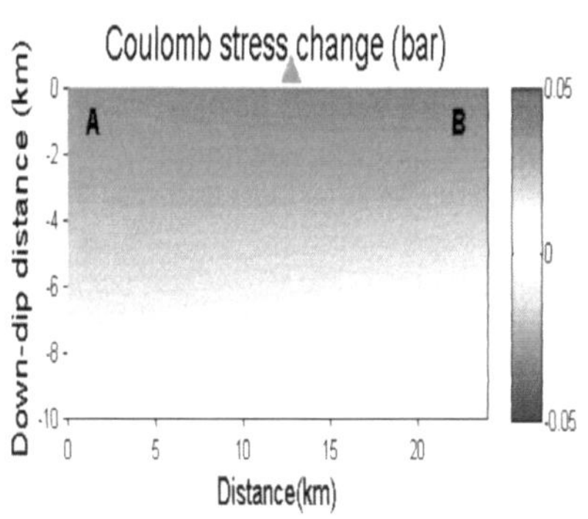

Figure 5. *Cross section* A-B *coulomb stress* change under mount Soputan

Figure 6. *Cross section* A-B *coulomb stress* change untuk mount Gamalama 2014 in depth 10 km

The cross section of the results obtained coulomb stress distribution changes under Soputan mountain to a depth of 10 km as shown in Figure 5. In the image cross section A-B in the area that increased stress working at a distance of 0 to 10 km. Rated positive Coulomb stress at a depth of 8 km is 0.023 bar.

The cross section of the results obtained coulomb stress distribution changes under Gamalama mountain to a depth of 10 km as shown in Figures 6. In the image cross section A-B in the area has increased coulomb stress. Value coulomb positive stress at a depth of 8 km is 0.007 bar.

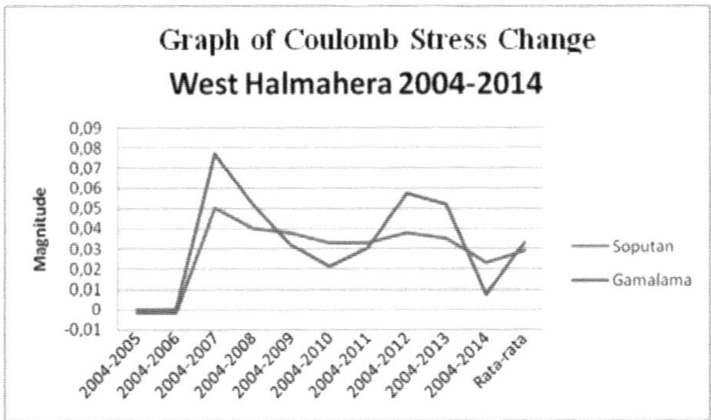

Figure 7. The coulomb stress Change graph on the mount Soputan and Mount Gamalama period 2004-2014

Based on Figure 7 it can be seen the value of coulomb stress on Mount Soputan and Mount Gamalama at 2004-2014. At the time the mountain Soputan increased coulomb stress the Gamalama mountain also increased coulomb stress. Likewise, when Mount Soputan decreased coulomb stress the Gamalama mountain coulomb stress also decreased. While the decline in stress coulomb lowest occurred in 2004-2005 due to the number of earthquakes is less than in later years.

For value coulomb stress Changes at a depth of 8 km on the mountain Soputan and Mount Gamalama in 2004-2014 respectively 0.023 bar and 0.007 bar. The value of coulomb stress is caused by the massive earthquake of Mw 7.2 that hit West Halmahera with a strike 42°, dip 35° and rake 110° in the opposite direction to that mountain areas and mountain Soputan Gamalama coulomb stress decreased to 0.023 bar and 0.007 bar.

Based on data, Mount Gamalama have increasing in volcanic activity 34 days after the mainshocks of West Halmahera Earthquake Mw 7.2, while Mount Soputan have increasing in volcanic activity 41 days after the mainshocks of West Halmahera Earthquake Mw 7.2. The Epicenter of west halmahera earthquake mainshocks Mw 7.2 was located ~156 km from mount Gamalama while mount Soputan located ~226 km from the epicenter of west halmahera earthquake mainshocks Mw 7.2.

Although the mountain Soputan and Mount Gamalama coulomb stress has value only 0.023 bar and 0.007 bar but indirectly affect volcanic activity in both the mountain. As a result of the increase in the value of the coulomb stress, physical conditions in the magma rising magma [6]. Among the increasing pressure of the gas bubbles of magma and magmatic system, sehungga causing cracks widen-fractures or the occurrence of new cracks in the rocks of the magma chamber wall and is characterized by volcano-tectonic earthquakes on Mount Soputan and Mount Gamalama. Based on data from PVMBG that when it mount Soputan and Mount Gamalama is experiencing the rainy season with moderate rainfall. Rainwater that seeps through the cracks-cracks of rocks and mixes with magma to form gas bubbles. The gas bubbles will increase the pressure in the magma chamber that previously had a change of dynamic stress. If the magma dome can not withstand the pressure from the inside due to the density of the walls of the dome was reduced because of rain, the magma dome had collapsed due to the gas pressure teperangkap in the magma chamber, causing phreatic eruption at Mount Soputan and phreatomagmatic on mountain Gamalama [6]. As a result of the earthquake that occurred in West Halmahera/Molucca sea or water it can indirectly affect volcanic activity is increasing to become status of alert on December 26th, 2014 on Mount Soputan and become status of alert on December 19th, 2014 on Mount Gamalama.

4. Conclusion

In accordance with the purpose of research, the results achieved are 1) Earthquake in West Halmahera indirectly affect mountain Soputan and Mount Gamalama due to long distances to be observed coulomb stress that can transfer stress and affect stress fracture in Soputan and Gamalama. 2) The method coulomb stress can see accumulate value of coulomb stress change correlation with adittion earthquake in surrounding West Halmahera. 3) Coulomb stress Changes that comes from the earthquake period November 2004-December 2014 including the west Halmahera 7.2 Mw earthquake happens Mount Soputan generate increased coulomb stress at a depth of 8 km is 0.023 bar while Gamalama mountain coulomb increased stress at a depth of 8 km is 0.007 bar resulted in a mountain of volcanic activity Soputan and Mount Gamalama increased caution status to become alert after ± 30 days a major earthquake occurs. Things were discovered from analysis used are 1) the results of this study can be concluded that the Molucca Sea and surrounding waters is a zone prone to earthquakes and volcanic activities, especially mountain Gamalama and Mount Soputan. 2) Based on the historical data from BMKG and Global CMT has earthquakes as much as 112 times the scale of magnitude 5.2 and 7.5 magnitude on the geographical area of 4.95 north latitude-1.05 south latitude and 123.49 east longitude-129.49 west longitude in the period November 2004 to December 2014. With this type of fault that occurred in the waters of West Halmahera dominant fault ride / reverse. 3) The distance and the large magnitude earthquake affecting the value and reach coulomb stress, and volcanic activity of Mount Soputan and Mount Gamalama.4) Coulomb stress Changes change on Mount Soputan has similarities with the coulomb stress changes on Gamalama mountain although the value is different. 5) The number of earthquakes that can be processed in coulomb 3.4 software just 127 so that only an earthquake with a magnitude 5.4 mw which can be processed.

5. Reference

[1] Anderson, E.M. 1951. The Dynamics of Faulting and Dyke Formation with Application to Britain. Oliver & Boyd, Edinburgh.

[2] Colin G Macpherson et ors, Geochemical Evolution of Magmatism in an Arc-Arc Collison: The Halmahera and Sangihe Arcs, Eastern Indonesia, in Robert D Larter, Ed(2003) Intra-Oceanic Subduction Systems, Geological Society of London, p208

[3]Dawid, S., Ferdy, Pasau G. 2015. Penentuan Lokasi Pergerakan Magma Gunung Api Soputan Berdasarkan Studi Sebaran Hiposenter Gempa Vulkanik Periode Mei 2013 – Mei 2014. Jurnal Ilmiah Sains Vol.15 No.2, oktober 2015

[4] Hall, R. 1998. The plate tectonics of Cenozoic SE Asia and the distribution of land and sea. in Hall, R. & J. D. Holloway (eds.). 1998. Biogeography and Geological Evolution of SE Asia. Backhuys Publisher, Leiden.

[5] Harris, R. A., and Simpson, R. W. (1992). Changes in static stress on southern California faults after the 1992 Landers earthquake, Nature, 360, 251-254.

[6] Hermawan Bagus Heru. Pengaruh Perubahan Stress Akibat Gempabumi di Selatan Pulau Jawa terhadap Peningkatan Aktivitas Vulkanik dan Erupsi Gunung Merapi periode Maret 2014. Universitas Gajah Mada. 2015

[7] Hudnut, K. W., Seeber, L., and Pacheco, J. (1989). Cross-fault triggering in the November 1987 Superstition Hills earthquake Sequence, southern California, Geophys. Res. Lett.,16, 199-202.

[8] Jaeger, J.C. and Cook, N.G.W., 1979. Fundamentals of rock mechanics. Chapman and Hall, London, 593 pp.

[9] Jaumé, S. C., and Sykes, L. R. (1992). Change in the state of stress on the southern San Andreas fault resulting from the California earthquake sequence of April to June 1992, Science, 258, 1325-1328.

[10] King G C P, Stein R S, and Lin J. 1994. Static Stress Changes and the Triggering of Earthquakes

: *Bulletin of the Seismological Society of America,* v.84(3) : 935-953.
[11] Larsen, S., Reilinger, R., Neugebauer, H., and Strange, W. (1992). Global Positioning System measurements of deformations associated with the 1987 Superstition Hills Earthquake: Evidence of conjugate faulting, J. Geophys. Res., 97, 4885-4902.
[12] Lay T, Terry C Wallace. 1995. *Modern Global Seismology*. Academic Press, California New York.
[13] Okada, Y. 1985. Surface deformation due to shear and tensile faults in a half-space. Bulletin of the Seismological Society of America 75, 1135Ð1154.
[14] Oppenheimer, D. H., Reasenberg, P. A., and Simpson, R. W. (1988). Fault plane solutionsfor the 1984 Morgan Hill, California, earthquake sequence: Evidence for the state of stress on the Calaveras fault, J. Geophys. Res., 93, 9007-9026.
[15] Reasenberg, P. A., and Simpson, R. W. (1992). Response of regional seismicity to the static stress change produced by the Loma Prieta earthquake, Science, 255, 1687-1690.
[16] Silver, E. A. & Moore, J. C. 1978. The Molucca Sea collision zone, Indonesia. Journal of Geophysical Research, 83, 1681-1691.
[17] Simpson, R. W., and Reasenberg, P. A. (1994). Static stress changes on central California faults produced by the Loma Prieta earthquake, U.S. Geol. Surv. Prof. Pap., in press
[18] Stein, R. S., and Lisowski, M. (1983). The 1979 Homestead Valley earthquake sequence, California: Control of aftershocks and postseismic deformation, J. Geophys. Res., 88, 6477-6490.
[19]Stein, R. S., King, G. C. P., and Lin, J. (1992). Change in failure stress on the southern San Andreas fault system caused by the 1992 Magnitude = 7.4 Landers earthquake, Science, 258, 1328-1332.

IOP Conf. Series: Earth and Environmental Science 56(2017) 012006 doi:10.1088/1755-1315/56/1/01200

Building vulnerability and human loss assessment in different earthquake intensity and time: a case study of the University of the Philippines, Los Baños (UPLB) Campus

I Rusydy[1,3,4*], D V Faustino-Eslava[2], U Muksin[3,6], R Gallardo-Zafra[5], J J C Aguirre[5], N C Bantayan[7], L Alam[4,8], and S Dakey[4,9]

[1] Geological Engineering Department, Syiah Kuala University, Jl. Syech Abdurrauf No. 7, Banda Aceh, 23111, Indonesia
[2]School of Environmental Science and Management, University of the Philippines, Los Baños, Pedro R. Sandoval Ave, Los Baños, Laguna, Philippines
[3]Tsunami and Disaster Mitigation Research Center (TDMRC), Syiah Kuala University, Jl. Prof. Dr. Ibrahim Hasan, Gampong Pie, Banda Aceh, 23233, Indonesia
[4]PARR Research Fellows at University of Philippines Los Baños, Pedro R. Sandoval Ave, Los Baños, Laguna, Philippines
[5]Department of Civil Engineering, University of Philippines Los Baños, Pedro R. Sandoval Ave, Los Baños, Laguna, Philippines
[6]Physic Department, Syiah Kuala University, Jl. Syech Abdurrauf No. 7, Banda Aceh, 23111, Indonesia
[7]College of Forestry and Natural Resources, University of the Philippines, Los Baños, Pedro R. Sandoval Ave, Los Baños, Laguna, Philippines
[8]LESTARI, Universiti Kebangsaan Malaysia (UKM), Kampung Bangi, 43600 Bangi, Selangor, Malaysia
[9]Visvesvaraya National Institute of Technology, South Ambazari Road, Nagpur, Maharashtra 440010, India

*email: ibnu@tdmrc.org

Abstract. Study on seismic hazard, building vulnerability and human loss assessment become substantial for building education institutions since the building are used by a lot of students, lecturers, researchers, and guests. The University of the Philippines, Los Banos (UPLB) located in an earthquake prone area. The earthquake could cause structural damage and injury of the UPLB community. We have conducted earthquake assessment in different magnitude and time to predict the posibility of ground shaking, building vulnerability and estimated the number of casualty of the UPLB community. The data preparation in this study includes the earthquake scenario modeling using Intensity Prediction Equations (IPEs) for shallow crustal shaking attenuation to produce intensity map of bedrock and surface. Earthquake model was generated from the segment IV and the segment X of the Valley Fault System (VFS). Building vulnerability of different type of building was calculated using fragility curve of the Philippines building. The population data for each building in various occupancy time, damage ratio, and injury ratio data were used to compute the number of casualties. The result reveals

IOP Conf. Series: Earth and Environmental Science 56 (2017) 012006 doi:10.1088/1755-1315/56/1/C

that earthquake model from the segment IV and the segment X of the VFS could generate earthquake intensity between 7.6 – 8.1 MMI in the UPLB campus. The 7.7 Mw earthquake (scenario I) from the segment IV could cause 32% - 51% damage of building and 6.5 Mw earthquake (scenario II) occurring in the segment X could cause 18% - 39% structural damage of UPLB buildings. If the earthquake occurs at 2 PM (day-time), it could injure 10.2% - 18.8% for the scenario I and could injure 7.2% - 15.6% of UPLB population in scenario II. The 5 Pm event, predicted will injure 5.1%-9.4% in the scenario I, and 3.6%-7.8% in scenario II. A nighttime event (2 Am) cause injury to students and guests who stay in dormitories. The earthquake is predicted to injure 13 - 66 students and guests in the scenario I and 9 - 47 people in the scenario II. To reduce the number of injuries, the authority of UPLB have to carry out the buildings restoration, set up earthquake early warning system in all dormitories, and evacuation sign board.

1. Introduction

The number of human losses when an earthquake occurred is related to the vulnerability of buildings to earthquakes. Another factor that could increase the number of losses is the number of people living or working in the buildings. Buildings that mostly inhibited by people are university or school buildings therefore it is important to study the vulnerability of buildings at university or schools. In this paper we analyzed the building vulnerability and human loss of the buildings of The University of The Philipines, Los Banos (UPLB) that used by students and faculty members of the university.

The University of The Philipines, Los Banos (UPLB) has several types of buildings and different use (academic, service, administration, and dormitories) and some of the buildings are more vulnerable than the other buildings when the earthquake occur. The Philippines has a lot of experiences with earthquakes which caused building damages and injured many people in various parts of the country. One of the greatest earthquakes is The 1990 Central Luzon earthquake in the magnitude of 7.6 Ms [1].

In order to estimate the number of injured people due to different magnitudes and time of earthquake occurrences, in this study we produce the seismic hazard map and building damage of the study area. The seismic hazard map was produced from the fault line which is close to study area. The Deterministic Seismic Hazard Analysis (DSHA) method was used generate earthquake model intensity or ground shaking model. Ground shaking is the primary factor that causes building damage compared to liquefaction, landslide, and seismic bearing capacity [2]. DSHA method used in this study is considered to be accurate since the tectonic setting of area is well studied [3]. The second process is to evaluate the building vulnerability based on the seismic performance in MMI (Modified Mercalli Intensity) of each building and fragility curve of different types of structure of structure [4]. The third step of the analysis is to estimate the number of injury of people working in the UPLB from the building vulnerability and the building inventory data (Detailed information of building and population) at University of The Philippines, Los Banos.

Previously in the Philippines, there is numerous study about earthquake assessment including, seismic hazards, damage assessments, simulations/scenarios, and site responds carried out by several researchers. The ground motion simulation technique was developed based on seismic hazard assessment from historical earthquakes and active faults in the Philippines [1]. The study produced the PGA (peak ground acceleration) map of the Philippines in expected event of 475 years recurrence (10% in 50 years) and occasional events of 100 years recurrence (39% in 50 years). The developed PGA map is a macro-zonation map and the acceleration in bedrock [1]. To develop the micro-zonation map, the site coefficient is needed. In Metro Manila, the earthquake damage estimation was conducted based on earthquake scenario and the seismic performance of the building was determined based on expert's judgment [5]. The building vulnerability calculated from the fragility curve which is varied in the different types of the buildings.

In this study, we develop the seismic hazard map in ground shaking model at the different sources of earthquakes, calculate the building vulnerability based on the fragility curve and estimate of injury number of UPLB communities for various magnitudes of earthquakes and occurrence time. The occurrence time of earthquakes used in this study is 2 Am for night-time scenario, 2 Pm and 5 Pm for the day-time scenario as suggested by FEMA [6].

2. Tectonic Setting of The Philippines

The Philippines located in one of the most active tectonic setting in Asia. Several numbers of destructive earthquakes have occurred in every part of the country. The Central Luzon earthquake in 1990 with a magnitude of M_S7.6 was the largest seismic event for Philippines people. Earthquakes in the Philippines could occur along the Philippines trench, the Manila trench, the Philippine Fault Zone (PFZ), and the Valley Fault System (VFS).

The Philippine trench, built by subduction of the western edge of the Philippine Plate penetrating beneath the Eurasia plate. This trench is the primary sources of the earthquakes [1]. The very seismically active Philippines trench extends from south of the Mindanao Island to the Luzon Island in the north for a distance of about 1400 [7]. The Manila Trench is a 1200-km long subduction zone between the South China Sea and the Northern Philippines plate which produced several mega earthquakes [8, 9]. Nguyen, et al., [9], developed the scenarios of the earthquakes for tsunami modeling and the seismic event were divided into six fault zones (the earthquake sources), and the worst case is the earthquake with magnitude of Mw 9.3.

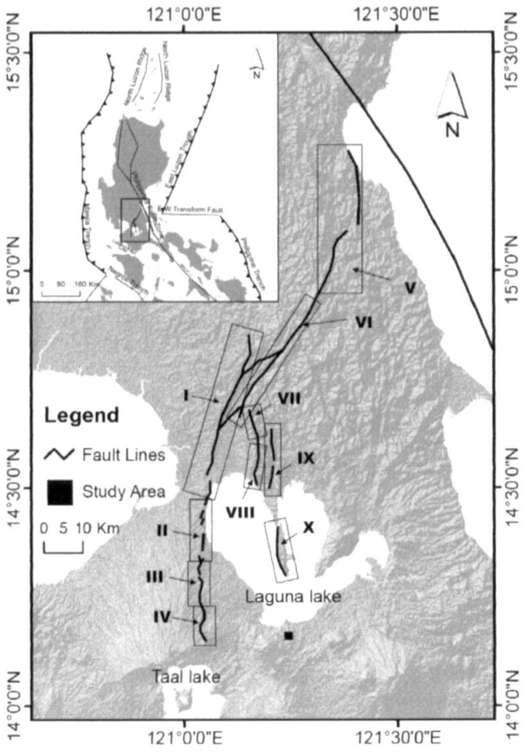

Figure 1. The fault system in the Philippines and the study area in the UPLB campus (black rectangle) and the tectonic map of Philippines, modified from [24, 25, 12], digital elevation model dataset is from NASA 2007

There are two on land sources of the earthquakes for the Philippines which are the Philippine Fault Zone (PFZ) and the Valley Fault System (VFS). The Philippine Fault Zone (PFZ) formed because of oblique motion of the Philippines Sea plate relative to the Eurasia plate [10, 11, 12]. PFZ is an active left lateral strike-slip fault system extending 1600-km from the Mindanao to the Luzon Island [12]. The 1990 Central Luzon earthquake was the most destructive earthquake occurred in the PFZ. The Valley Fault System (VFS) is the nearest fault to study area which is around 20-50 km. The VFS is a 130 km right lateral strike-slip fault system beginning from south of Canlubang, Calamba, Laguna to the Umiray and the Dingalan Bay in the north in NNE (North North East) of the strike [12]. The VFS is divided into two major fault systems, the East Valley Fault System (EVFS) and the West Valley Fault System (WVFS). The WVFS is divided into four minor segments, and the EVFS is divided into six minor segments (see Figure 1).

The Pittland–Sungay segment IV of the WVFS and the Talim Island segment of the EVFS are the imminent fault to UPLB with the length of about 20 – 30 km and both segments is recognized as the active faults. The magnitude of earthquake along this fault system have been estimated based on empirical relationships among surface displacement, rupture length, and magnitude within a range of 6.1 to 7.5 Mw (empirical relationship based on length) and in the range of 7.4 to 7.7 Mw (empirical relationship based on displacement) [12]. The earthquake estimation of each segment is shown in Table 1.

Table 1. Magnitude estimates in EVFS and WVFS based on length of segments and maximum displacement [12], see Figure 1 for their location.

Fault Segments	Length (Km)	Est. Mag. based on length (Mw)	Est. Mag. based on displacement (Mw)
Segment I (Rodriguez–Taguig)	45	7.0	7.7
Segment II (Sucat–Binan)	14	6.4	-
Segment III (Binan–Sto. Domingo)	12	6.4	7.6
Segment IV (Pittland–Sungay)	7.5	6.1	7.7
Segment V (Umiray)	35	6.9	-
Segment VI (San Mateo-Rodriguez)	37	6.9	7.4
Segment VII (Antipolo)	9	6.2	-
Segment VIII (Angono)	10	6.3	-
Segment IX (Binangonan)	13	6.4	-
Segment X (Talim Island)	14.5	6.5	-

The sources of the earthquake models were generated from Segment IV and Segment X and ground shaking models were produced in MMI. The ground shaking or intensity models were highly influenced by local geology condition or site coefficient. Different site coefficient has different site respond to the earthquake model and it is also influenced by soil properties or geotechnical condition underlying the buildings [13, 14, 15].

3. Methodology

3.1. Data preparation

In this study, the earthquake model was generated into two scenarios in a precise way and the model refers to the previous study about the seismicity and the tectonic setting [12]. The ground shaking or intensity scenario was developed using the Intensity Prediction Equations (IPEs) for global application.

This new global shaking attenuation was developed for shallow crustal and near earthquake [16]. The building vulnerability or damage ratio estimation in different types of building was calculated based on fragility curve [4]. The fragility curve is a function of the intensity of the earthquake in MMI. The flow chart of this study is shown in Figure 2.

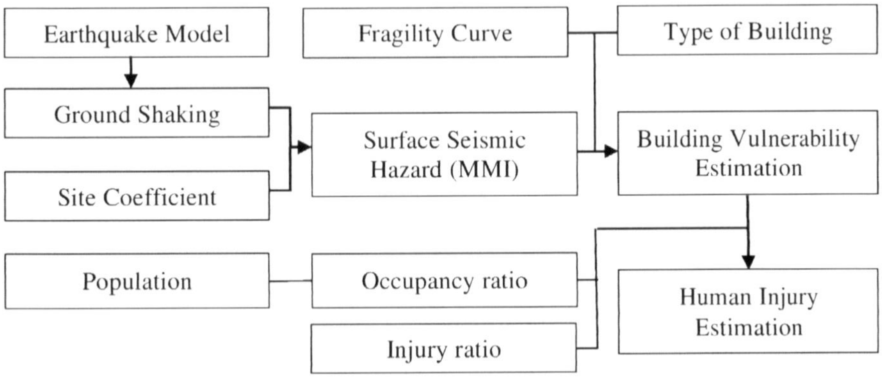

Figure 2. The Flowchart of earhquaake hazard, building vulnerability and human injury estimation in several earthquake scenarios

3.2. Earthquake Model

The generated earthquake model depends on the shaking attenuation that is influenced by physical properties of structures and experimental data [16]. Shaking attenuation is an empiric approach to determine ground motion in PGA (peak ground acceleration) or intensity of the earthquake in MMI (Modified Mercalli Intensity). The latest seismological approach employs the intensity values for efficiently communicating seismic hazard and risk information to the public and media [16]. The different tectonic setting has different shaking attenuation and the choice of what shaking attenuation will be applied in the study based on the similarity of tectonic setting and the source of the earthquake. In this study we used the shaking attenuation [16] to produce earthquake model from the segment IV and the segment X. The shaking attenuation particular designed [16] to compute the intensity in the shallow active tectonic crust in the range of magnitude of Mw 5 - 7.9. This shaking attenuation named as the Intensity Prediction Equations (IPEs) for global application.

The IPEs method was developed using 13,077 earthquakes around the world since 1973 and 1,613 from Asia countries and some of the data taken from the historical earthquake in the Philippines. The the intensity of the earthquake in every distance of rupture zone could be determined by:

$$I\left(M, R_{rup}\right) = c_0 + c_1 M + c_2 \ln \sqrt{R_{rup}^2 + [1 + c_3 e^{(M-5)}]^2} + S \tag{1}$$

where I is intensity in MMI, M is magnitude of the earthquake in Mw, R_{rup} is the distance to source of the earthquake in Km, c_0 is a coefficient number of 3.950, c_1: 0.913, c_2: −1.107, c_3: 0.813. The site correction S is related to the site amplification representing the amplification factor of the local soil expressed by:

$$S = 3.18 \log (F) \tag{2}$$

The earthquake intensity models that are generated from the segment IV and X are the intensity in bedrock without any consideration of site coefficient or site amplification. The site coefficient and the site characteristic of the soil can be determined accurately from shear wave velocity (Vs) and Horizontal Vertical Spectral Ratio (HVSR). However, those parameters also can be estimated from Standard Penetrating Test (SPT), Cone Penetrating Test (CPT), Dilatometer Test (DMT), and Pressure Meter Test (PMT) [13]. In UPLB campus, the Vs data is not available thus we used the N-SPT data and classified it based on the soil classification as suggested by ASCE/SEI 7-10 [17]. In this study we used the previously calculated PGA map for the University of Philippines, Los Banos from Torregosa, et al [1] and Aguirre [18] in the range between 0.2 to 0.4 g. Based on N-SPT data, the type of soil in UPLB campus is considered as soft soil type E (SE).

Table 2. Site coefficient (F_a) for five different soil classifications based on ASCE/SEI 7-10 [17] for PGA map of 0.4 g.

Site Class	S_A	S_B	S_C	S_D	S_E
N-SPT	NA	NA	>50	15 to 50	<15
F_a	0.80	1.00	1.20	1.40	1.72

3.3. Human Injury Assessment

The UPLB campus is located 16 km from the segment X and 24 km from the segment IV (see Figure.1). The nearest and shallow crustal earthquake could cause the highest impact to the building in term of damage and casualty [16]. Different type of building has different resistance to ground shaking thus will increase the value of building vulnerability and injury ratio. We refer to fragility curve of the Philippine building in MMI determined previously [4]. The type of building in determination of the fragility curve is similar to that proposed by FEMA [6].

The fragility curve shows the damage ratio or building vulnerability of the structure on a scale of 0 to 1 toward ground shaking [4]. In this study, we used this fragility curve for different types of buildings in the UPLB campus. The main reason we used this fragility curve is because it was developed for buildings in the Philippines (Figure 3).

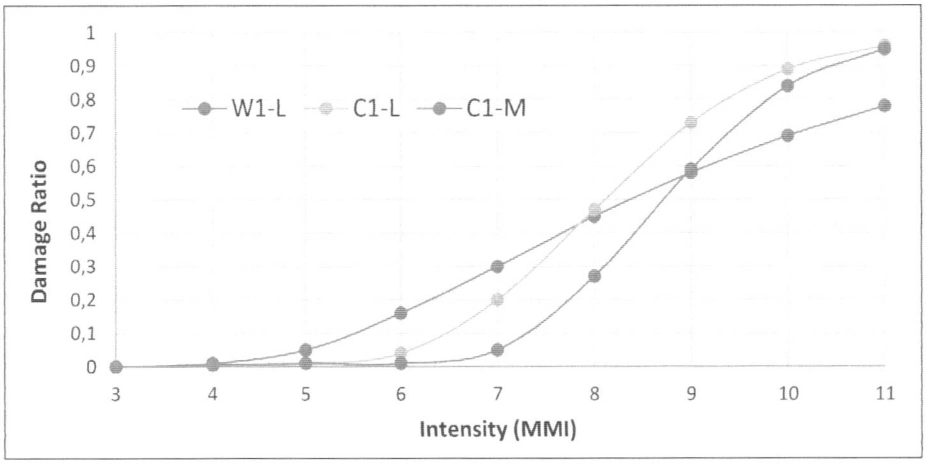

Figure 3. Fragility Curve of 3 types of buildings in the UPLB campus as a function of intensity in MMI [4]

Valino [19] and Arellano [20] had conducted the building inventory classification of the lower-campus and the upper-campus of the UPLB. They classified 67 buildings in UPLB Lower Campus into two types of building: (1) Low-Rise Concrete moment frame (C1-L) and (2) Medium Rise Concrete Moment Frame (C1-M). We updated the inventory building data from 67 to 145 buildings based on a previous study conducted by Valino [19] and Arellano [20] but we only focused on UPLB lower-campus. The types of buildings in the UPLB lower campus were divided into three type: (1) Low-Rise Concrete moment frame (C1-L), (2) Medium Rise Concrete Moment Frame (C1-M), and (3) Low-Rise Light wood-frame (W1-L). To determine the building vulnerability in different types of buildings we refer to the fragility curve in Figure (3). Building vulnerability of each building results in different number of casualties because the people is considered to be trapped in collapsed or heavily damaged buildings [21]. There are many equations were used to calculate the casualty ratio. In this study, we used equation (3) which was proposed by Hashemi & Alesheikh [21]. The equation computes the injured ratio separately based on occupancy of the buildings at the earthquake occurrence time, damage ratio or building vulnerability, population of each building, and injury ratio as follows:

$$Injured \quad = Occupancy\ ratio\ x\ population\ x\ damage\ ratio\ x\ injury\ ratio \qquad (3)$$

At the night-time (2 Am), the occupancy ratio was assumed as 1 for dormitories because at this time all students are sleeping in the dormitories but considered as 0 for offices and classroom building in the UPLB campus. For the day-time (2 Pm) scenario, the occupancy ratio in offices, classroom and dormitories is 1. At 5 pm, most of the class were dismissed, and the staff in the offices would already be out of the office thusthe occupancy ratio at this time is predicted of 0.5. The population data of the offices and the classrooms were uncalculated and we had no access to the data. The data of the student population in dormitories was obtained from the authority of dormitories and the Programmatic Environmental Performance Report and Management Plan [22] as shown in Table 3. The avoid the unavailability of population data, we used percentage for the population of all office buildings (instead of number) thus the casualty or injury estimation result is also shown in percentage (%). The worst scenario of the injury ratio refers to Hashemi & Alesheikh [21] which was developed for the Teheran city. Although the injury ratio developed by Hashemi & Alesheikh [21] as shown in Table 4 is considered to be relatively high, but the final result of injury could be adjusted by the damage ratio and the occupancy ratio.

Table 3. The student population at dormitories in the UPLB campus. The data was obtained from Programmatic Environmental Performance Report and Management Plan [22] and building authority.

No.	Dormitories	Population	Source of Data
1.	Male Residence Hall	646	PEPRMP 2014
2.	Female Residence Hall	444	PEPRMP 2014
3.	Veterinary Medicine Residence Hall	376	Authority of Dormitory
4.	New Dormitory	294	Authority of Dormitory
5.	ACCI	90	PEPRMP 2014
6.	SEARCA Residence Hotel	90	Authority of SEARCA
7.	ATI-NTC	126	Authority of Dormitory
8.	International House	124	PEPRMP 2014

Table 4. The worse scenario of injury ratio for different type of building proposed by Hashemi & Alesheikh [21]

No.	**Type of Building**	**Injury Ratio**
1.	RC Building > 4 stories (C1-M)	0
2.	RC Building < 4 stories (C1-L)	0.2
3.	Timber Frame Building (W1-L)	0.4

4. Result

4.1. Earthquake hazard model

This study develops two earthquake model at three different times of events: 2 pm, 5 pm, and 2 am. The earthquake models which were calculated using IPEs equation is shown in Figure 3. This result has no consideration of site coefficient. Equation (2) was used to calculate the site coefficient and used as the input of equation (1) to determine the surface seismic intensity. According to ASCE /SEI 7-10 [17] in Table 2, the type of soil E produces the site coefficient of 1.72. Table 5 shows the total intensity on the surface after calculated by using formula (1) and (2) in two different magnitudes and sources of the earthquakes. The site coefficient of 1.72 was assumed to have the same effect to all buildings in the UPLB campus since the UPLB campus was built on nearly homogenously pyroclastic deposits.

The earthquake in scenario I occur along the segment IV (Pittland–Sungay) with magnitude of Mw 7.7 and the earthquake in scenario II occur along the segment X (the Talim Island) with magnitude of 6.5 Mw. The scenario I is the worst scenario for UPLB campus because it is possible to produce the intensity of about 8.1 on the surface. It is noted that the intensity of 8 able to destroy masonry building type B [23]. The Masonry building (type B) was built with excellent workmanship and mortar; reinforced, but was not designed for lateral forces resistance. The intensity in scenario II have similarity of destructive to the Masonry building type B.

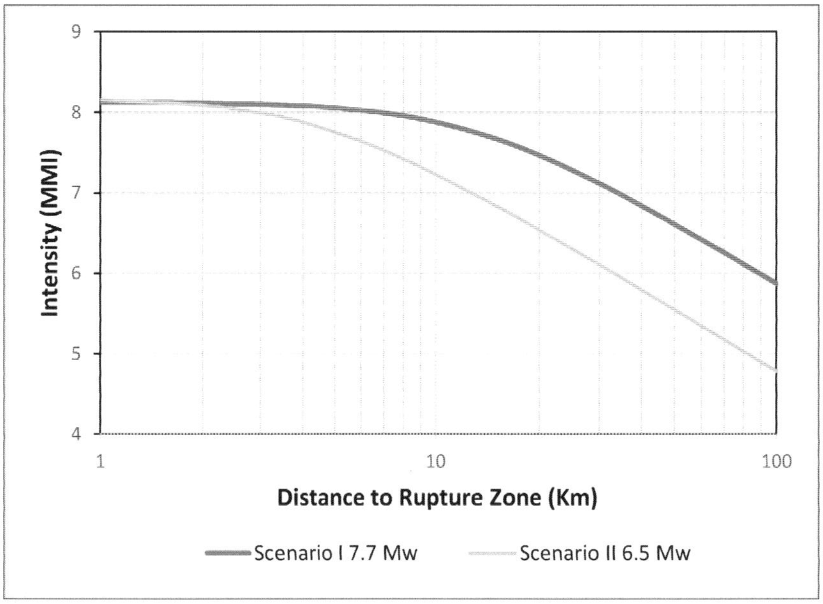

Figure 4. Predicted intensity of the R_{rup} using IPEs for Mw 7.1 and Mw 6.5 earthquakes and without consideration of site coefficient.

Table 5. Earthquake scenarios and ground shaking respond in bedrock and surface

Earthquake Scenario	Magnitude (M)	Distance (R_{rup})	Intensity at Bedrock	S	Intensity at Surface
I	7.7 Mw	24 Km	7.3	0.82	8.1

Earthquake Scenario	Magnitude (M)	Distance (R_{rup})	Intensity at Bedrock	S	Intensity at Surface
II	6.5 Mw	16 Km	6.8	0.82	7.6

4.2. Building Vulnerability

Using the fragility curve in Figure 3, we were able to calculate the damage ratio in a different types of buildings which depent on the surface intensity as shown in Table 5. The magnitude of Mw 7.7 in the scenario I tear the building in a ratio range of 0.32-0.51. This ratio mean, around 32% to 51% of the UPLB buildings in Lower-campus was estimated to be destroyed due to Mw 7.7 earthquake from the segment IV of the Valley Fault System (VFS). This earthquake damages around 51% of the Low-Rise Concrete moment frame (C1-L) building, 47% of the building types of Low-Rise Light wood-frame (W1-L) and 32% of the building types of Medium Rise Concrete Moment Frame (C1-M) as shown in Table 6.

Table 6. Damage ratio in a different type of building and scenarios.

Type of building	Number of Buildings	Damage Percent (%)	
		Scenario I	Scenario II
C1-M	17	32	18
C1-L	126	51	36
W1-L	12	47	39

The building with Low-Rise Concrete Moment Frame (C1-L) is the most vulnerable building in UPLB lower-campus where 87% of total building in UPLB lower-campus is C1-L. An earthquake with magnitude of Mw 7.7 from the segment IV is worst scenario. The earthquake could damage 51% of almost all building in the UPLB lower-campus. Most of C1-L buildings in the UPBL campus are old buildings and constructed before 1977 and between 1978 and 2000 [19]. Therefore, it is necessary to revitalize the buildings in the UPBL campus. The damage ratio of C1-L and W1-L types of buildings in scenario II are considered to be lower than that of scenario I.

4.3. Human Casualty Analysis

The first step in human casualty analysis is to determine precisenumber of population, damage ratio, and intensity of the earthquake model. In the case of intensity map, without any detailed knowledge of how the building responds to the ground shaking at different periods, this method could potentially add more uncertainty to loss estimation [16]. In this study at the first stage, we develop the intensity map model using IPEs equation, the population in the building, and damage ratio as precise as possible. The human casualty analysis caused by two earthquake scenarios was applied into three times: 2 pm, 5 pm and 2 am as suggested by FEMA [6].

4.3.1. Daytime (2 Pm) Earthquake Event

Daytime at 2 pm is a busiest time during a day and at this particular time many students are in classrooms and administrative staff are in their office. The number casualty is calculated by using equation (3). The result shown in Table 7 describes that if the students and staffs have no adequate access to evacuation zone, around 10.2% students or officers will be injured in C1-L buildings, 0% in C1-M building, and 18.8% in W1-L buildings for scenario I. For scenario II, the intensity of earthquake is predicted of 7.6 MMI and it would effect to 7.2% injuries in C1-L buildings, 0% in C1-M buildings, and 15.6% in W1-L buildings.

Table 7. The percentage of injury in different type of building at 2 Pm of earthquake event

Type of building	Injury in Percent (%)	
	Scenario I	Scenario II
C1-M	0	0
C1-L	10.2	7.2
W1-L	18.8	15.6

4.3.2. Daytime (5 Pm) Earthquake Event

At 5 pm, the students are mostly outside the building and most of the staff about to going out from the office. At this time period, a half of students stay in dormitories and half of staff are working in the office or the occupancy ratio is 0.5. If an earthquake occurs at this time, the estimated injuries would be 5.1% of population in C1-L building, 0% of population of C1-M building and 9.4% of population in W1-L building for scenario I. For scenario II, the percentange population will injure are 3.6% in C1-L building, 0% in C1-M building and 7.8% in W1-L building. The detailed estimation of injuries is shown in Table 8.

Table 8. The percentage of injury in different type of building at 5 Pm of earthquake event

Type of building	Injury in Percent (%)	
	Scenario I	Scenario II
C1-M	0	0
C1-L	5.1	3.6
W1-L	9.4	7.8

4.3.3. Nighttime (2 Am) Earthquake Event

If earthquake occur at 2 am, all student are expected to stay in dormitories and none of students are in the office. In this time scenario, we are able to calculate the number of injury since the population data of residences is available as shown in Table 3. The occupancy ratio at 2 am is 1 for all residences and 0 for offices. There are 8 residential accommodated by the students and guests of the UPLB. In all two scenarios of the earthquakes, the Male Residence Hall is highly vulnerable and we estimate the number of injuries of 66 students in scenario I, and 47 students in scenario II. SEARCA Residence Hotel and New Dormitory is less vulnerable. The number of injured people in UPLB at night-time event is shown in Table 9.

Table 9. The number of injury at university residential at 2 Am of earthquake event

No.	Dormitories	Type of Building	Peak Population	Number of Injury in different scenarios	
				Scen. I	Scen. II
1.	Man Residence Hall	C1-L	646	66	47
2.	Women Residence Hall	C1-L	444	45	32
3.	Veterinary Medicine Residence Hall	C1-L	376	38	27
4.	New Dormitory	C1-M	294	0	0
5.	ACCI	C1-L	90	9	6
6.	SEARCA Residence Hotel	C1-M	90	0	0

No.	Dormitories	Type of Building	Peak Population	Number of Injury in different scenarios	
				Scen. I	Scen. II
7.	ATI-NTC	C1-L	126	13	9
8.	International House	C1-L	124	13	9

5. Conclusion

This study has successfully built earthquake model and predicted the worst scenario for UPLB communities during a different time when an earthquake occurred. Earthquake shaking model in MMI was developed using global applicable macro-seismic Intensity Prediction Equations (IPEs). Even though it is a global shaking attenuation, but the equation was purposed for a shallow crustal earthquake or inland earthquake. In this study, the earthquake model was produced along the Valley Fault System (VFS) which is a shallow crustal earthquake.

The UPLB campus has different types of buildings and all buildings in the UPLB campus are divided into three types: C1-L, C1-M and W1-L. Different type of building and ground shaking will respond in various damage ratios. In a worst scenario in scenario I, 126 of C1-L buildings have 51% of damage and cause 10.2% population injury at day-time. The earthquake occurring at night-time (2 Am) is expected to injure 184 students and guests in worst scenario. To avoid this condition, the authority of UPLB have to conduct the restoration of the buildings and set up earthquake early warning system all dormitories, evacuation sign board and road.

In future, some suggestions have to put on top to improve this study in the UPLB in particular and the Laguna province in general. The respond of the building toward the ground shaking depends on the site coefficient of soil beneath the building. The site coefficient plays a crucial factor in ground shaking intensity and has to be determined carefully. To improve the result of this study, the GIS analysis is recommended to be performed in future to determine the distribution of damage building.

Acknowledgments

Authors would like to acknowledge the assistance and the encouragement to the research assistant (RA) of UPLB, for technical support during the stay at the University of The Philippines, Los Baños (UPLB). This study will be impossible to conduct without the funding support from the START International Inc., in Pan-Asia Risk Reduction (PARR) fellowship program 2016 – 2017.

6. References

[1] R. F. Torregosa, M. Sugito and N. Nobuoto, "Strong Motion Simulation for the Philippines Based on Seismic Hazard Assessment," *Journal of Natural Disaster Science*, vol. 23, pp. 35-51, 2001.

[2] S. F. Cinicioglu, I. Bozbey, S. Oztoprak and M. K. Kelesoglu, "An integrated earthquake damage assessment methodology and its application for two districts in Istanbul, Turkey," *Engineering Geology*, vol. 94, p. 145–165, 2007.

[3] S. Karimzadeh, M. Miyajima, R. Hassanzadeh, R. Amiraslanzadeh and B. Kamel, "A GIS-based Seismic Hazard, Building Vulnerability and Human Loss Assessment for The Earthquake Scenario in Tabriz," *Soil Dynamics and Earthquake Engineering*, vol. 66, p. 263–280, 2014.

[4] E. Tingatinga, B. Pacheco, J. Hernandez Jr, U. Ignacio, P. Castro, F. Germar, R. Longalong, W. Mata, M. Pascua, R. Suiza and L. Tan, "Development of Vulnerability Curves of Key Building types to Different Hazard in Philippines," 1 July 2013. [Online]. Available: http://icrm.ntu.edu.sg/NewsnEvents/Doc/Documents/GEM-SEA/[Tingatinga]%20GEM%20Singapore%20workshop%202013.07.01.pdf. [Accessed 27 September 2016].

[5] H. Miura, S. Midorikawa, K. Fujimoto, B. M. Pacheco and H. Yamanaka, "Earthquake damage

estimation in Metro Manila, Philippines based on seismic performance of buildings evaluated by local experts' judgments," *Soil Dynamics and Earthquake Engineering,* vol. 28, p. 764–777, 2008.

[6] FEMA, "Multi-hazard Loss estimation Methodology, Earthquake Model (HAZUS-MH)," Department of Homeland Security Emergency Preparedness and Response Directorate, Washington, D.C, 2010.

[7] G. Galgana, M. Hamburger, R. McCaffrey, E. Corpuz and Q. Chen, "Analysis of Crustal Deformation in Luzon, Philippines using Geodetic Observations and Earthquake Focal Mechanisms," *Tectonophysics,* vol. 432, p. 63–87, 2007.

[8] D. E. Hayes and S. D. Lewis, "Structure and Tectonics of The Manila Trench System, Western Luzon, Philippines," *Energy,* vol. 10, pp. 263-279, 1985.

[9] P. H. Nguyen, Q. C. Bui, P. H. Vu and T. T. Pham, "Scenario-based Tsunami Hazard Assessment for The Coast of Vietnam from The Manila Trench Source," *Physics of the Earth and Planetary Interiors,* vol. 236, p. 95–108, 2014.

[10] H. Rowlett and J. A. Kelleher, "Evolving seismic and tectonic patterns along the Western margin of the Philippine Sea Plate," *J. Geophys. Res.,* vol. 81, p. 3518–3524, 1976.

[11] G. M. Besana and M. Ando, "The central Philippine Fault Zone: Location of Great Earthquakes, Slow Events and Creep Activity," *Earth Planets Space,* vol. 57, p. 987–994, 2005.

[12] R. E. Rimando and P. L. Knuepfer, "Neotectonics of the Marikina Valley Fault System (MVFS) and Tectonic Framework of Structures in Northern and Central Luzon, Philippines," *Tectonophysics,* vol. 415, p. 17 – 38, 2006.

[13] M. K. Akin, S. L. Kramer and T. Topal, "Empirical correlations of shear wave velocity (Vs) and penetration resistance (SPT-N) for different soils in an earthquake-prone area (Erbaa-Turkey)," *Engineering Geology,* vol. 119, p. 1–17, 2011.

[14] A. M. Mohamed, H. A. Hafiez and M. Taha, "Estimating the near-surface site response to mitigate earthquake disasters at the October 6th city, Egypt, using HVSR and seismic techniques," *NRIAG Journal of Astronomy and Geophysics,* vol. 2, p. 146–165, 2013.

[15] S. Fabbrocino, G. Lanzano, G. Forte, F. S. d. Magistris and G. Fabbrocino, "SPT blow count vs. shear wave velocity relationship in the structurally complex formations of the Molise Region (Italy)," *Engineering Geology,* vol. 187, p. 84–97, 2015.

[16] T. I. Allen, D. J. Wald and C. B. Worden, "Intensity attenuation for active crustal regions," *J Seismol,* vol. 16, p. 409–433, 2012.

[17] ASCE/SEI 7-10, "Minimum Design Loads for Buildings and Other Structures," American Society of Civil Engineers, Virginia, 2010.

[18] J. J. C. Aguirre, "Probabilistic Seismic Hazard Analysis of Laguna, Philippines," University Of The Philippines – Los Baños, Laguna, 2013.

[19] M. Q. M. Valino, "Rapid Visual Seismic Vulnerability Assessment of University of the Philippines Los Baños Lower Campus School and Administration Buildings of Selected Colleges," University of the Philippines, Los Baños, Laguna, 2014.

[20] C. R. P. Arellano, "Earthquake Loss Estimation For Buildings Of UPLB," University Of The Philippines – Los Baños, Laguna, 2016.

[21] M. Hashemi and A. A. Alesheikh, "A GIS-based Earthquake Damage Assessment and Settlement Methodology," *Soil Dynamics and Earthquake Engineering,* vol. 31, p. 1607–1617, 2011.

[22] PEPRMP, Programmatic Environmental Performance Report and Management Plan, Los Banos: University of The Philippines, Los Banos, 2014, p. 133.

[23] C. F. Richter, Elementary Seismology, London: Bailey Bros & Swinfen Ltd, 1958.

[24] M. L. P. Bautista and K. Oike, "Estimation of the magnitudes and epicenters of Philippine historical earthquakes," *Tectonophysics,* vol. 317, p. 137–169, 2000.

[25] JICA, MMDA, PHIVOLCS, "Earthquake Impact Reduction Study for Metropolitan Manila, Republic of Philippines," JICA, Tokyo, 2004.

Advantages of Realistic Model Based on computational method: NDSHA versus Standard PSHA

I Irwandi[1,2,3]

[1]Tsunami and Disaster Mitigation Research Center (TDMRC), Syiah Kuala University, Jl. Prof. Dr. Ibrahim Hasan, Gampong Pie, Banda Aceh, 23233, Indonesia
[2]Physics Department, Syiah Kuala University, Jl. Syech Abdurrauf, Banda Aceh, 23111, Indonesia
[3]Geophysics Engineering Department, Syiah Kuala University, Jl. Syech Abdurrauf, Banda Aceh, 23111, Indonesia

*email: irwandi@unsyiah.ac.id

Abstract. Procedure of standard Seismic Hazard Assessment (PSHA) has a problem with over simplifying recurrence since being represented by a linear relationship. However, the relationship will be satisfied only if the size of the study area is large enough with respect to linear dimensions of sources.. PSHA lies in attenuation relations are usually not translation invariant in the phase space. Regarding the problem of completeness data, instead of using recurrence relationship DSHA select the most credible earthquakes. However, the DSHA remain lies in attenuation relations, assume the same propagation model for all the events, but such a hypothesis is not very realistic. On the contrary, NDSHA procedure has advantages in calculation strong ground motion from the realistic model of synthetic seismograms from source specific properties and cooperates with the available structural model. Additionally, The NDSHA produced more information such as PGD, PGV, PGA for horizontal and vertical components each. Using realistic computation for Banda Aceh, NDSHA provides an accurate value for each of those components, achieving probability of exceedance in the range between 10% to 2% probability of exceedance PGA from PSHA computation. Regarding some limitation from PSHA, Indonesia needs to establish research on NDSHA for the area has critical infrastructures to face the seismic hazard.

1. Introduction

Recent destructive earthquakes in Sumatra, such as Sumatra Andaman (2004), Nias (2005), Mentawai (2007), Padang (2009), Indian Ocean Strike Slip (2012), and Takengon (2013), provided new challenges for urgent action to improve a procedure for seismic hazard assessment (SHA). Realistic and accurate SHA will benefit the community and scientific studies. Moreover, this research may improve the SHA results in previous studies (Panza, Irikura, et al., 2011; Stein, Geller, & Liu, 2011, Panza, et al. 2012).

The mapping of the expected earthquake ground motion that accounts for event's recurrence – e.g., the standard Probabilistic Seismic Hazard Assessment (PSHA) – may be suitable for insurances in dealing with economical profit. When dealing with critical structures, where it is necessary to consider extremely long time intervals, the standard PSHA estimates are by far unsuitable, due to their basic heuristic limitations (Panza ,et al. 2012).

A thorough discussion of different approaches available for SHA is behind the purposes of the present review. In fact, the PSHA method (Bommer & Abrahamson, 2006 and Cornell, 1968) and its applications are widely described in the literature, and have been the subject of intensive debate in recent years (e.g., Bilham, 2009; Klugel, 2007a). Criticisms have been expressed on the probabilistic method by many authors (e.g., Castanos & Lomnitz, 2002; Klugel, 2007b; Krinitzsky, 1998; Wang, 2011, Peresan & Nekrasova 2014), who evidenced some essential limits in the physical and mathematical models, as well as in other basic assumptions. Several issues related to the PSHA approach and possible alternatives are discussed in the recent Topical Volume of Pure and Applied Geophysics (Panza, Irikura, et al., 2011). A posteriori quantitative evaluation of the performances of the PSHA is given by Kossobokov and Nekrasova (2010, 2011). They showed that the worldwide maps resulting from the Global Seismic Hazard Assessment Program, GSHAP (Giardini, Grunthal, Shedlock, & Zhang, 1999), are grossly misleading, as proved by fatal evidence of the deadliest earthquakes occurred since the year 2000, selected event for Sumatra are shown in Table 1.

Why are the standard probabilistic methods of estimating seismic hazard and risks too often wrong? A simple answer exists to the question in the title of this chapter: most, if not all, the standard probabilistic methods to assess seismic hazard, namely PSHA, and associated risks are based on subjective, commonly unrealistic, and even erroneous assumptions about seismic recurrence (Panza et al., 2014).

Table 1. List of the deadliest earthquakes occurred during the period 2000-2011, and the corresponding intensity differences, $\Delta I_0 = I_0M - I_0(mPGA)$, among the observed values and predicted by GSHAP. $I_0(M)$ and $I_0(mPGA)$ are computed from the observed magnitude M and the maximum GSHAP PGA around the observed epicenter, see (Panza et al., 2013, Kossobokov and Nekrasova, 2010).

Region	Date	M	Fatalities	Intensity difference ΔI_0
Sumatra-Andaman "Indian Ocean Disaster"	26.12.2004	9.0	227898	4.0
Nias (Sumatra, Indonesia)	28.03.2005	8.6	1313	3.3
Padang (Southern Sumatra, Indonesia)	30.09.2009	7.5	1117	1.8

2. Method

Seismic hazard assessment procedures generally can be divide into four steps as shown in Figure 1. Each method (PSHA, DSHA, and Neo-DSHA) is starting from scenario of seismic source, earthquake locations in space lead to a distribution of location. Figure 1a schematically shows the procedure for conduction a PSHA. The figure illustrates the specific hazard of ground motion, but it could be redrawn in like fashion for any other earthquake hazard. The step divides the earthquake threat into sources (these may be identified faults or geographical areas) that produce earthquakes and earthquake characteristics. It is assumed that earthquakes act independently – that is, the occurrence of an event at one source does not affect the occurrence of events at other source (McGuire, 2004).

The second step of the PSHA method is recurrence, which can be represented by a linear relation, and this will produce useful information for risk analysis. However, the relation will be satisfied only if the size of the study area is large with respect to linear dimensions of the sources. The third step of the PSHA method lies in attenuation relations, which are usually not translation invariant in the phase space (M, R, S), M – magnitude; R – source distance; S - local soil conditions, i.e. the relative decay is independent from (M, R, S). Even when the attenuation relations are translation invariant, they are not a conditional probability density function, but they are those that represent the functional dependency of the random spectral acceleration on the random variates, magnitude, distance and measurement error (Klügel, ENGEO 90, 186-192, 2007).

Regarding the issue with the data completeness, instead of using recurrence relations as mentioned in the second step of the DSHA method, this study selects the most credible earthquakes fixed source distance R and magnitude M as shown in Figure 1b. However, the DSHA method remains in attenuation relations, which often emphatically called laws, by assuming the same propagation model for all of the events, but such a hypothesis is not very realistic (Panza, 2010). Therefore, this study produces DSHA for Sumatra after Natawidjaja & Triyoso (2007).

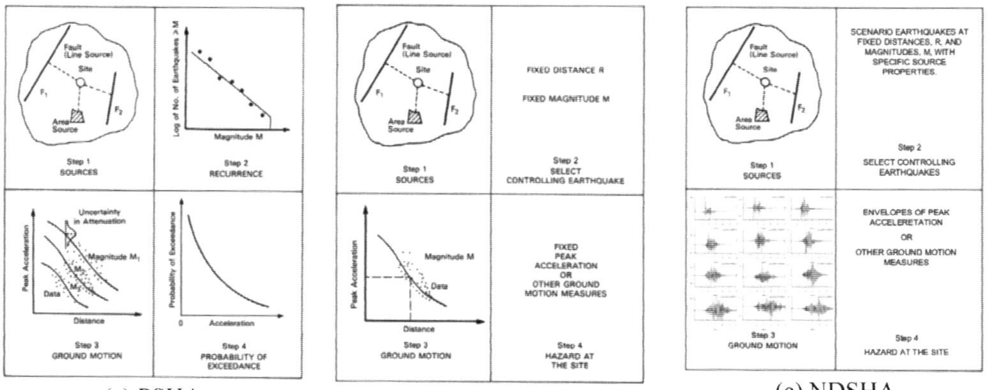

(a) PSHA (b) DSHA (c) NDSHA

Figure 1. The comparison steps in performing PSHA, DSHA, and NDSHA Panza (2010)

Regarding the issue with the estimation ground shaking lies in generic attenuation relations for all types of sources, NDSHA (Neo-Deterministic Seismic Hazard Assessment) procedure has advantages in calculation strong ground motion from realistic model of synthetic seismograms from source specific properties, see Figure 1c. The NDSHA procedure, which represents a drastic enhancement of DSHA, permits incorporating, as they become available, new geophysical and geological data, as well as the information from the different Morphostructural Zonation (MZ) developed for the space-time identification of strong earthquakes. All this leads to the natural definition of a set of scenarios of expected ground shaking at the bedrock (Panza et al., 2013).

Recent advances in the physical knowledge of seismic waves generation and propagation processes, along with the improving computational tools, make it feasible the realistic modeling of the ground shaking caused by an earthquake, taking into account due consideration the complexities of the source and of the propagation path. A neo-deterministic scenario that bases its approach on the seismic hazard assessment (NDSHA) has been developed to naturally supply realistic time series of ground shaking,

including reliable estimation of ground displacement readily applicable to seismic isolation techniques (Panza et al., 2013).
The NDSHA approach has already been applied in several regions worldwide, including a number of local scale studies accounting for two-dimensional and three-dimensional lateral heterogeneities in inelastic media. NDSHA is an innovative, but already well consolidated, procedure that supplies realistic time histories, with very solid physics roots, from which it is natural to retrieve peak values for ground displacement, velocity and design acceleration in correspondence of earthquake scenarios (e.g., Parvez et al., 2010; Paskaleva et al., 2010). Synthetic seismograms can be efficiently constructed with the modal summation technique (e.g., Panza et al., 2001; La Mura et al. 2011) to model ground motion at sites of interest, using the available knowledge of the physical process of earthquake generation and wave propagation in realistic media and this makes it possible to easily perform detailed parametric analysis that permit to account for the uncertainty in input information.

Estimates of seismic hazard obtained using NDSHA and PSHA are compared for the Italian territory (Fig. 2). NDSHA provides values larger than those given by PSHA in areas where large earthquakes are observed and in areas identified as being prone to large earthquakes but provides lower values in low-seismicity areas. These differences suggest the adoption of the flexible, robust, and physically sound NDSHA approach for reliable seismic hazard estimation. NDSHA allows us to overcome the shortcomings of PSHA, including its proven tendency to underestimate ground shaking for the largest earthquakes, especially for those areas characterized by a prolonged quiescence, i.e. in tectonically active sites where events of only moderate size have occurred in historical times (Panza et al. 2011, Panza and Romanelli 2014).

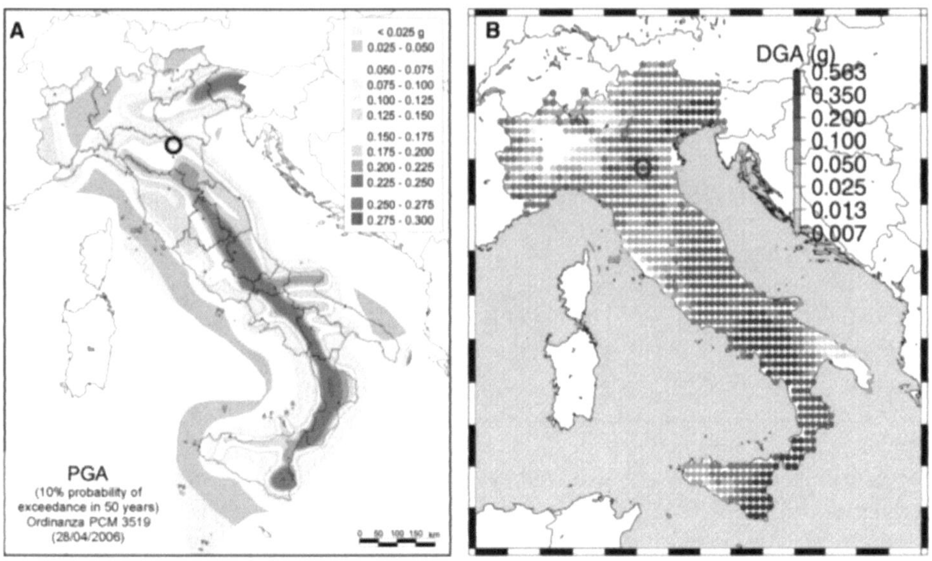

Figure 2. Seismic hazard maps developed using a PSHA and b NDSHA. The PSHA map is currently the official reference seismic hazard map for Italy. The epicenter of the Emilia earthquake is marked (black dot in (a); blue circle in (b). a Puts the Emilia epicenter in an area expected to have low ground shaking. Contrary to this, b assesses the seismic hazard in the area near the epicenter to be higher (from Peresan and Panza 2012) .

3. Official Seismic Hazard Map for Indonesia

Probabilistic seismic hazard map of Indonesia expressed in terms of expected PGA (g) with a probability of exceedance of 10% in 50 years (in a return period of 475 years), see Figure 3. In 2009 the Ministry of Public Work decided to establish a team to revise and update the SNI 03-1726-2002 seismic hazard map of Indonesia. There are three seismic source models used in their assessment: fault zone, subduction zone, and gridded seismicity. The earthquake source models are derived using all available information from earthquake catalogs, tectonic setting, regional and geological data and focal mechanisms. Fault source is treated as a plane in 3-D space for calculation of distance from a site to a certain point at the plane. The main input fault parameters to develop the seismotectonic sources are fault traces, focal mechanism, slip-rate, dip, length and width of the fault (Irsyam et al., 2010 and 2013).

Due to insufficient ground motion records needed to develop an attenuation relationship (GMPE) in Indonesia region, the relations developed from similar regions are used. The selection is based upon the similarity on source characteristic, geologic and tectonic conditions where the attenuation functions are developed. Most of the attenuation relations used in this study are Next Generation Attenuation (NGA), which are derived using worldwide observed earthquake data.
The new Indonesian code for building, SNI 1726-2012, follows the concept of MCEG used by ASCE 7-10 for the purpose of geotechnical calculation. It combines both the results from PSHA for 2% probability of exceedance in 50 years (2,500 years earthquake) and DSHA for area located near active fault. Both approaches are utilized according to the procedure proposed by Leyendecker et al. (2000), (Irsyam et al., 2010 and 2013).

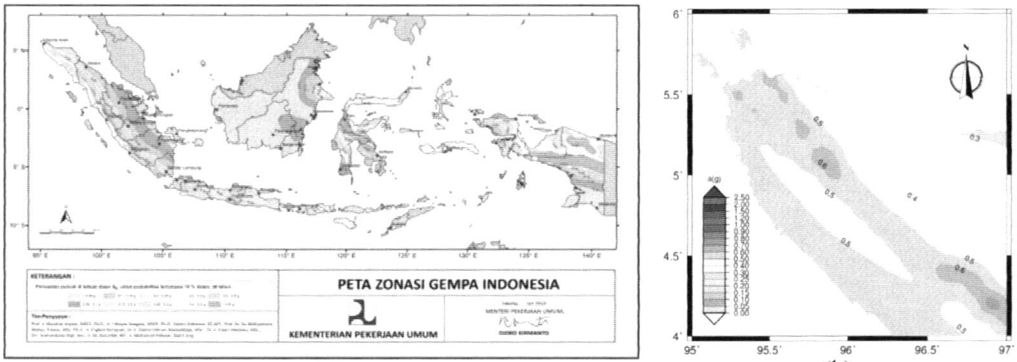

(a) (b)

Figure 3. (a) Seismic Hazard of Indonesia, Ministry General Work (PU) 10% probability of exceedance in 50 years SNI 1726-2012. http://loketpeta.pu.go.id/peta/zonasi-gempa-indonesia-4/. (b) PGA scale to Banda Aceh and surrounding area (Fauzi et al. 2013).

4. Comparison PSHA and NDSHA for a site on Banda Aceh

Regional scale NDSHA will compute synthetic seismogram from several sources defined for most significant historical or expectation sources, (Figure 4a Seismicity around Banda Aceh) for each area in the cell area, namely source definition. At the regional scale, a set of sources is defined in the tectonically active areas of the considered region. From these sources, and once the physical properties of average structural models have been defined, wave propagation is efficiently modeled with the modal summation technique and broadband synthetic seismograms are generated at the free surface on a predefined grid of points covering the study region (Florsch, Fah, Suhadolc, & Panza, 1991; Panza, 1985; Panza et al., 2012) .

In fact, recent destructive earthquakes demonstrate that a single hazard map cannot meet the requirements from different end-users. Therefore, the computation strong ground motion will calculate not only PGA but also PGD and PGV from seismograms generated by all sources. In order to investigate the most significant sources for hazard at the Banda Aceh city, we have calculated synthetic seismograms from all earthquake sources M>4 (Figure 4a) and selecting the most significant earthquake sources for contribution to strong ground motion PGD, PGV, and PGA at horizontal and vertical dimensions as produced in Figure 4b.

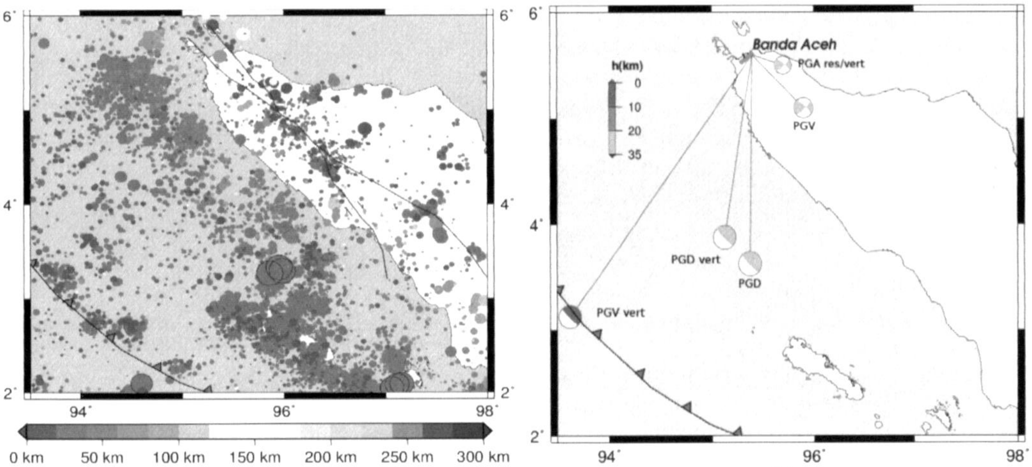

(a) (b)

Figure 4. (a) Seismicity around Banda Aceh. (b) The source location which contributes to significant strong ground motion PGD, PGV, and PGA at maximum frequency 10Hz. The PGD and PGV for the vertical and horizontal component have different sources. Only PGA for the horizontal and vertical components have the same sources.

Table 2. Analysis of the peak values of seismograms shows most significant sources for each component, located near the Banda Aceh basin (longitude=95.4°, latitude=5.6°).

	Source location (degree and km)				Focal mechanic (°)			magnitude	peak value	note
Component	longitude	latitude	distance	depth	strike	dip	rake			
PGD Horizontal	95.375	3.625	218	31	325	7	100	8.83	45.55 m	
PGV Horizontal	95.9	5.1	78	20	313	72	168	7.41	38.83 m/s	
PGA Horizontal	95.7	5.5	35	20	313	72	168	6.5	0.314 g	
PGD Vertical	95.125	3.875	193	28	325	7	100	8.73	38.33 m	
PGV Vertical	93.625	3.125	337	5	325	7	100	8.31	21.86 m/s	
PGA Vertical	95.7	5.5	35	20	313	72	168	6.5	0.123 g	

The result, calculated at 10hz maximum frequency, shows the source of PGD and PGV coming from a subduction zone and for PGA coming from strike-slip fault as PGD in the city is the dominant contributor in a big earthquake at a subduction zone, whereas PGA is related to a near source. However, for vertical components, almost all vertical PGD are contributed by a subduction zone because it has a vertical movement caused by the Sunda megathrust mechanic. The PGA for vertical components remains dominant by a near field source as the strikes slip is the nearest source in the land of Sumatra.

The NDSHA produced several sets of information as shown in Table 2, which contains six PGD, PGV, and PGA in horizontal and vertical components. The table shows vertical components that have lower values compared with those in the horizontal components. Therefore, to reduce a computing cost, the vertical component is ignored. The results provide more sets of information compared with the results that PSHA provides as shown in Figure 3. The available data from PSHA (Fauzi et al. 2013) are the PGA values at bedrock with 10% probability of exceedance in 50 years. The official PSHA map for PGA at bedrock with 10% and 2% probability of exceedance in 50 years are 0.298 g and 0.552 g. The PSHA analysis can give you various PGA values depending on your insurance analysis, and this is one of the reason why PSHA more popular. The NDSHA provides an exact value for each component, naturally, with realistic computation and the value in Banda Aceh in the range between 2% and 10% probability of PGA exceedance compared to the PSHA calculation.

5. Conclusions and Suggestion

The limitations that PSHA estimates are due not only to the data scarcity but also to the invalid physical model and mathematical formulation employed (Castanos H. e Lomnitz C., (2002), PSHA: is It Science?, Eng. Geo., 66, 315-317. Wang, 2011; Paskaleva et al., 2007).The limitations become unacceptable when considering the number of casualties and injured people (Wyss et al., 2012). The evolving situation makes it compulsory for any national or international regulation to be open to accommodate the most important new results, as they are produced and validated by the scientific community.

NDSHA has the capability to solve some issues, which have largely been neglected in traditional hazard analysis. For example, how crustal properties affect attenuation as ground motion parameters are not derived from simplified attenuation relations but rather from synthetic time histories. Starting from the available information on the Earth's structure (mechanical properties), seismic sources, and the level of seismicity of the investigated area, it is possible to estimate PGA, PGV, and PGD for horizontal and vertical components or any other parameters relevant to seismic engineering, which can be extracted from the synthetic seismograms (Panza 1985, Panza et al., 2013).

NDSHA is an innovative, and already well consolidated, procedure that supplies realistic time histories, with very solid physics roots, from which it is natural to retrieve peak values for ground displacement, velocity, and design acceleration in correspondence of earthquake scenarios (e.g., Parvez et al., 2010; Paskaleva et al., 2010). As PSHA is limited, Indonesia needs to establish further research on the uses of NDSHA for areas, whose critical infrastructures are at risk of seismic hazard.

6. References

[1] Bilham, R. (2009). The seismic future of cities. Bulletin of Earthquake Engineering. doi:10.1007/s10518-009-9147-0. 2009.

[2] Bommer, J.J., and Abrahamson, N.A., 2006, Why do modern probabilistic seismic-hazard analyses often lead to increased hazard estimates? Bull. Seismo. Soc. Am.,96, 1,976–1,977.

[3] Castanos, H., & Lomnitz, C. (2002). PSHA: is it science? Engineering Geology, 66, 315–317.

[4] Cornell, C.A., 1968, Engineering seismic risk analysis, Bull. Seismo. Soc. Am., 58, 1,583-1,606.

[5] Fauzi, U.J., Fauzi, A., Irsyam, M., and Toha, F.X. (2011). Proposed Long Period Transition Map for New Indonesia Earthquake Resistant Building Code Based on Indonesia Seismic Hazard Map 2010. In Proceedings of the Annual International Conference, Syiah Kuala University-Life Sciences & Engineering Chapter (Vol. 1, No. 2).
[6] Fauzi, U.J. (2013). (Personal Communication).
[7] Florsch, N., Fah, D., Suhadolc, P., & Panza, G. F. (1991). Complete synthetic seismograms for high-frequency multimode SH-waves. Pure and Applied Geophysics, 136, 529–560.
[8] Giardini, D., Grunthal, G., Shedlock, K. M., & Zhang, P. (1999). The GSHAP global seismic hazard map. Annali di Geofisica, 42, 1225–1230.
[9] Irsyam, M., Sengara, W., Aldimar, F., Widiyantoro, S., Triyoso, W., Natawidjaja, D.H., Kertapati, E., Meilano, I., Asrurifak, M., Suhardjono, and Ridwan, M. (2010). Summary of Study: Development of Seismic Hazard Maps of Indonesia for Revision of Hazard Map in SNI 03-1726-2002. Indonesia Seismic Hazard Map Revision Team.
[10] Irsyam, M., Asrurifak, M., Ridwan, M., Aldiamar, F., Sengara, I.W., Widiyantoro, S., and Firmanti, A. (2013). Past Earthquakes in Indonesia and New Seismic Hazard Maps for Earthquake Design of Buildings and Infrastructures. In Geotechnical Predictions and Practice in Dealing with Geohazards (pp. 33-46). Springer Netherlands.
[11] Klugel, J. U. (2007a). Comment on "why do modern probabilistic seismic-hazard analyses often lead to increased hazard estimates" by Julian J. Bommer and Norman A. Abrahamson. Bulletin of the Seismological Society of America, 97, 2198–2207.
[12] Klugel, J. U. (2007b). Error inflation in probabilistic seismic hazard analysis. Engineering Geology, 90, 186–192.
[13] Krinitzsky, E.L. (1998). The hazard in using probabilistic seismic hazard analysis for engineering. Environmental & Engineering Geoscience. IV(4), 425–443.
[14] Kossobokov, V. G., & Nekrasova, A. K. (2010). Global seismic hazard assessment program maps are misleading. Eos Transaction American Geophysical Union, 91(52). Fall Meet. Suppl., Abstract U13A-0020.
[15] Kossobokov, V. G., & Nekrasova, A. K. (2011). Global seismic hazard assessment program (GSHAP) maps are misleading. Problems of Engineering Seismology, 38(1), 65–76, (in Russian).
[16] La Mura, C., Yanovskaya, T.B., Romanelli, F., and Panza, G.F. (2011). Three-Dimensional Seismic Wave Propagation by Modal Summation: Method and Validation. Pure and Applied Geophysics, 168(1-2), 201-216. Special Issue: Advanced Seismic Hazard Assessment. Part I: Seismic Hazard Assessment.
[17] Leyendecker, E.V., Hunt, R.J., Frankel, A.D. and Rukstales, K.S. (2000). Development of Maximum Considered Earthquake Ground Motion Maps. Earthquake Spectra, 16, 21–40.
[18] McGuire, R. K. (2004). Seismic Hazard and Risk Analysis, Earthquake Engineering Research Institute, Oakland, CA
[19] Natawidjaja, Hilman, D., and Triyoso, W. (2007). The Sumatran Fault Zone-From Source To Hazard. Journal Of Earthquake and Tsunami, Vol. 1, No. 1 (2007) 21–47.
[20] Panza, G. F. (1985). Synthetic seismograms: The Rayleigh waves modal summation. Journal of Geophysics, 58, 125–145.
[21] Panza, G.F., (2010). PSHA and NDSHA in Italy (Presentation). Advanced Conference on Seismic Risk Mitigation and Sustainable Development, ICTP, Trieste 10-14 May 2010.
[22] Panza, G. F., Irikura, K., Kouteva, M., Peresan, A., Wang, Z., & Saragoni, R. (2011). Introduction to the special issue on advanced seismic hazard assessments. Pure and Applied Geophysics, Vol. 168, 1–2. doi:10.1007/s00024-010-0179-9.
[23] Panza, G.F., Mura, C.L., Peresan, A., Romanelli, F., and Vaccari, F. (2012). Seismic Hazard Scenarios as Preventive Tools for a Disaster Resilient Society. Chapter Three. Advances in Geophysics. Vol. 53, 2012, pp. 93–165.
[24] Panza, G.F., Peresan, A., and Mura, C.L. (2013). Seismic Hazard and Strong Ground Motion: an Operational Neo-Deterministic Approach from National to Local Scale. In UNESCO-

EOLSS Joint Commitee (Ed.), Encyclopedia of Life Support Systems (EOLSS), Geophysics and Geochemistry. Developed under the Auspices of the UNESCO, Eolss Publishers, Oxford ,UK, pp. 1E49.

[25] Panza, G. F., & Romanelli, F. (2014). Seismic waves in 3-D: from mantle asymmetries to reliable seismic hazard assessment. Earthquake Science, 27(5), 567-576.

[26] Parvez, I.A., Romanelli, F., and Panza, G.F. (2010). Long Period Ground Motion at Bedrock Level in Delhi City from Himalayan Earthquake Scenarios. Pure and Applied Geophysics.

[27] Paskaleva, I., Dimova, S., Panza, G. F., & Vaccari, F. (2007). An earthquake scenario for the microzonation of Sofia and the vulnerability of structures designed by use of the Eurocodes. Soil Dynamics and Earthquake Engineering, 27, 1028–1041.

[28] Peresan A, Panza GF (2012) Improving earthquake hazard assess-ments in Italy: an alternative to ''Texas Sharpshooting''. EOS 93:538–539.

[29] Peresan, A., & Nekrasova, A. (2014). Why are the standard probabilistic methods of estimating seismic hazard and risks too often wrong. Earthquake Hazard, Risk and Disasters, 309.

[30] Standar Nasional Indonesia (SNI) (2002). Tata Cara Perencanaan Ketahanan Gempa untuk Bangunan Gedung (SNI 03-1726-2002), Badan Standardisasi Nasional.

[31] Stein, S., Geller, R., & Liu, M. (2011). Bad assumptions or bad luck: why earthquake hazard maps need objective testing. Seismological Research Letters, 82, 5.

[32] Wang, Z. (2011). Seismic hazard assessment: Issues and alternatives. Pure and Applied Geophysics, Vol. 168, 11–25. doi:10.1007/s00024-010-0148-3.

[33] Wyss, M., Nekrasova, A., & Kossobokov, V. (2012). Errors in expected human losses due to incorrect seismic hazard estimates. Natural Hazards, 62(3), 927–935. doi: 10.1007/s11069-012-0125-5.

Acknowledgments
I wish to acknowledge assistance from Department of Geoscience, Trieste University, and International Centre for Theoretical Physics ICTP and Dikti for financial support from Hibah Bersaing. I wish to encouragement from Physics Department, Geophysics Engineering Department, and TDMRC.

Simulation of surface heat fluxes of Typhoon Songda (Chedeng) 2011 using WRF-ARW model

Muhammad[1], R I Lestari[1], F Mulia[1], Y Ilhamsyah[1,3], Z Jalil[2] and Rizwan[1]

[1]Dept. of Marine Sciences, Faculty of Marine and Fisheries, Syiah Kuala University, Banda Aceh 23111, Indonesia
[2]Dept. of Physics, Faculty of Math. And Natural Sciences, Syiah Kuala University, Banda Aceh 23111, Indonesia
[3]Tsunami and Disaster Mitigation Research Center (TDMRC), Syiah Kuala University, Banda Aceh 23111, Indonesia

E-mail: y.ilhamsyah@gmail.com

Abstract. Heat fluxes particularly latent heat is important to drive the development, formation, and intensification of Typhoon Songda (Chedeng). The research was carried out by performing WRF ARW. Three domains with finest resolution at 3.2-km in domain three were utilized in the model. The model involved significant physics parameters, e.g., Kain-Fritsch in the cumulus scheme, Yonsei university in the PBL scheme, and WRF Single-Moment 3-class in the microphysics scheme . The analysis focused on May 26th upon mature stage of Songda (Chedeng). The result showed that the simulation of the eye, three-dimensional structure of internal wind flow, and surface heat fluxes were well-performed. The intensity of Songda (Chedeng) was represented by azimuthal velocity. It showed that the maximum wind was 72 ms^{-1} occurred at the eye wall at critical radius of 20-km from the eye center where large portion of latent heat available in the area. Significant variation of surface sensible and latent heat fluxes were occurred between the inner and outer core. Thus, it affected to develop a strong horizontal temperature gradient which further intensify the cyclonic inward penetration into the inner core. In terms of disaster risk reduction, this study bring benefit to assist operational weather forecaster to produce good short-range forecasts of the Typhoon intensities. If the surface heat fluxes increase gradually, early warning system on typhoon intensities that will affect over particular region is then released.

1. Introduction

Heat exchange between ocean and atmosphere is often characterized by sensible and latent heat transfer. Sensible heat usually contributes to either increasing or decreasing atmospheric temperature. Latent heat, on the other hand, takes form as evaporation and is a source of continuous hidden heat transfer as a result of water vapor condensation. Latent heat then becomes among primary energy to drive the intensification of Tropical Cyclone (TC) [1]. The mature TC intensity is typically described by axisymmetric distribution of azimuthal velocity from inner to outer structure [2, 3]. The stronger circulation is mostly found in the eye wall where latent heat is abundant in the area. Thus, the eye wall is known as the most intense zone in TC. Circulation in the eye wall might influence a positive feedback to strengthen TC since it provides moisture supply into the system.

An effort to study the relationship between TC intensity and heat fluxes had been done [4]. He constructed an idealized, steady-state model of mature TC to investigate the maintenance of TC associated with oceanic heat transfer. In recent published paper, he developed the latest theoretical concept based on numerical simulation to study the intensification of TC controlled by surface fluxes [5, 6]. Another numerical study has been done by using a coupled hurricane-ocean model to examine the ocean effect (involving the calculation of surface heat fluxes) on TC intensity [7]. The utilization of a numerical weather prediction model such as earlier version of Weather and Research Forecasting (WRF) to study surface heat fluxes on TC intensity was done by testing two Planetary Boundary Layer (PBL) scheme to obtain a better simulation of latent heat fluxes. Meanwhile, in the present paper, Advanced Research WRF (hereafter refer to WRF ARW) with state-of-the-art of atmospheric simulation is applied in the model to obtain a better simulation of Typhoon Songda (Chedeng) (refer to TSC) surface heat fluxes.

Typhoon Songda (called Chedeng in the Philippines) occurrence over western North Pacific Ocean was chosen as a case study in the present paper. TSC, local name of TC, was the fourth named TC of the 2011 Northern Hemisphere tropical season. It was the most devastating storm to strike western North Pacific Ocean for the period of 2011. TSC lasted for almost two weeks from May 18th to 30th, 2011 and enhanced to reach peak activity of powerful category 5 Super Typhoon between May 24th and 26th. Low pressure area was centered nearby off west coast of Manila. The exhibition of mature Songda (Chedeng) as well as the fact that Songda (Chedeng) is not a subject of exploration yet soon motivates the study. Therefore, the present study attempt to show the roles of heat fluxes to maintain mature TC intensity by utilizing WRF model.

2. Material and Methods

The domain of study areas is between 10°N - 20°N and 120°E - 130°E for period of mature stage of TSC. The research was carried out by performing WRF ARW model version 3.3. WRF ARW is a non-hydrostatic model with terrain-following in the vertical-sigma coordinate. A review of mathematics, physics, dynamics, and thermodynamics of the model can be found in Skamarock et al. [8]. In the meantime, three domains were employed in the model as shown in figure 1. Large-scale domain 1 and 2 consist of 100 x 107 and 187 x 241 horizontal grids and cover spatial resolution of 28.9–km and 9.6-km, respectively. Domain 3 consists of 247 x 259 horizontal grids at finest 3.2-km resolution. Domain 3 was utilized to analyze the result. Besides, 28 pressure levels ranging from 1000-100 hPa were employed in the model. The model was also supported by following physics parameterization: (a) surface and Planetary Boundary Layer (PBL) scheme, (b) cumulus scheme, and (c) microphysics scheme. List of physics parameterization used in the model was given in Table 1. Similar parameterization was also adapted by Osuri et al. [9] to simulate different TC over north Indian Ocean. The subroutine to calculate heat and moisture fluxes from the surface was included in the model. The initial boundary condition in the model are : (a) 30 second (0.925-km) resolution of USGS terrain height data, (b) 30 second (0.925-km) resolution of global 24-category USGS land use/cover data, and (c) 1.0° latitude x 1.0° longitude grids NCEP Final Analysis (FNL) data with grib2 format. The description of the data are as follows: (a) temporal resolution of 6-hourly (0000, 0600, 1200, and 1800 UTC) dataset, (b) pressure levels are available from 1000 to 10 hPa, and (c) variables include surface pressure, sea-level pressure, geopotential height, temperature, sea surface temperature, soil values, ice cover, relative humidity, zonal and meridional velocity, vertical velocity, vorticity, and ozone concentration. The selected simulation days are from May 25th to 27th, 2011 when Songda (Chedeng) turn into category 5 Super Typhoon. Analysis will focus on second day on May 26th.

Table 1. Physics of WRF ARW model used in the study.

Physics categories	Scheme
Surface layer	Monin-Obukhov with Carslon-Boland viscous sub-layer
Land surface	Noah land surface model
PBL	Yonsei university PBL
Cumulus parameterization	Kain-Fritsch scheme
Longwave radiation	Rapid Radiative Transfer Model (RRTM) scheme

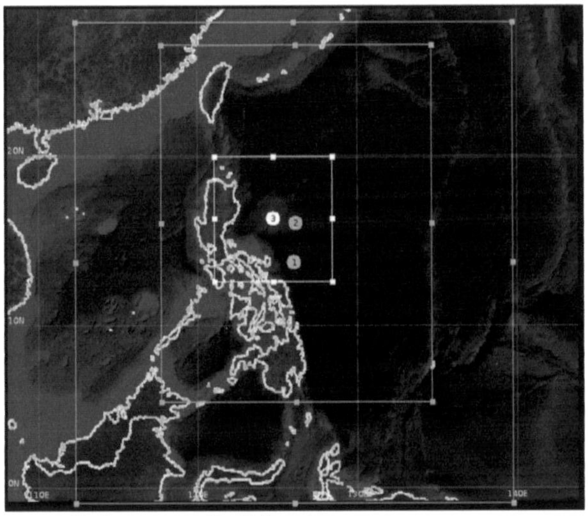

Shortwave radiation	Dudhia scheme

Figure 1. Domains of WRF ARW simulation.

Microphysics	WRF Single-Moment 3-class scheme

3. Results and Discussion

The simulation of the spiral clouds and the internal structure of winds flow of TSC were shown in figure 2. The whirling clouds were counter-clockwise since it established in the Northern Hemisphere. The eye featuring the core of TSC was well-simulated as shown in figure 2a. The eye of TSC was reported to have minimum Sea-Level Pressure (SLP) of 920 hPa. The relative calm condition at the center was main feature of the eye. Another feature was typically recognized by subsidence of winds as shown in figure 2b. In the eye, the air gradually sinked and heated by compression. Thus, it was responsible for warming the zone. As a result, the eye is also well-known as the warm core.

Halverson et al. [10] in their observational study found that maximum temperature anomaly in the warm core was 11°C near 500 hPa with minimum SLP and wind speeds of 969 hPa and 54 m s-1, respectively. However, their finding was lower than energetic TC shown in Hawkins and Imbembo [11]. In field study conducted by them to investigate TC Inez in 1966, the maximum temperature anomaly at the same level was 16°C with central SLP of 927 hPa. These differences might refer to the differences of size distribution on two TCs. On the other hand, the subsidence in the eye led to cloudless condition characterized by clear appearance of the sky.

Surrounding the eye was the eye wall of TSC, here the typical TC characteristics appeared. The eye wall was the most devastating zone in Songda (Chedeng) since the strongest winds usually occurred in the zone. Songda (Chedeng) had maximum sustained winds of 72 ms^{-1}. Willoughby et al. [12] and Black and Willoughby [13] found that the eye wall was a well-defined wind maximum area. It can be explained that the decreasing pressure toward the core produced a steep pressure gradient. Consequently, it caused the velocity of inward curving winds to increase. The conservation of angular momentum was behind the acceleration. As the rotation of Songda (Chedeng) was identified by convergence of wind into low pressure center, angular momentum associated with Earth's rotation was concentrated into angular momentum associated with Songda (Chedeng) wind. When the inward of warm, moist surface air moved toward the core, it turned upward and rise in a circle of Cumulonimbus towers as shown in figure 2b.

In addition, the eye wall of Songda (Chedeng) implied the occurrence of the intense convective activity. McNoldy [14] found that strong convection might even form three eye walls during the intensification of TC. Meanwhile, outer structures of TSC were associated with spiral rainbands as shown in figure 2a. Spiral rainbands consisted of convective clouds with updrafts were found on the inward side of the band axis and downdrafts were located just outside the inward. At the top of the Typhoon, the flow was outward, leaving the center of the storm. Thus, it allowed the air at the surface to flow inward. Upper and lower level flow were in the opposite direction as shown in figure 2b. Both eye wall and spiral rainbands shown in figure 2a were called zone of deep convection. Houze [15] explained that in addition to contributing TC genesis, deep convection has intense updrafts but weak downdrafts as proved by the present study. Furthermore, the spinning flow of rainbands was essential to provide fuel to develop the storm since it provided moisture to get into the eye wall of Songda (Chedeng). Nevertheless, according Hack and Schubert [16], it also contributed to the release of latent heat away from the center, reducing TC intensity. Wang [17] found that the existence of strong spiral rainbands might limit TC intensity. He also concluded that cooling in the spiral rainbands maintained the intensity and the size of TC while heating maked the intensity to decrease but to increase the size of TC.

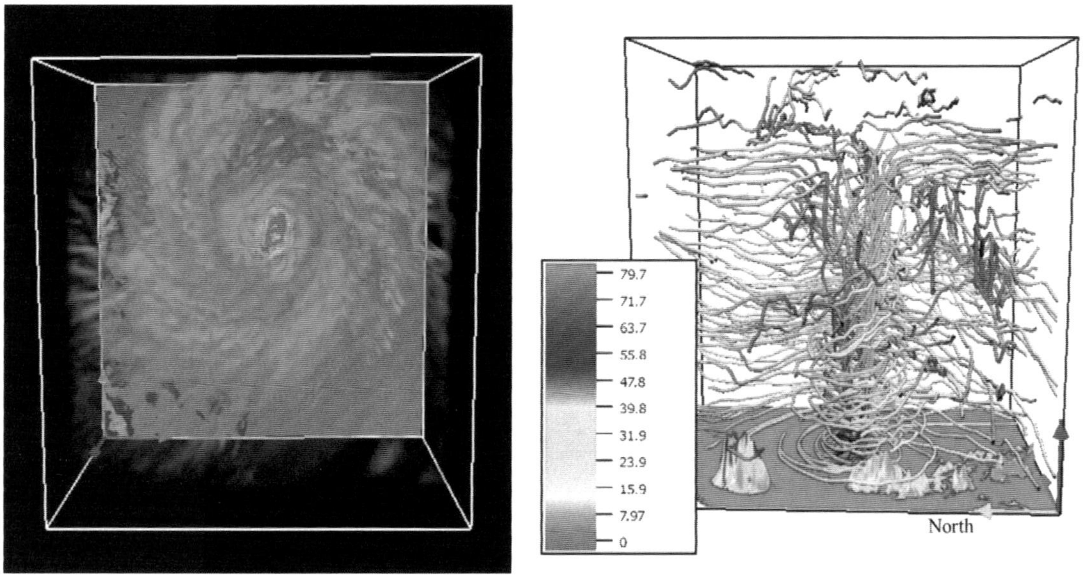

(a) **(b)**

Figure 2. Simulation of Category 5 Super Typhoon Songda (Chedeng) at 1700 UTC on May 26th, 2011 , a) swirling shape of clouds, b) three-dimensional structure of wind circulation inside the storm. Bar chart in the left in figure 2b indicates the magnitudes of wind speeds in ms^{-1} (not the terrain profile in the lowest background).

Figure 3 showed the distribution of azimuthal velocity of TSC in the boundary layer. The diameter of the eye was about 200-km in width. The velocity sharply decreased leading to light wind in the area. The decrease of azimuthal wind at the eye during mature stage of TC is often characterized by V-shaped of axisymmetric structure. As already discussed above that the maximum velocity of 72 ms^{-1} occurred at the outside edge of Songda (Chedeng) eye wall at critical radius of 20-km as shown in figure 3. The velocity gradually decreased towards spiral rainbands and outer structure of Songda (Chedeng). Stern and Nolan [18] inspected some TCs and found different critical radius of maximum wind nearby the inner core. From figure 3, it was shown that Songda (Chedeng) had approximately 200-km in radial distance from the center of the eye to outer structure of Songda (Chedeng).

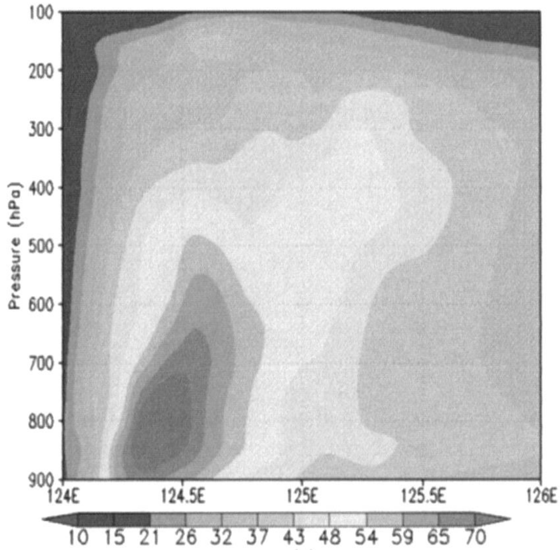

Figure 3. Simulation of West-East vertical structures of azimuthal velocity (ms^{-1}) at Category 5 Super TSC at 1700 UTC on May 26th, 2011.

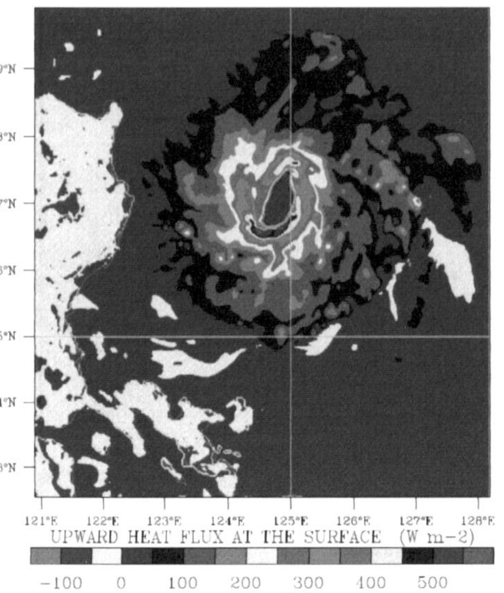

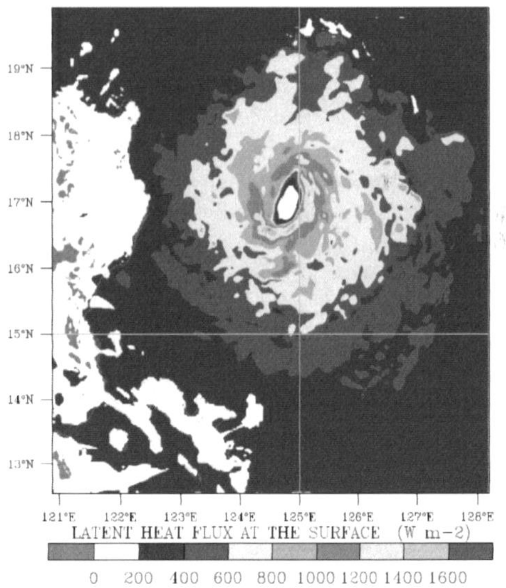

Figure 4. Simulation of surface heat fluxes of TSC at 1700 UTC on May 26th, 2011, a) sensible (or upward heat flux), b) latent heat flux. Both Units in Wm^{-2}.

The greatest wind speeds in the eye wall related to higher amount of surface latent heat flux in the area as shown in figure 4b. Mark [19] found that high quantities of latent heat release were concentrated in the eye wall. From figure 4b, it is demonstrated that just outside of the eye, surface latent heat flux reached 1400 Wm^{-2}. Meanwhile, at the core and the edge of the eye wall, latent heat flux at the surface slowly decreased from 1200 down to 1000 Wm^{-2}. In addition, surface latent heat flux in spiral rainbands and outer structure fell to 800 Wm^{-2} and 600 Wm^{-2}, respectively. Compared to sensible heat (or upward heat flux at surface) as shown in figure 4a, latent heat flux from the ocean was three times greater than sensible heat flux. Principally, the reason for this is that most of energy transfer released from the ocean to the atmosphere is exchanged to latent heat. It can bring to a close understanding that latent heat flux plays a major role on Songda (Chedeng) intensity than sensible heat flux. Sensible heat flux at the surface is defined as conductive heat flux from the Earth's surface to the atmosphere indicated by positive values. Moreover, heat energy transfer from the atmosphere to the ocean is indicated by negative values. The highest energy transfer occurred at the center of the eye wall reached 350 Wm^{-2}. Additionally, surface sensible heat flux directly influenced Songda (Chedeng) temperature. With the exception of the eye, it further implied that higher sensible heat flux might reflect higher temperature in the same occasion. For that reason, the warm and cool region of Songda (Chedeng) occurred at the eye wall and spiral rainbands, respectively. Consequently, it increased temperature gradient and strengthened Songda (Chedeng). In conclusion, from figure 4a b, it was depicted that surface sensible and latent heat fluxes between the eye wall and spiral rainbands showed much higher differences. Therefore, as discussed before, it might have led to the strong development of horizontal temperature gradient which further drives the formation and intensification of cyclonic circulation of into the inner core Songda (Chedeng) as shown in figure 2b. Black and Holland [20] provided a thorough aircraft study and found that strong horizontal advection might enhance the inward penetration. Overall, figure 4a and b showed that fluxes production was maximized at the eye wall. Hence, the reason behind the destructive part of Songda (Chedeng) becomes obvious since moisture transport from the ocean was greater in the area. In terms of disaster risk reduction, this study bring benefit to assist operational weather forecaster to produce good short-range forecasts of the Typhoon intensities. If the surface heat fluxes increase gradually, early warning system on typhoon intensities that will affect over particular region is then released.

4. Concluding remarks

WRF ARW model was performed to simulate the surface heat fluxes of TSC. The counter-clockwise spinning clouds featuring the eye as well as wind circulation structure inside Songda (Chedeng) were well-simulated. The simulation of subsidence of wind at the core of Typhoon Songda was well-observed. The greatest winds of 72 ms^{-1} arise at the eye wall of Songda (Chedeng). The reason behind this was the sharp pressure gradient as the decrease of pressure into the core led to accelerate the inward rotational wind. The entering wind at the eye wall was deepened by conservation of angular momentum associated with Earth's coriolis force. As the innermost of warm, moist surface air circulates headed for the core, it rises aloft at the top of anvil clouds. Strong updrafts were observed in the spiral rainbands whereas weak downdrafts occurred just outside of updrafts in the spiral rainbands. At the top of Songda (Chedeng) boundary layer, the flow travelled away from storm center and provided a sufficient space to the inward surface flow. The illustration of azimuthal velocity obviously showed the emergence of maximum velocity at critical radius of 20-km from the eye center. However, the slope of velocity decline exponentially as the radius away. In accordance with category 5 Super Typhoon, it is also reported that the Songda (Chedeng) had 200-km in radius. From the heat fluxes simulation, it is shown that latent heat flux at the surface was three times larger than sensible heat flux. The greatest proportion of latent heat was deposited at the eye wall of Songda (Chedeng). Large in number of latent heat flux is among critical factor that trigger the strong cyclonic velocity at the eye wall. On the other hand, sensible heat flux at the surface organized Songda (Chedeng) temperature, sufficient energy transport from the ocean to the atmosphere was also maximized at the eye wall. Variation in number of heat fluxes observed between inner and outer structure stimulate to enlarge a strong temperature gradient which further strengthen the rotational of inward penetration in Songda (Chedeng). For that reason, the role of surface heat fluxes in determining Songda (Chedeng) intensity become more apparent to address.

Acknowledgments

The author expresses gratitude to Prof. Dr. Ahmad Bey for the discussion, Mr. A Fachri Radjab for giving an opportunity to carry out the study at Tropical Cyclone Warning Center BMKG, Mr. Zainal Abidin for kind assistance throughout WRF simulation, and anonymous reviewers for comments. Thankful was also addressed to DIKTI for providing financial support to the study.

5. References

[1] Ooyama K 1969 *J. Atmos. Sci.* **26** 3-40
[2] Wu CC and Cheng HJ 1999 *Mon. Wea. Rev.* **127** 3003-3031
[3] Nguyen MC *et al.* 2011 *Quart. J. Roy. Meteor. Soc.* **137** 829-844
[4] Emanuel KA 1986 *J. Atmos. Sci.* **43** 585-604
[5] Rotunno R and Emanuel KA 1987 *J. Atmos. Sci.* **44** 542–561
[6] Emanuel K 2012 *J. Atmos. Sci.* **69** 988-996
[7] Schade LR and Emanuel KA 1999 *J. Atmos. Sci.* **56** 642–651
[8] Skamarock WC *et al.* 2008. *NCAR Technical Note*
[9] Osuri KK *et al.* 2012 *Nat. Hazards*. **63** 1337–1359
[10] Halverson JB *et al.* 2006 *J. Atmos. Sci.* **63** 309-324
[11] Hawkins HF and Imbembo SM 1976 *Mon. Wea. Rev.* **104** 418–442
[12] Willoughby HE, Clos J and Shoribah M 1982 *J. Atmos. Sci.* **39** 395-411
[13] Black LM and Willoughby HE *Mon. Wea. Rev.* **120** 947-957
[14] McNoldy BD 2004 *Bull. Amer. Meteor. Soc.* **85**1663-1666
[15] Houze RA 2010 *Mon. Wea. Rev.* **138** 293-344
[16] Hack JJ and Schubert WH 1986 *J. Atmos. Sci.* **43** 1559-1573
[17] Wang Y 2009 *J. Atmos. Sci.* **66** 1260-1273
[18] Stern DP and Nolan DS 2009 *J. Atmos. Sci.* **66** 3579-3600
[19] Mark FD 1985 *Mon. Wea. Rev.* **113** 909-930
[20] Black PG and Holland GJ 1995 *Mon. Wea. Rev.* **123** 2007-2028

IOP Conf. Series: Earth and Environmental Science **56** (2017) 012009 doi:10.1088/1755-1315/56/1/012009

Numerical simulations of land cover roughness influence on tsunami inundation in Ulee Lheue Bay, Aceh-Indonesia

Tursina[1,*] **and Syamsidik**[1,2]

[1]Tsunami Computation and Visualization Laboratory, Tsunami and Disaster Mitigation Research Center (TDMRC), Syiah Kuala University, Jl. Prof. Ibrahim Hasan, Gampong Pie, Banda Aceh, 23233, Indonesia.
[2]Lecture at Civil Engineering Department, Syiah Kuala University, Jl. Syeh Abdurrauf, No. 7, Banda Aceh, 23111, Indonesia.

*email: tursina@tdmrc.org

Abstract. Most of the flat area at Ulee Lheue Bay of Aceh Besar is affected by tsunami wave during the 2004 Indian Ocean Tsunami. The coastal area with no significant barrier caused tsunami wave inundating further inland. Coastal forest is known as an effective way to reduce tsunami impact in the coastal environment. In this paper, we used 2 Dimensional Horizontal (2DH) numerical model Cornell Multi-grid Coupled Tsunami (COMCOT) and Delft3D to evaluate the coastal forest influence which interpreted by variable Manning's coefficient of tsunami inundation. We also discussed the hydrodynamic characteristic of tsunami wave such as maximum water elevation and flow velocities to evaluate how effective is the coastal forest in reducing tsunami impacts at Ulee Lheue Bay. The results showed that the coastal forest effect is not significant difference both water elevation and tsunami inundation. Meanwhile, the coastal forest with 100 m width can reduce tsunami wave velocities about 40 %. It is important because high tsunami velocities caused severely eroded in the coastal area.

1. Introduction

Twelve years ago, the great Sumatera earthquake had shaken the Aceh Province with 9.1 Mw. A giant tsunami attacked the coastal area around the India Ocean that arrived in Sumatra's shores about 30 minutes after the earthquake [11]. Due to the tsunami waves, low lying areas of Aceh coastal zones were inundated by the tsunami waves. The tsunami caused major losses on human casualties and properties. After the tsunami, people started to realize how vulnerable Aceh coastal area toward tsunami. The dense of Aceh population around the coast amplified impacts on human casualties. Hence, we need to develope ways to reduce the impacts of the tsunami. Concept of coastal forest is known as an ecological method to reduce the impact of the tsunami on the coastal area. Unfortunately, the infrequent and unpredictable natural disaster of tsunami make it difficult to collect field research data of tsunami physical processes. Therefore, a numerical simulation is an important tool to better understand tsunami processes [3;9]. The information of tsunami hydrodynamics such as inundation area, flow depth and velocities will help coastal planners in improving tsunami hazard assessment to mitigate tsunami disaster.

Several studies have been conducted to focus on numerical simulation of tsunami hydrodynamic aspects including tsunami inundation, the effect of coastal forest, and land cover. For example, [4] used numerical simulation to investigated tsunami inundation and influence of land cover on flood patterns on the Andaman Sea coast of Thailand. Their study revealed flow velocities reduced up to 50

% by mangroves. Another study evaluated the reduction of tsunami inundation by coastal forest in Yogyakarta [7]. They found that the coastal forest with 100 m width reduced inundations flux, depth, and area by 17.6, 7.0 and 5.7 %, respectively. A study of the role of coastal forest on tsunami mitigation also has been conducted with field investigation. [2] argued that the coastal forest reduce tsunami disaster by stopping drifting materials or ships carried by a tsunami, to reduce tsunami energy, to form sand dune protecting tsunamis as well as high waves, and to catch persons brought back by a tsunami to the sea. All of the study areas mentioned above were located outside of Aceh Province. These motivate this study to particularly investigate the effect of land cover in Aceh where the 2004 Indian Ocean tsunami occurred.

This study is aimed at evaluating the damping effects of coastal forest width to reduce tsunami inundation that useful for mitigating future tsunamis in Ulee Lheue Bay. This study produced a detailed map of land cover in the study area derived from IKONOS imagery from January 2003 (before the tsunami). We set up another map of land cover for two scenarios with various width of coastal forest (mangrove) and fishpond. Scenario 1 and scenario 2 were set to have 50 m and 100 m of forest and fishpond, respectively. We focused on the coastal forest area. In the Delft3D numerical model, land cover information is interpreted by Manning coefficient (n) that was adopted from [6]. The computational domain of the study area was set to have a grid size of 6 m x 6 m. We have some limitations in this study. The numerical models used in this study did not simulate sediment transport. In addition, this study also did not consider debris of the collapse buildings. The trees were consider to sustain the tsunami wave. We placed observation points and cross sections to observe the hydrodynamics aspects of the tsunami waves.

2. Study Area

The Ulee Lheue Bay was chosen as the study area in this research. The area was severely damaged by the 2004 Indian Ocean tsunami. It is located on the western coast of Banda Aceh, the capital city of Aceh Province. The Ulee Lheue Bay is dominated by two types of coastal morphology. At the west part, the coastal has one headland named as Ujong Pancu. Meanwhile, at the eastern part, the area is relatively flat. According to pre-tsunami satellite imagery (IKONOS 2003), the land cover of Ulee Lheue Bay was dominated by fishponds (tambak). Beside the fishponds, this area had also beach area, forest area, settlement, and vegetation area. Before the 2004 tsunami, no hard structures for coastal protection along the coastline along this bay. The computational area of Ulee Lheu Bay ranges from 5.22°E to 5.579°E and from 95.225°N to 95.280°N. The study area is shown in Figure 1.

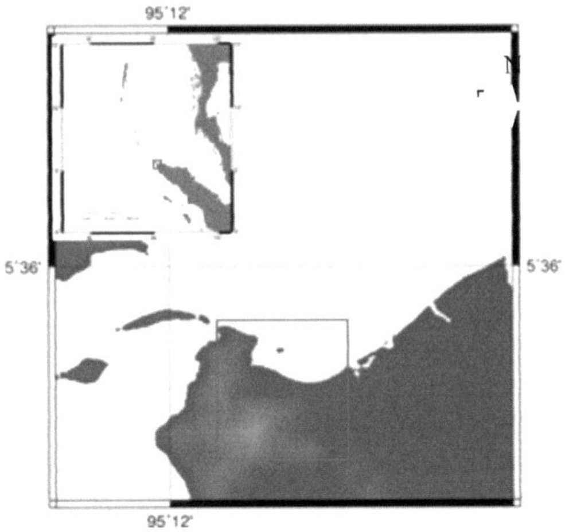

Figure 1. The map of Ulee Lheue Bay. The study area is marked by the red rectangle

3. Methods

Numerical simulations were performed using Cornell Multi-grid Coupled Tsunami Model (COMCOT). COMCOT model is used to simulate tsunami generation and propagation. Meanwhile, for tsunami inundation and observation of tsunami hydrodynamics, Delft3D model is used. Both models were connected at an open boundary (north boundary) using time series data of tsunami water level produced in COMCOT simulation. The methodological approach is shown in Figure 2.

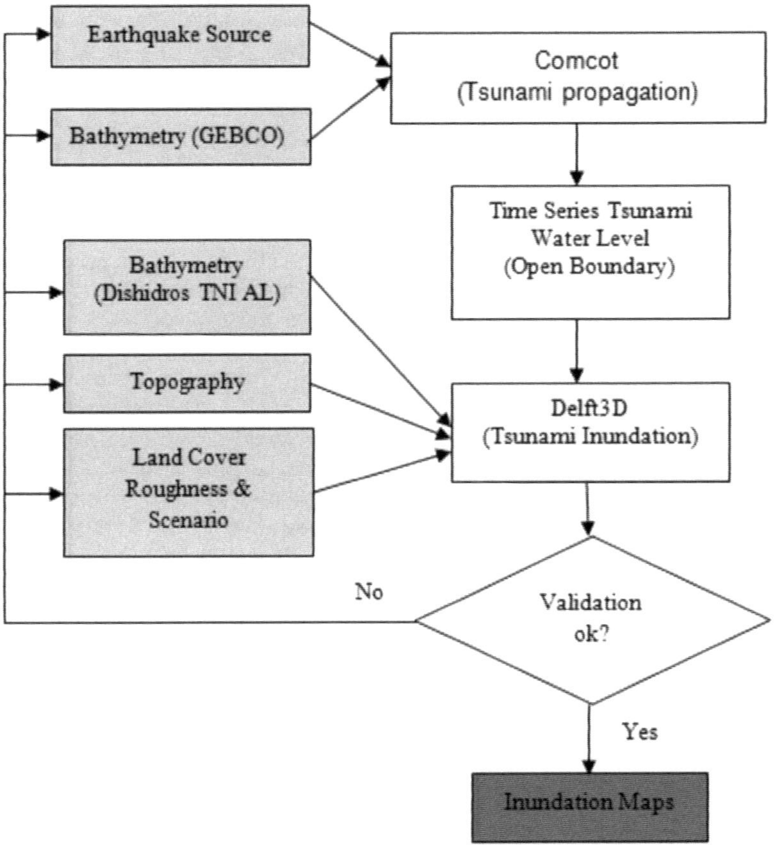

Figure 2. Methodological approach of the numerical model in this study

3.1 Earthquake source mechanisms and tsunami generation

The COMCOT model requires several data input to simulate tsunami generation. The data are bathymetry data and the parameter of the 9.1 Mw earthwquake that triggered the 2004 Indian Ocean Tsunami. A list of mechanisms of the 2004 earthquake (see Table 1) were adopted from [8] and [10] have validated this model. The initial sea surface elevation map according to the source mechanisms model used in this study is shown in Figure 3.

Table 1. Multi-fault Scenario of the earthquake parameter in the 2004 tsunami event [8]

	Fault segmen	Coordinat		Parameter *Fault*			*Depth* (km)	*Dis* (m)	*Width W* (km)	*Length L* (km)
		Long (°)	*Lat* (°)	*Strike*	*Dip*	*Slip*				
9.1 Mw	1	98.845	2.671	301.8	13.22	90	10.1	5	113.728	137.154
	1	95.21	1.915	301.8	5.2	90	0.1	4	110.273	137.211
	2	94.964	3.312	309.14	13.23	90	10	16	113.624	109.662
	2	94.3	2.583	309.5	5.21	90	0.1	15	110.137	123.126
	3	94.183	4.206	328.16	13.48	90	10	5	111.504	125.25
	3	93.4	3.667	329.9	5.4	90	0.1	5	106.331	158.965
	4	93.582	5.473	336.16	14.48	130	10	14	103.985	182.4
	4	92.85	5.133	341.05	6.34	130	0.1	16	90.568	183.497
	5	93.084	7.051	342.91	14.88	130	10	5	101.266	163.41
	5	92.35	6.75	341.39	6.42	130	0.1	5	89.367	172.928
	6	92.524	8.416	334.56	14.83	130	10	15	101.569	176.108
	6	91.733	8.1667	334.66	6.14	130	0.1	16	93.561	181.21
	7	92.089	10.085	357.33	15.25	90	10	4	98.863	167.488
	7	91.312	10.075	0.77	6.5	90	0.1	5	88.3	204.321
	8	92.259	11.637	10.71	15.71	90	10	14	96.999	156.887
	8	91.562	11.75	12.08	7.12	90	0.1	16	80.698	166.563
	9	92.602	13.067	15.94	16.36	90	10	6	92.323	167.403
	9	91.988	13.25	17.23	7.79	90	0.1	16	73.763	172.847

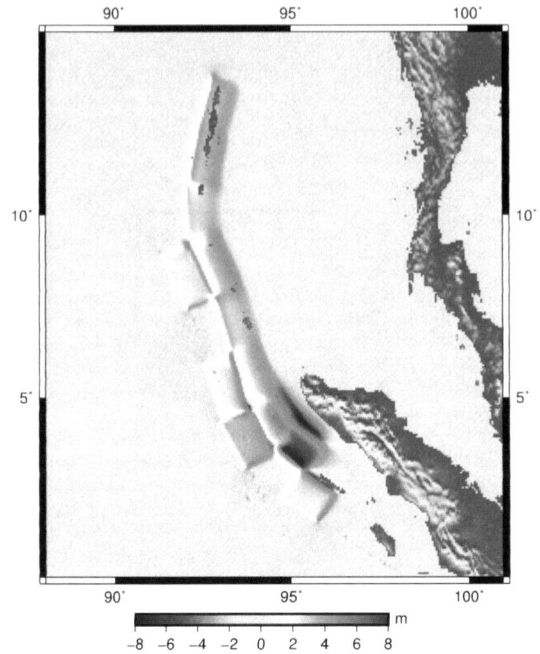

Figure 3. The map of the initial surface elevation derived from multi fault model by [8]

IOP Conf. Series: Earth and Environmental Science **56** (2017) 012009 doi:10.1088/1755-1315/56/1/012009

3.2 Numerical simulation of tsunami propagation

The nonlinear Shallow Water Equations (SWE) in COMCOT are given by

$$\frac{\partial \eta}{\partial t} + \frac{\partial hu}{\partial x} + \frac{\partial hv}{\partial y} = 0 \quad (1)$$

$$\frac{\partial u}{\partial t} + u\frac{\partial u}{\partial x} + v\frac{\partial u}{\partial y} + \frac{gn^2}{h^{4/3}} u\sqrt{u^2 + v^2} = -g\frac{\partial \eta}{\partial x} \quad (2)$$

$$\frac{\partial v}{\partial t} + u\frac{\partial v}{\partial x} + v\frac{\partial v}{\partial y} + \frac{gn^2}{h^{4/3}} v\sqrt{u^2 + v^2} = -g\frac{\partial \eta}{\partial y} \quad (3)$$

Where η = surface elevation (m); h = the total water depth (m); u dan v = the depth average velocities in the x dan y directions, respectively (m/s); ρ = the water density (kg/m^3); g = the gravitational acceleration (m/s^2); n = the Manning's coefficient for bottom roughness.

In this study, COMCOT model was applied to five simulation layers as given in Figure 4. To compute the wave propagation for the first four layer, the linear shallow water equatio and the spherical coordinate system were used. The bathymetry data for layers 1-4 were generated from the GEBCO data. The layer 5 that covered only the Ulee Lheue Bay used nonlinear shallow water equation and it was solved in a cartesian coordinate system. For the nearshore zone (Layer 5), the bathymetry data were digitized from DISHIDROS of the Indonesian Navy. The roughness coefficient in COMCOT simulation was set to n = 0.02 for the whole model domain simulation. In Layer 5, which is the focus of this study, the land cover roughness coefficients before tsunami were transferred into Delft3D model as Manning coefficients. The grid size of Layer 5 Delft3D model was set to 6 m x 6 m. For evaluating the coastal forest effect, we used two scenarios with coastal forest widths of 50 m and 100 m for scenario 1 and scenario 2, respectively. The details of the five layer simulations used in this study are summarized in Table 1.

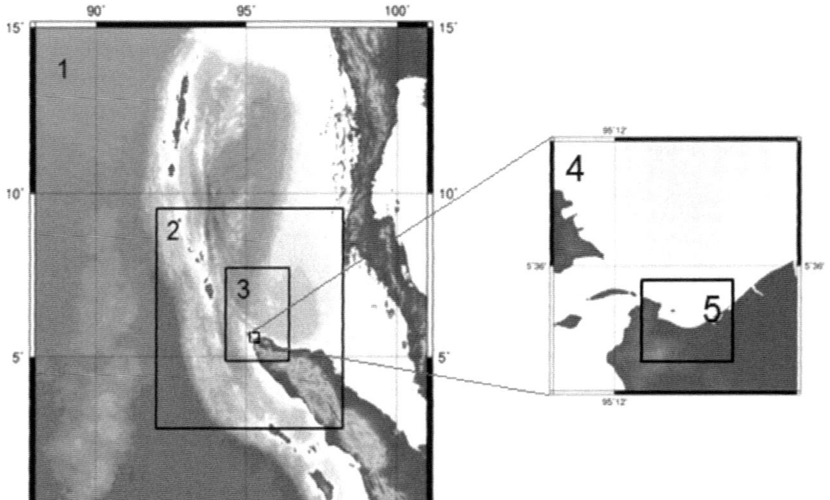

Figure 4. Simulation layer for COMCOT model

Table 1. Information on the setup of the five layer for COMCOT simulations

	Layer 1	Layer 2	Layer 3	Layer 4	Layer 5
Grid size (m)	1 min = 1851 m	0.5 min = 928 m	0.167 min = 309.33 m	0.056 min = 103.11 m	0.014 min = 25.78 m
Number of grids	nx = 777 ny = 898	nx = 716 ny = 596	nx = 537 ny = 519	nx = 213 ny = 210	nx = 184 ny = 188
Time step (sec)	1	1	1	1	1
Coordinate System	Spherical	Spherical	Spherical	Spherical	Cartesian
SWE	Linear	Linear	Linear	Linear	Nonlinear

3.3 Land cover roughness

In this study, land cover roughness was interpreted using Manning's roughness coefficient (n), which variation of land cover classification. The detailed maps of land cover on the innermost layer before tsunami digitize using Quantum GIS from IKONOS satellite imagery captured in January 2003. The Ulee Lheue Bay's land cover (before the tsunami) has classified into the beach, fishpond, vegetations area, settlements, paddy field, and forest. For each area, the Manning coefficient was adopted from [6]. The land cover classification around the Ulee Lheue Bay before tsunami are shown in Figure 5 (a). The smallest roughness value was specified for the sea (n = 0.013). Meanwhile, the smallest roughness value on land was specified for the fishpond area (n = 0.017). The value roughness (n = 0.04) is a higher roughness in this study that specified for the settlement area.

The land cover scenarios in the simulation were used to investigate the role of coastal forest on reducing tsunami wave impacts. See Figure 5 (b) and Figure 5 (c) for more details. For scenario 1, the coastal forest width is 50 m and fishpond area was set at the same width. For scenario 2, we set 100 m width for coastal forest and 100 m for fishpond area. Simulation of tsunami inundation with various Manning coefficients were performed using Delft3D model simulation. Coastal forest is assumed to be able to reduce tsunami impact. Therefore, in the simulations, we focused the observation of the results in coastal forest area.

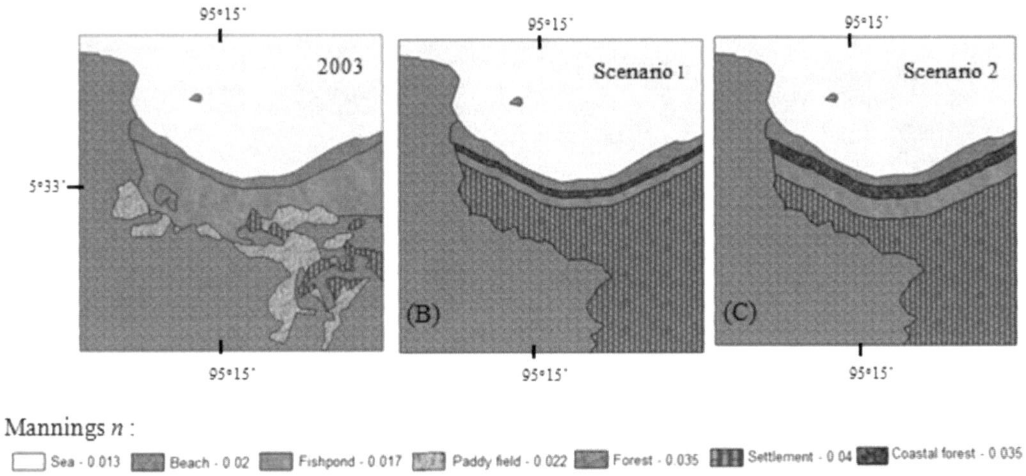

Figure 5. Roughness maps on land cover in Ulee Lheue Bay (a) 2003, (b) Coastal forest 50 m width and (c) Coastal forest 100 m width

4. Results

The numerical simulation results were analyzed to evaluate the land cover influence on the tsunami inundation area. The results are presented in the spreading of maximum tsunami inundation for three types of land cover, maximum water elevation, and flow velocities.

4.1 Influence of land cover roughness on tsunami inundation

Based on the land cover condition before the tsunami, our simulation shows that the tsunami waves inundated all of the flat area in the Ulee Lheue Bay. Only at the mountain side, tsunami waves were halted (the mountain side is indicated in dark blue in Figure 6). The steeper slope of the foothill stops tsunami propagation for inundating farther inland. The results were found similar at two models scenarios of land cover. For both scenarios (Figures 6 (c) and 6 (d)), we found a minor change in term of inundation area. The results were indicated by the small difference of color scale at the southern part of inundation limit of simulation area. For more detail of the tsunami maximum water level, the results were presented at observation points A and B. Points A and B are located in front of the mangrove area. The tsunami maximum water levels at observation points A and B are presented in Figure 7. There is no significant difference in maximum water level between the simulation using land cover condition in 2003 and with scenarios using two coastal forest widths.

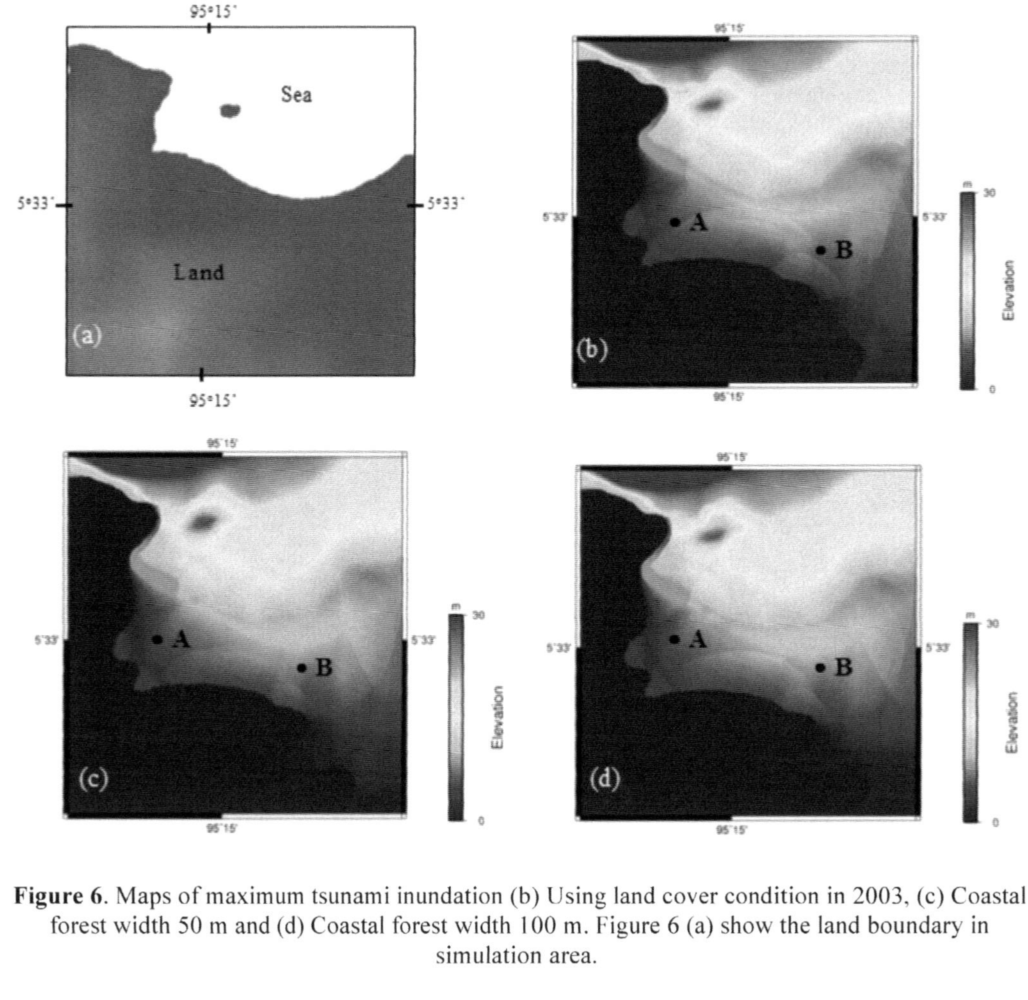

Figure 6. Maps of maximum tsunami inundation (b) Using land cover condition in 2003, (c) Coastal forest width 50 m and (d) Coastal forest width 100 m. Figure 6 (a) show the land boundary in simulation area.

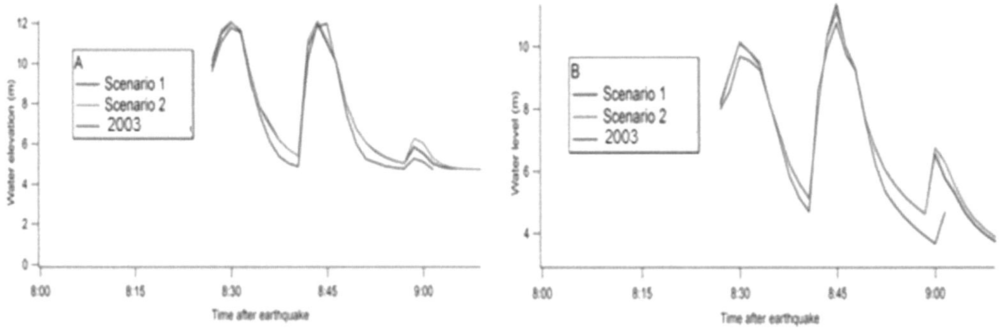

Figure 7. Differences in maximum tsunami water level

4.2 Influence of land cover roughness on tsunami velocity

On the contrary, the comparison of the velocity maximum shows more significant differences. The comparison results are shown in Figure 8. Based on observation points A and B, the purple solid line indicates the maximum tsunami velocity found in the simulation using land cover types/condition 2003 is larger than the velocity results in simulations that considered coastal forest. The first wave with the existing land cover hit the Ulee Lheue Bay with 5 m/s of velocity. On the other hand, the velocity decreased to 3.1 m/s in coastal forest with the width of 100 m and decreased to 3.0 m/s in coastal forest with the width of 50 m. The results show the velocity of the tsunami waves can be reduced approximately 40 % compared to the existing land cover in the study site.

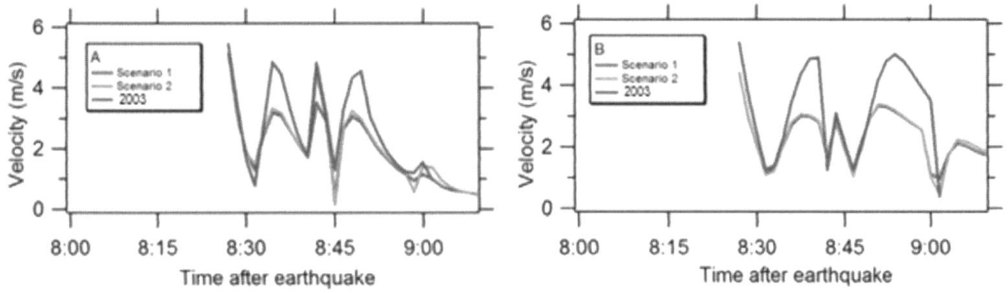

Figure 8. Differences in maximum tsunami velocity

5. Discussion

In this paper, the Manning coefficient is a sensitive parameter to response the velocity of the tsunami wave. Figure 9 presents the maximum velocity in a north-south direction for profile 3. The solid line represents the simulation across the coastline from the offshore until inundation limit. The coastal forest is located about 250 m distance from the shoreline. There is 250 m width of beach area in this cross section. The purple solid line represents the actual maximum velocity of the tsunami wave. The maximum velocity is increased with a rather consistent trend, started around the coastline until the limit of inundation. About 500 m to the land ward area, the maximum velocity in both of scenarios with coastal forest decreased with a steeper slope than the simulation results based on the land cover condition in 2003. The result shows that tsunami waves velocity decreased after being reduced by coastal forest. The 50 m width and 100 m width of coastal forests have a similar pattern of the velocity cross-profile. The coastal slope also influences to decrease of the tsunami waves velocity. These simulations gave a similar results to a study conducted in Phang Nga and Phuket, Thailand [4].

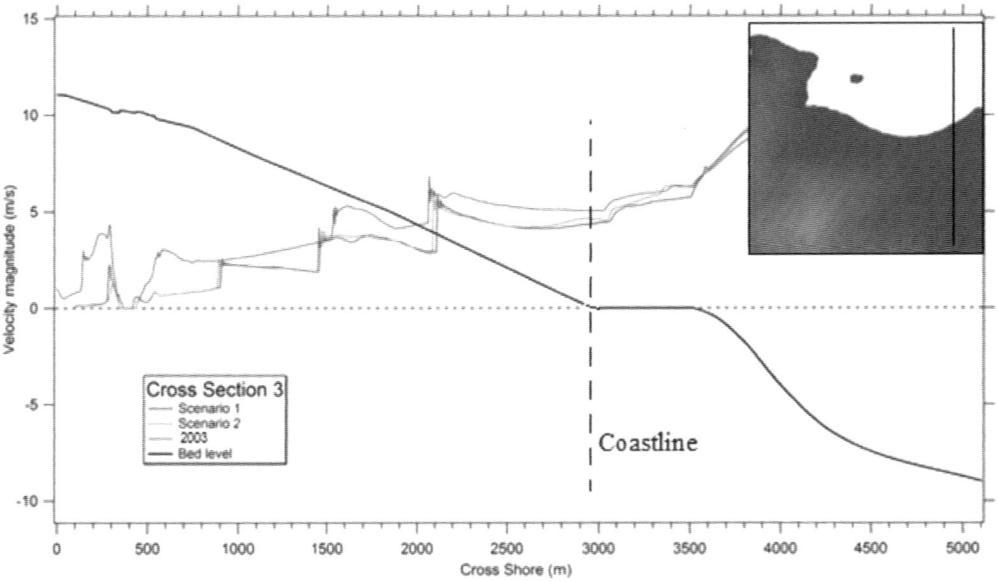

Figure 9. Maximum tsunami velocities in north-south direction for cross section 3

6. Conclusion

This paper report the numerical simulation results to investigate the role of coastal forest to reduce tsunami waves energy, in term of tsunami inundation area, the tsunami wave heights, and the wave velocity. The role of coastal forest to reduce the tsunami inundation was minor. It was shown by small differences in the area of tsunami inundation using a number of variations of coastal forest widths. However, a significant difference was seen in term of the velocity of the tsunami waves. With the land cover change to the coastal forest, the velocity of the tsunami waves can be reduced to almost 40% compared to the existing land cover found in the study site. The reducing velocity will reduce adverse impacts of future/potential tsunamis in this area. Although the topography is the most critical factor to influence the tsunami inundation pattern, the land cover roughness also needs to be considered.

7. References

[1] Gayer G, Leschka S, Nohren I, Larsen O and Gunther H 2010 Tsunami Inundation Modelling Based on Detailed Roughness Maps of Densely Populated Areas *Nat Hazards* **10** 1679-2010

[2] Harada K and Imamura F 2005 Effect of Coastal Forest on Tsunami Hazard Mitigation – A Preliminary Investigation Chapter Tsunamis Advances in *Natural and Technological Hazards Research* **23** pp 279-292

[3] Ioualalen M, Asavanant J, Kaewbanjak N, Grilli S T, Kirby J T and Watts P 2007 Modelling the 26 December 2004 Indian Ocean Tsunami: Case Study of Impact in Thailand *J. Geophys. Res* **112** C07024 doi:10.1029/2006JC003850

[4] Kaiser G, Scheele L, Kortenhaus A, Lovholt F, Romer H and Leschka S 2011 The Influence of Land Cover Roughness on the Results of High Resolution Tsunami Inundation Modelling *Nat. Hazard Earth Syst. Sci* **11** 2521-2540.

[5] Leschka S, Pedersen C and Larsen O 2009 On the Requirements for Data and Methods in Tsunami Inundation Modelling – Roughness Map and Uncertainties *Proc. of the South China Sea Tsunami Workshop*, Penang, Malaysia

[6] Li L, Qiu Q dan Huang Z 2012 Numerical Modeling of the Morphological Change in LhokNga, west Banda Aceh, during the 2004 Indian Ocean Tsunami : Understanding Tsunami Deposits Using a Forward Modeling Method, *Nat Hazard* **64** 1549-1574

[7] Ohira W, Honda K and Harada K 2012 Reduction of Tsunami Inundation by Coastal Forest in Yogyakarta, Indonesia: A Numerical Study *Nat. Hazards Earth Syst. Sci* **12** : 85-95

[8] Romano F 2009 *The Rupture Process of Resent Tsunamigenic Earthquake by Geophisical Data Inversion* Disertasi Universitas Degli Bologna

[9] Sugawara D, Goto K dan Jaffe B E 2014 Numerical Models of Tsunami Sediment Transport – Current Understanding and Future Directions *Marine Geologi* **352** 295-320

[10] Syamsidik, Rasyif TM dan Kato S 2015 Development of Accurate Tsunami Estimated Times of Arrival for tsunami-prone cities in Aceh Indonesia. *Int. J. of Disaster Risk Reduction*, **14(4)** 403-410

[11] Syamsidik, Tursina, Meutia A, Al'ala M, Fahmi M and Meilianda E 2016 Numerical Simulations of Impact of the 2004 Indian Ocean Tsunami on Coastal Morphological Changes Around the Ulee Lheue Bay of Aceh, Indonesia *J. of Earthquake and Tsunami* Vol.**11** No.2

Acknowledgments
Authors would like to thank the Partnership Enhanced Engagement in Research (PEER) promote by USAID by granting a research grant in Cycle 3 Sponsor Grant Award Number: AID-OAAA-A-11-00012 and Sub Grant Number PGA-2000004893 which funded this research with the project title" Tsunami Wave Impact on Coastal Morphological Change Based on Sediment Transport Simulation". Authors are also grateful for the member of Coastal Research Group and TDMRC Syiah Kuala University for support and help.

Lightning hazard region over the maritime continent observed from satellite and climate change threat

Y Ilhamsyah[1,2,5], Y Koesmaryono[1], R Hidayat[1], J Murjaya[3], I W Nurjaya[4] and Rizwan[1]

[1]Applied Climatology, Graduate School of Bogor Agricultural University, Bogor 16680, Indonesia
[2]Dept. of Marine Sciences, Faculty of Marine and Fisheries, Syiah Kuala University, Banda Aceh 23111, Indonesia
[3]Head of Centre for Engineering Seismology, Potential Geophysics, and Standard Time, Agency for Meteorology, Climatology, and Geophysics (BMKG), Jakarta Indonesia 10720
[4]Dept. of Marine Sciences and Technology, Faculty of Fisheries and Marine Sciences, Bogor Agricultural University, Bogor 16680, Indonesia
[5]Tsunami and Disaster Mitigation Research Center (TDMRC), Syiah Kuala University, Banda Aceh 23111, Indonesia

Abstract. Climate change would lead to such hydrometeorological disaster as: flash-flood, landslide, hailstone, lightning, and twister become more likely to happen in the future. In terms of lightning event, one research question arise of where lightning would be mostly to strike over the Maritime Continent (MC)?. The objective of the research is to investigate region with high-density of lightning activity over MC by mapping climatological features of lightning flashes derived from onboard NASA-TRMM Satellite, i.e. Optical Transient Detector/Lightning Imaging Sensor (OTD/LIS). Based on data retrieved since 1995-2013, it is seasonally observed that during transition season March to May, region with high vulnerability of lightning flashes cover the entire Sumatra Island, the Malacca Strait, and Peninsular Malaysia as well as Java Island. High-frequent of lightning activity over the Malacca Strait is unique since it is the only sea-region in the world where lightning flashes are denser. As previously mentioned that strong lightning activity over the strait is driven by mesoscale convective system of Sumatra Squalls due to convergences of land breeze between Sumatra and Peninsular Malaysia. Lightning activity over the strait is continuously observed throughout season despite the intensity reduced. Java Island, most populated island, receive high-density of lightning flashes during rainy season (December to February) but small part in the northwestern of Java Island, e.g., Bogor and surrounding areas, the density of lightning flashes are high throughout season. Northern and southern parts of Kalimantan and Central part of Sulawesi are also prone to lightning activity particularly during transition season March to May and September to November. In the eastern part of MC, Papua receive denser lightning flashes during September to November. It is found that lightning activity are mostly concentrated over land instead of ocean which is in accordance with diurnal convective precipitation event due to the existence of numerous mountainous island in MC. The malacca strait however is the only exception and turn into a unique characteristic of convective system over MC and the only sea-region in the world where lightning activity is the greatest.

1. Introduction

Lightning event often correspond to hazard, it is true as many reports in the community in which lightning is responsible for lethal impact, even the strike was also allegedly to have linked with early

2015 crash of Air Asia flight number QZ8501 which was later denied. The lightning turn into the deadliest weather phenomena over the Maritime Continent (MC), situated between 18.5N-15.5S and 90.5-162.5E.

As it is located over the tropical region with strong diurnal convection, MC grows into the most vulnerable region of lightning activity in the world [1]. Many areas over Indonesia for instance, which experience numerous amount of lightning flashes. One of the greatest is Cibinong Bogor, West Java - Indonesia, situated at 106.8E and 6.43S, is known to have the most severe lightning occurrences which was also documented in the 1988 Guiness Book of World Records. It was recorded that Cibinong Bogor experienced 322 thunderstorm days annually of only 80 days in normal. Following huge number of lightning flashes over MC, it then raise a question of which region over MC where flashes are denser mainly during transition season when it becomes more frequent to occur in this period. Whether the transition season is the highest, it also remains question since there is no intensive investigation of climatological features of lightning that cover a whole MC yet. If regions with intense lightning activity are identified, in the forthcoming years, spots of lightning hazard over MC could be widely known and well-predicted and focus could be given for anticipating future climate change threat over the regions [2].

Back to the regional density, lightning detection provided by Indonesian Agency for Meteorology, Climatology, and Geophysics (BMKG) might be employed to detect regions with denser flashes. However, detection devices installed at BMKG stations are yet in the form of lightning detection network, but still only a single-operational system which has limited coverages of 300 km away from the source. The single-system also has weakness in determining the precise location of lightning strokes which require improvement using triangulation method with stations nearby. In addition to strokes location, BMKG's lightning detection is yet to provide long-term measurement. Some ground-based stations has gathered more than ten years of lightning data particularly stations stretched in Java and North Sumatra but still many are newly-operated leading to difficulties in determining full climatology of lightning activity over MC. From space, lightning activity has been intensively monitored since 1995 using onboard optical sensor installed at NASA-Tropical Rainfall Measuring Mission (TRMM) under Optical Transient Detector/Lightning Imaging Sensor (OTD/LIS). Several research have employed one and half-years of satellite data (December 1997 to May 1999) to observe lightning distributions over Indonesian Archipelago [3, 4]. Meanwhile, from ground measurement network, since December 1994 to January 1996 lightning activity has been observed over Java Island [5]. Some research to observe diurnal variations of lightning had also been done using both LIS and ground measurement from World Wide Lightning Location Network as comparison [6, 7]. But still previous studies are yet to be sufficient to describe full climatology of lightning activity over MC. Since MC cover vast marine-continental region supported by the availability of long-term observation, thus, the utilization of lightning data derived from satellite is urgently required to identify regions with intense lightning activity over MC. On the other hand, the added value of detectable lightning flash density is that such information is increasingly demanded not only for lightning safety reason but also to be considered as part of cost-benefit analysis by leading national or private companies, such as: oil, electrical, and even insurance companies. The objectives of the research are to map lightning climatology over MC and to determine detection efficiency (DE) between satellite and local lightning detection system.

2. Materials and Methods

Lightning data are retrieved from OTD/LIS, onboard optical sensor installed at NASA-TRMM satellite [8, 9]. The device captured lightning flashes (both Intra-cloud and cloud-to-ground) since May 1995-December 2013. The data are in the gridded HDF format, available online at http://lightning.nsstc.nasa.gov/data/index.html [10, 11]. Spatial resolutions of the data are 0.5^0 longitude x 0.5^0 latitude, cover the entire MC between 18.5N-15.5S and 90.5-162.5E. Meanwhile, temporal resolution is monthly lightning flash density data. For validation, DE determination is carried out by comparing between flashes and lightning strokes density derived from BMKG's cloud-to-ground lightning detection single-system over Bogor District in 2012. DE is calculated using the following expression:

$$DE = \frac{\rho_{SAT_{an}}}{\rho_{BMKG_{an}}} \qquad (1)$$

Meanwhile, stroke density is calculated using the following expression:

$$\rho_{BMKG_{an}} = \frac{Str_{tot}}{L} \qquad (2)$$

Di mana : $\rho_{BMKG_{an}}$ = annual density of BMKG (flashes/km^2/year).
$\rho_{SAT_{an}}$ = annual density of satellite (strokes/km^2/year).
Str_{tot} = Total annual strokes (stokes/year).
L = Areas (km^2).

The gridded lightning data are read and stored in text format using HDFView. The data processing and mapping utilize MATLAB-mapping toolbox with the following script:

```
clear all
clf
load HRMC_COM_FR3.txt
flash1=HRMC_COM_FR3(100:220,400:700);
load HRMC_COM_FR4.txt
flash2=HRMC_COM_FR4(100:220,400:700);
load HRMC_COM_FR5.txt
flash3=HRMC_COM_FR5(100:220,400:700);
flashtotal=(flash1+flash2+flash3);
%------------------------
load Latitude_monthly.txt
lat=Latitude_monthly(100:220);
load Longitude_monthly.txt
lon=Longitude_monthly(400:700);
max(max(flashtotal))
axesm('MapProjection','mercator','MapLatLimit',[-18 19],'MapLonLimit',[88 165])
framem;
plabel('PlabelLocation',5); mlabel('MlabelLocation',10);
contourfm(lat,lon,flashtotal,0.05:0.01:0.4)
colorbar
load coast
geoshow(lat, long,'Color','red');
axis off
```

3. Results and Discussion

According to 2013 Intergovernmental Panel on Climate Change (IPCC) report, it is projected that in the near-future tropical region will be "very likely" to "suffer from" weather and climate extreme events. There is a curious statement that wet areas become wetter, why so?. It then certainly implies the tropical region including the MC. MC would tend to be wetter because the accumulation of heat would concentrate over the region which have an effect on the increase of sea-surface temperature. As temperature increase, the release of water vapor in the atmosphere would be higher and convective system would grow stronger and spread widely, cover a whole MC. These processes becomes vulnerable to the development of deep cumulonimbus clouds that initiate the growth of multicell to supercell thunderstorms. The event is often regarded as local disaster where torrential rain resulted from intense thunderstorms could lead to flash-flood, landslide and frequently accompanied by

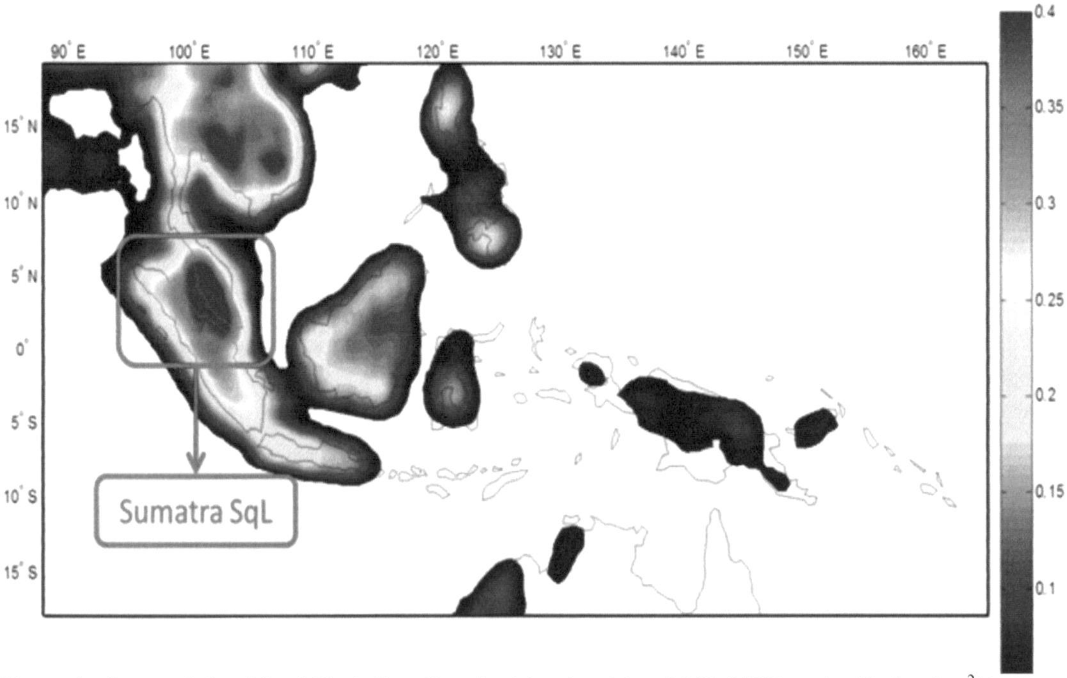

Figure 1 . Seasonal Combined Flash Rate Density Mar-Apr-May (1995-2013), units (flashes/km^2/day). Sumatra SqL refer to Sumatra Squalls

vulnerable on intense thunderstorms particularly supercell thunderstorms. Similarly, the release of latent heat which is continuing to increase also trigger the development of deep convective clouds over the ocean, and when merge with convective clouds from the land, it might drive the formation of mesoscale convective system accompanied with lines of thunderstorm (refer to squall-line thunderstorm), distributed widely and cover most parts of MC. One of the famous squall line thunderstorms that is Sumatra Squall-lines that occurred over the Malacca Strait due to convergences of land breeze coming from Sumatra Island and Peninsular Malaysia that produce heavy rainfall and frequent lightning activity in the morning along eastern coast of the island. This is one example, and of course, MC will be vulnerable to future hidrometeorological disaster as a result of the changing climate, e.g., heavy rainfall, hailstone, twister, lightning and perhaps these events has been underway in the present time. In order to anticipate such worst events, it is necessary to map regions with high vulnerability of hydrometeorological disaster over MC. Thus, adaptation could be implemented to cope with future climate change impact.

In relation to lightning occurrence, based on data retrieved since 1995-2013, it is seasonally observed that during transition season March to May, region with high vulnerability of lightning flashes cover a whole Sumatra Island, the Malacca Strait, and Peninsular Malaysia as well as Java Island (Figure 1). Strong intensity of lightning activity over the Malacca Strait is unique since it is the only sea-region in the world where lightning flashes are the greatest. As previously mentioned that strong lightning activity over the strait is driven by mesoscale convective system of Sumatra Squalls due to convergences of land breeze between Sumatra Island and Peninsular Malaysia. Lightning activity over the strait is continuously observed throughout season despite the intensity reduce.

Meanwhile, Java Island, most populated island, receive high-density of lightning flashes during rainy season (December to February) but small part in the northwestern of Java Island, e.g., Bogor and surrounding areas, the density of lightning flashes are high throughout season (Figure 2).

Northern and southern parts of Kalimantan and Central part of Sulawesi are also vulnerable to lightning activity particularly during transition season March to May and September to November. In

IOP Conf. Series: Earth and Environmental Science **56** (2017) 012010 doi:10.1088/1755-1315/56/1/012010

the eastern part of MC, Papua receive denser lightning flashes during September to November (Figure 3).

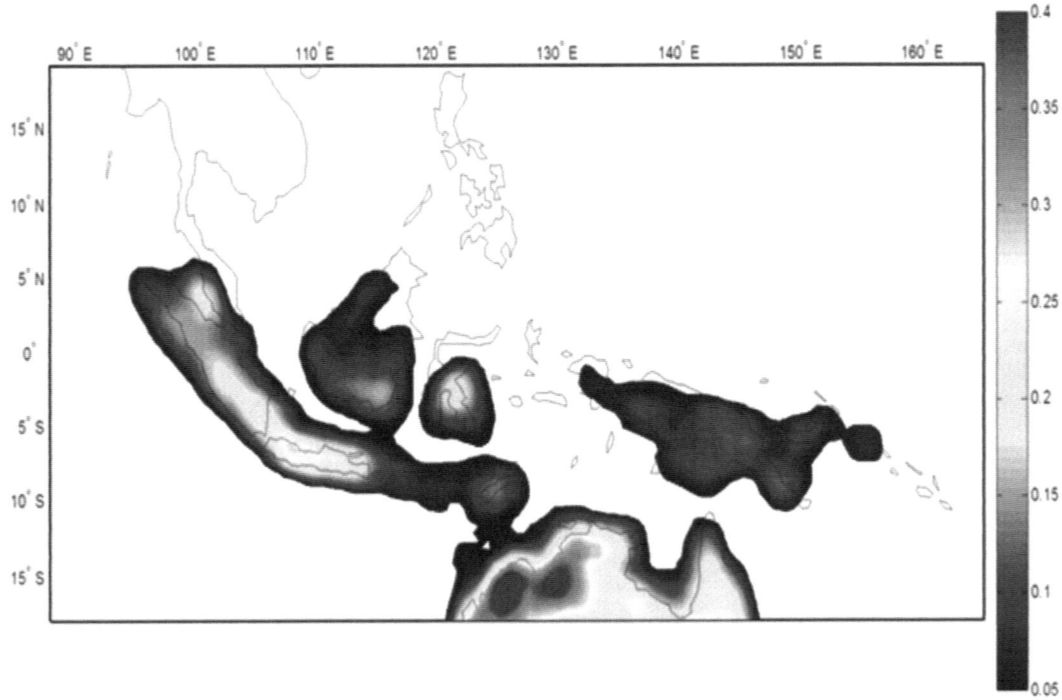

Figure 2. Seasonal Combined Flash Rate Density Dec-Jan-Feb (1995-2013), units (flashes/km^2/day).

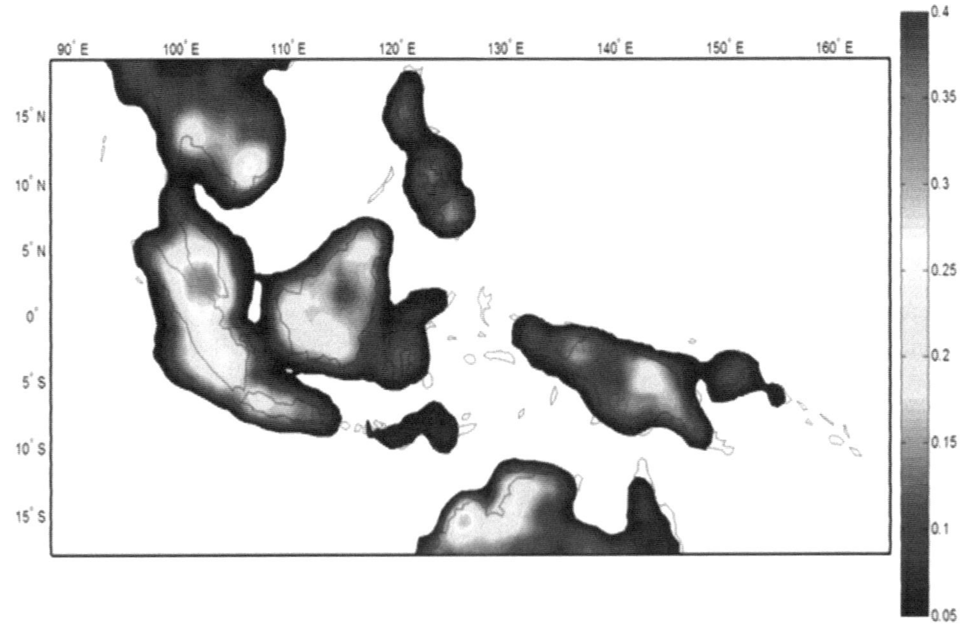

Figure 3. Seasonal Combined Flash Rate Density Sept-Oct-Nov (1995-2013), units (flashes/km^2/day).

It is interesting to observe that during dry season, lightning achieve higher density in the western part of MC (e.g., Sumatra Island, the Malacca Strait, and Peninsular Malaysia). Dry season particularly over Sumatra often relate to forestfire, whether high-density of lightning activity over the region has strong correlation on forestfire?, it remains interesting subject to be investigated further. One thing that could be taken into account that convective system over the region is very active, even during dry season it might lead to atmospheric disturbances over the region (Figure 4).

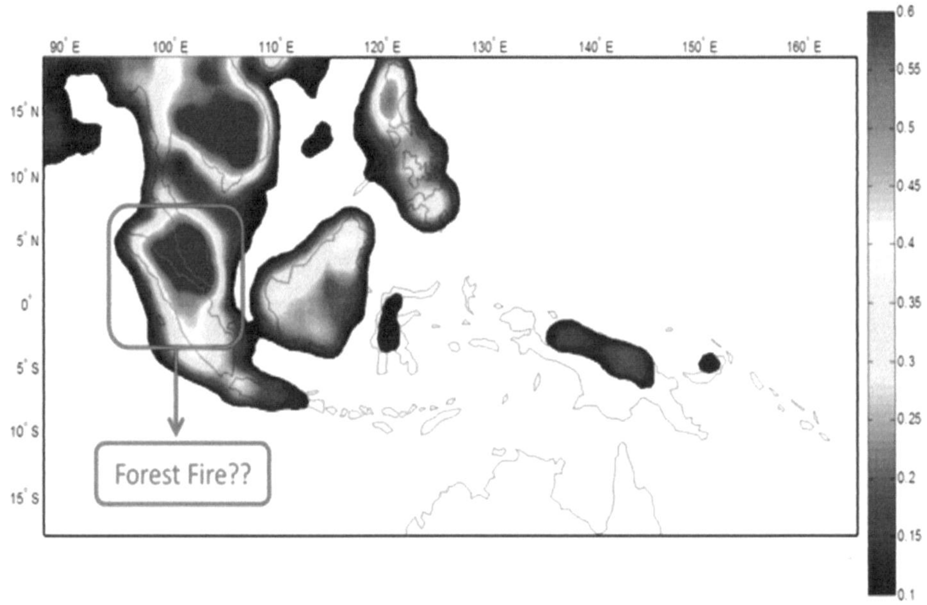

Figure 4. Seasonal Combined Flash Rate Density Jun-Jul-Aug (1995-2013), units (flashes/km^2/day).

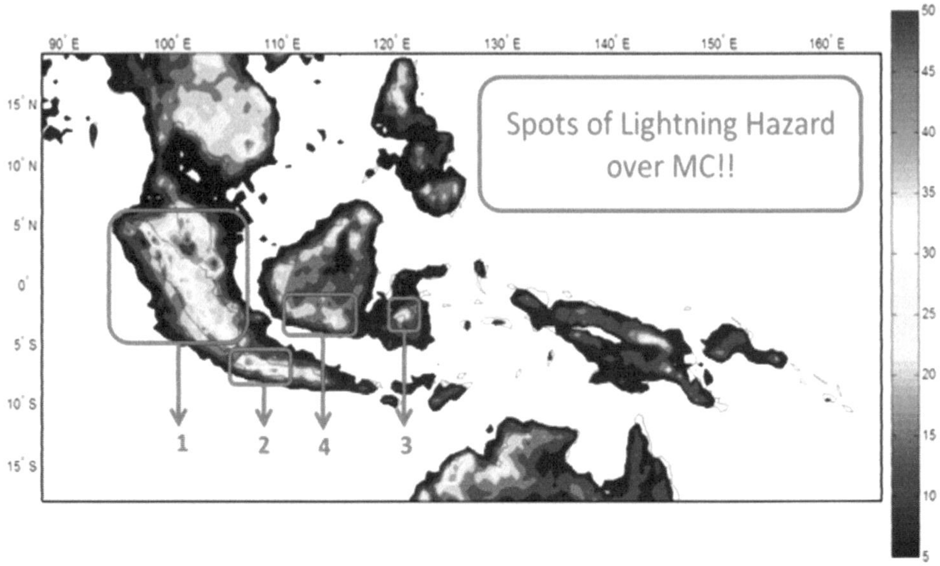

Figure 5. Combined Flash Rate Density Climatology (1995-2013), units (flashes/km^2/year).

Based on seasonal and full climatology observation, it is well-identified that there are four spots of intense lightning activity which later correspond to lightning hazard region over MC. Sumatra remains at the top of lightning hazard region, while Java is the second vulnerable region of lightning hazard. Central Sulawesi and South Kalimantan are the third and fourth lightning hazard region over MC, respectively (Figure 5). In the near-future where climate change becomes unequivocal and hydrometeorological hazard is very likely to occur, the regions require crucial attention to anticipate severe lightning disaster. In addition, The mapping of lightning climatology would benefit to assist the development of lightning protection around industrial areas and transmission lines, to harvest lightning for future renewable energy, to describe the development of potential agricultural areas, describe areas of potential mining.

It is found that lightning activity are mostly concentrated over land instead of ocean which is in accordance with diurnal convective precipitation event due to the existence of numerous mountainous island in MC [12]. The malacca strait however is the only exception and turn into a unique characteristic of convective system over MC and the only sea-region in the world where lightning activity is the greatest.

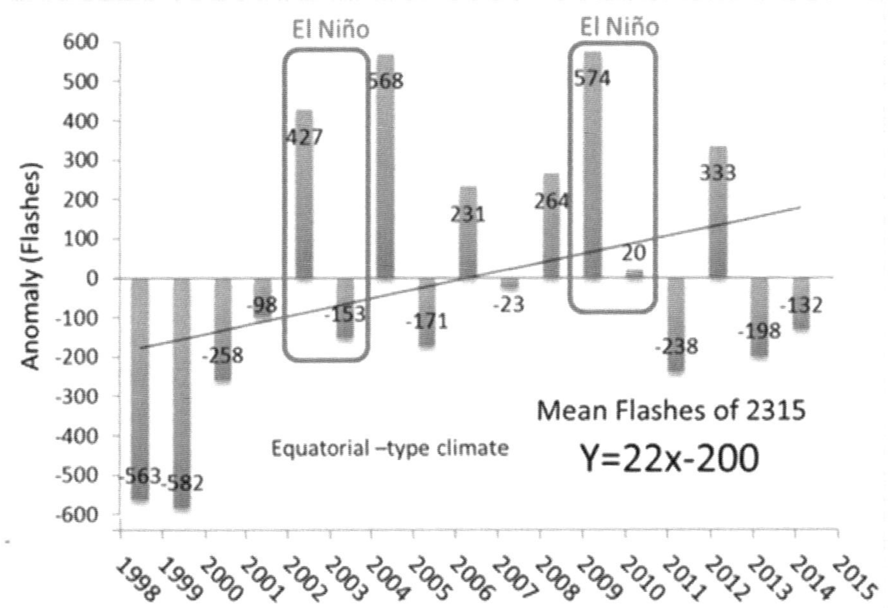

Figure 6. Anomaly of Annual Trends of Lightning Climatology over the Malacca Strait (1998 2014), units (flashes).

In relation to climate change, it is shown that anomaly of annual trends of lightning flashes over the Malacca Strait and Greater Jakarta Areas (Jabodetabek) have significantly increased from year to year (Figure 6 and 7). The Malacca Strait, equatorial-type climate, showed that in the period of El Niño 2002-2003 and 2009-2010, lightning is higher in the beginning of the event. On the contrary, over Jabodetabek which is monsoonal-type climate, lightning is lower at first. It is interesting to investigate more details of the relationship between lightning flashes and El Niño event particularly in mosoonal-type climate region which is largely affected by El Niño.

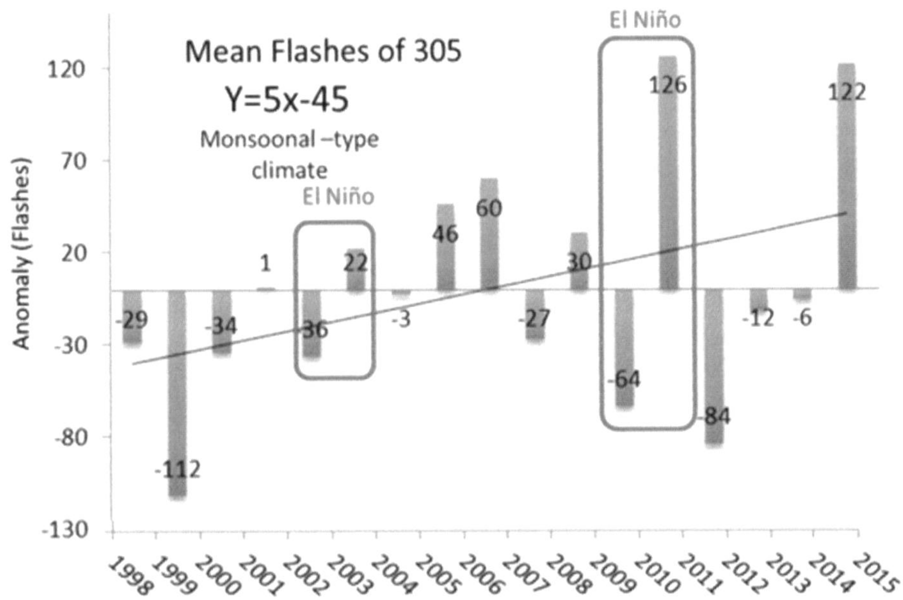

Figure 7. Anomaly of Annual Trends of Lightning Climatology over Greater Jakarta Areas, Jabodetabek (1998 2014), units (flashes).

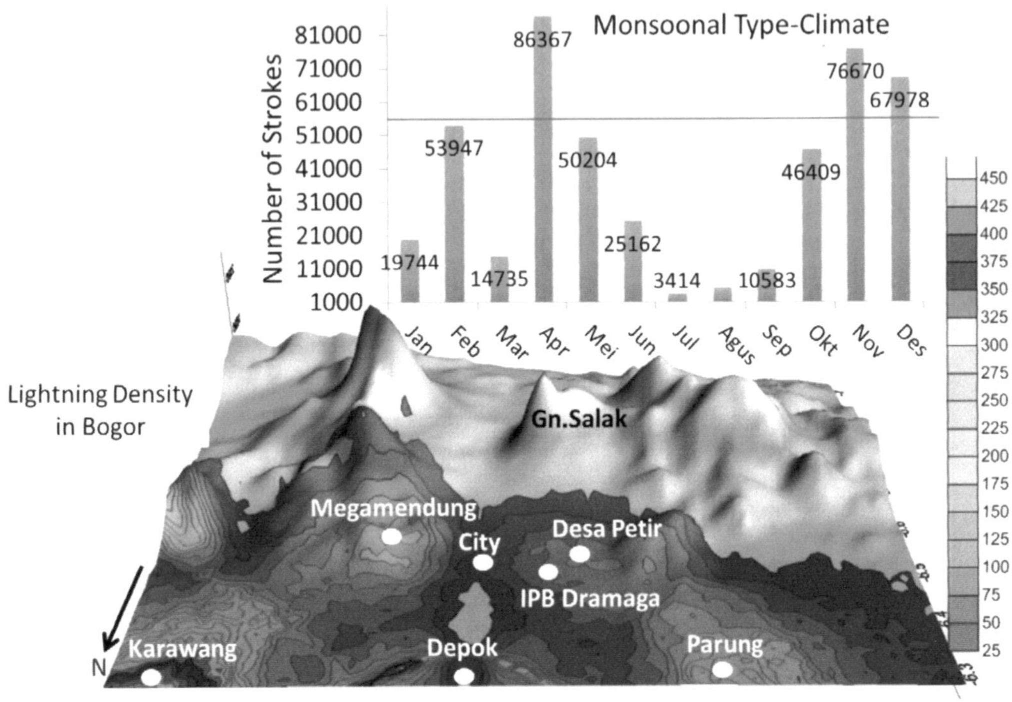

Figure 8. Spots of lightning hazard in Bogor District.

For DE, Bogor District, 1988 Guiness book of world records of thunderstorm days, is chosen for lightning validation between flashes derived from satellite and strokes detected by BMKG's lightning detection. It is obtained from equation (2) that lightning density over Bogor District is 50 strokes/km^2/year (Figure 8). According to Figure 5, flash density over Bogor District observed from satellite is 45 flashes/km^2/year. By using equation (1), we arrive at 90 % of DE, indicating that lightning activities are well-observed from space. Thus, satellite service is superior in its application on lightning investigation [13].

4. Concluding remarks

Specifically, regions with high-density of lightning activity are well-identified seasonally and annually. Most of them are stretched in the eastern coast of Sumatera, the Malacca Strait, Peninsular Malaya, West Java, South Kalimantan and Central Sulawesi. DE is nearly-perfect. Generally, lightning activity are mostly concentrated over land instead of ocean. Climate change threat become likely to occur near-future as indicated by the increase of annual trends of lightning activity over MC.

Acknowledgments

Authors express gratitude to Ministry of Research, Technology, and Higher Education Indonesia under Postgraduate Scholarship Program for financial support of the research.

5. References

[1] Kodama YM, Tokuda M and Murata F 2006 *J. Meteor. Soc. Japan* **84A** 133-149
[2] IPCC 2013 *The Physical Science Basis* (Geneva: Working group I contribution to the IPCC fifth assessment report climate change 2013)
[3] Hamid EY, Kawasaki ZI and Matsuura K 1999 *Journal of Atmospheric Electricity* **19** 153-164
[4] Hamid EY, Kawasaki ZI and Mardiana R 2000 *Power Engineering Society Winter Meeting IEEE* **3** 2077-2081.
[5] Hidayat S and Ishii M 1998 *J. Geophys. Res.-Atmos.* **103(D12)** 14001–14009
[6] Virts KS, Wallace JM, Hutchins ML and Holzworth RH 2013 *J. Atmos. Sci.* **70** 3128-3146
[7] Virts KS, Wallace JM, Hutchins ML and Holzworth RH 2013 *Bull. Amer. Meteor. Soc.* **94** 1381-1391.
[8] Buechler DE, Koshak, WJ, Christian HJ and Goodman SJ 2014 *Atmos. Res.* **135-136** 397-403
[9] Boccippio DJ, Koshak WJ and Blakeslee RJ 2002 *J. Atmos. Oceanic Technol.* **19** 1318-1332
[10] Cecil DJ 2006 *LIS/OTD 0.5 Degree High Resolution Monthly Climatology (HRMC)* (Alabama: NASA Global Hydrology Resource Center DAAC)
[11] Cecil DJ, Buechler DE and Blakeslee RJ 2014 *Atmos. Res.* **135-136** 404-414
[12] Qian J-H 2008 *J. Atmos. Sci.* **65** 1428-1441.
[13] Rudlosky SD and Shea DT 2013 *Geophys.Res. Lett.* **40** 1–5

Developing estimating equations of fatality ratio based on surveyed data of the 2011 Great East Japan Tsunami

P Latcharote[1], N Leelawat[1], A Suppasri[1] and F Imamura[1]

[1]Tsunami Engineering Laboratory, International Research Institute of Disaster Science, Tohoku University, 468-1, Aoba, Aramaki-Aza, Aoba, Sendai, 980-0845, Japan

Abstract. The 2011 Great East Japan tsunami caused a wide range of devastating tsunami with maximum tsunami height of 40 m and 19,000 casualties especially along the Tohoku coast of Japan. The purpose of this study is to develop estimating equations of fatality ratio from tsunami arrival time for future tsunami loss assessment and investigate the effect of two coastal topography types namely, Sanriku-ria coast and Sendai plain. In this study, fatality ratio was defined as number of fatality divided by total number of people in a small scale of towns along the shoreline and tsunami arrival time was calculated from TUNAMI modelling with nesting-grids of 1350 m, 450 m, 150 m, and 50 m. Then, linear and nonlinear regression analysis were performed to develop a relationship model between fatality ratio and tsunami arrival time. Based on the results, a strong correlation that fatality ratio decreases with longer arrival time was found in both Sanriku-ria coast and Sendai plain. For different coastal types, different distributions of fatality ratio with tsunami arrival time are observed, in which fatality ratio of Sendai plain is higher than that of Sanriku ria-coast at the same arrival time generally.

1. Introduction

The 2011 Great East Japan tsunami generated by a magnitude 9.0 earthquake caused various kinds of damage, including a large number of fatalities, whereas fatalities caused by ground shaking were less severe [1]. The Tohoku coast comprises of two coastal topography types namely, Sanriku ria-coast and Sendai plain as shown in figure 1. Based on past studies about fatality ratio of people during the 2011 Great East Japan tsunami, influential factors on fatality ratio can be classified to tsunami characteristic of a focused area and personal characteristic of evacuation perspective [2]. Previous study found high correlation between coastal topography and inundation depth from the 2011 tsunami [3]. Difference in the coastal topography types lead to difference in tsunami characteristic and personal characteristic of evacuation. The tsunami characteristic is a factor of tsunami height, tsunami arrival time, coastal type, etc. and the personal characteristic is a factor of age, gender, preparedness, occupation, etc [2]. A relationship model between fatalities and influential factors should be proposed for contributing to local government and insurance society for future tsunami loss assessment, such as the Nankai tsunami. In general, tsunami height is a key factor which has a significant effect on a trend of fatality ratio. However, tsunami arrival time is also important for evacuation of people to safety areas.

Some past studies have discussed on fatality ratio in a large scale, which may include population outside of tsunami inundation zone and inconsistency due to such large scale. These assumptions may lead to error in developing the relationship between fatalities and influential factors. In addition, several past studies focused on tsunami height or inundation depth as the main factor affecting fatalities [2][4][5]. This study proposed a relationship model between fatality ratio and tsunami arrival time based on the surveyed data in each coastal affected area and investigate effect of two coastal topography types, which are related not only the tsunami characteristic but also experience of historical events. The surveyed data of fatalities in each small scale of towns were collected by the reconstruction support survey archive [6]. In this study, a statistical analysis was performed, and several estimating equations for fatality ratio from different definitions of tsunami arrival time were developed. The results of this study are expected to help the estimation of fatalities with other considerations and also to predict fatality ratio in future tsunamis, which have shorter arrival time than the 2011 Great East Japan tsunami.

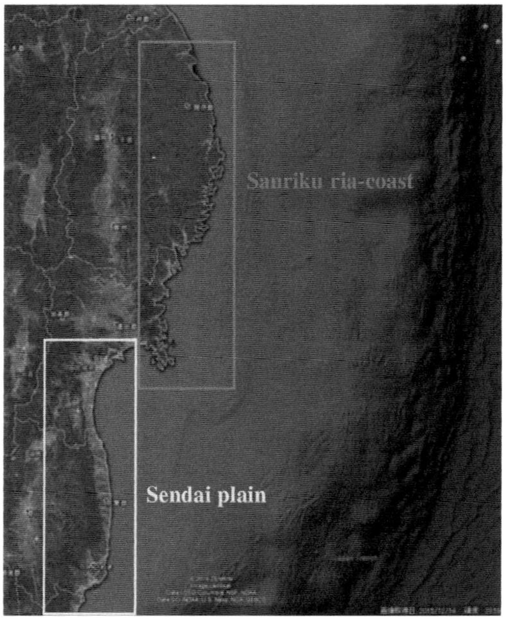

Figure 1. Coastal topography type along the Tohoku coast of Japan (Google Earth).

2. Methodology

2.1. Numerical tsunami simulation

In order to obtain tsunami arrival time at each town along the shoreline, numerical tsunami simulation was performed by TUNAMI modelling with nesting grids of 1350 m, 450 m, 150 m, and 50 m as shown in figure 2. Tsunami arrival time was calculated from the finest grid at the shoreline of each town.

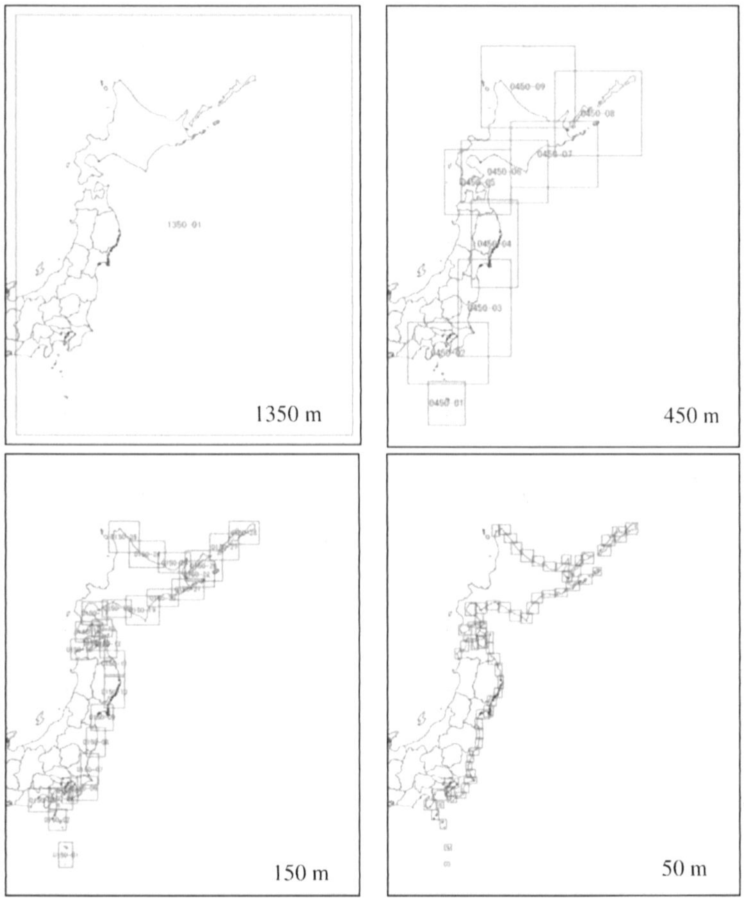

Figure 2. Tsunami modelling grids of 1350 m, 450 m, 150 m, and 50 m.

TUNAMI modelling was originally developed by Tohoku University aiming to simulate tsunami propagation and inundation. Numerical tsunami simulation was performed to collect tsunami arrival time at the shoreline inside the 50-m grid. The selected towns in this study are based on the location in insurance data, which cannot be disclosed. Tsunami arrival time was calculated in 46 towns along Sanriku ria-coast from Misawa city in Aomori prefecture to Iwate prefecture until Ishinomaki city in Miyagi prefecture and in 17 towns along Sendai plain from Ishinomaki city to Kitaibaraki city in Ibaraki prefecture. Initial seafloor deformation was calculated by a rectangular fault model [7] using the fault parameters of the 2011 Great East Japan earthquake with 55-subfault model [8] as shown in figure 3.

Figure 3. Initial seafloor deformation from 55-subfault model.

The results of numerical tsunami simulation are classified to:

- Initial: tsunami arrival time when inundation depth is equal or greater than 0.5 m [9]
- First: tsunami arrival time when the first peak of tsunami flow occurs
- Max: tsunami arrival time when maximum inundation depth occurs

2.2. Validation of tsunami arrival time

Table 1. Tsunami arrival time from surveyed data and numerical simulation

Coastal Type	Prefecture	City	Town	Tsunami arrival time (min)			
				Survey	Simulation		
					Initial	First	Max
Sanriku	Aomori	Misawa	Mikawame	-	54	55	65
Sanriku	Iwate	Kuji	Osanai	47	46	47	49
Sanriku	Iwate	Kunohe	Noda	-	42	46	48
Sanriku	Iwate	Shimohei	Fudai	-	37	37	37
Sanriku	Iwate	Shimohei	Shimanokoshi	-	31	32	36
Sanriku	Iwate	Shimohei	Wano	-	35	37	37
Sanriku	Iwate	Shimohei	Iwaizumi	-	28	29	34
Sanriku	Iwate	Miyako	Omoe	72	23	24	30
Sanriku	Iwate	Miyako	Taro	31	28	30	34
Sanriku	Iwate	Miyako	Koganji	72	31	32	36
Sanriku	Iwate	Miyako	Takahama	-	35	38	46
Sanriku	Iwate	Shimohei	Yamada	-	28	29	39
Sanriku	Iwate	Shimohei	Sakaida	-	29	30	35
Sanriku	Iwate	Shimohei	Dai13Chiwari	-	23	26	31
Sanriku	Iwate	Kamihei	Otsuchi	22	24	27	34
Sanriku	Iwate	Kamaishi	Heita	-	21	23	34
Sanriku	Iwate	Kamaishi	Hakozaki	24	27	30	36
Sanriku	Iwate	Ofunato	Massaki	27	24	27	38
Sanriku	Iwate	Ofunato	Yunagomae	25	26	37	40

Sanriku	Iwate	Ofunato	Takonoura	-	24	26	39
Sanriku	Iwate	Ofunato	Yamaguchi	-	29	33	43
Sanriku	Iwate	Ofunato	Yoshihama	27	14	16	32
Sanriku	Iwate	Ofunato	Okirai	27	20	21	31
Sanriku	Iwate	Ofunato	Ryori	27	20	21	36
Sanriku	Iwate	Rikuzen-Takata	Kesen	39	35	40	43
Sanriku	Iwate	Rikuzen-Takata	Otomo	41	40	47	47
Sanriku	Iwate	Rikuzen-Takata	Hirota	39	29	32	37
Sanriku	Iwate	Rikuzen-Takata	Yonesaki	40	33	37	42
Sanriku	Miyagi	Kesennuma	Karakuwa	38	23	26	38
Sanriku	Miyagi	Kesennuma	Asane	41	30	31	40
Sanriku	Miyagi	Kesennuma	Nagaisofunahara	41	31	31	43
Sanriku	Miyagi	Kesennuma	Motoyoshi	39	28	30	36
Sanriku	Miyagi	Kesennuma	Minato	55	43	48	50
Sanriku	Miyagi	Kesennuma	Akaiwaminato	55	41	46	46
Sanriku	Miyagi	Minami-sanriku	Natari	-	24	27	40
Sanriku	Miyagi	Minami-sanriku	Shizugawa	-	32	34	46
Sanriku	Miyagi	Ishinomaki	Kitakami	-	27	30	43
Sanriku	Miyagi	Ishinomaki	Ogatsu	-	31	40	40
Sanriku	Miyagi	Ishinomaki	Ojima	-	30	38	38
Sanriku	Miyagi	Ishinomaki	Ayukawahama	-	18	21	38
Sanriku	Miyagi	Ishinomaki	Futawatashihama	-	25	27	41
Sanriku	Miyagi	Ishinomaki	Yagawahama	-	28	32	58
Sanriku	Miyagi	Ishinomaki	Yoriisohama	-	24	29	60
Sanriku	Miyagi	Ishinomaki	Kyubunhama	-	26	31	72
Sanriku	Miyagi	Ishinomaki	Watanoha	-	52	55	55
Sanriku	Miyagi	Ishinomaki	Kitsunezakihama	-	38	41	53

Coastal Type	Prefecture	City	Town	Tsunami arrival time (min)			
				Survey	Simulation		
					Initial	First	Max
Sendai	Miyagi	Ishinomaki	Sakana	-	55	58	62
Sendai	Miyagi	Ishinomaki	Shiotomi	-	60	66	66
Sendai	Miyagi	Onagawa	Onagawahama	39	22	26	41
Sendai	Miyagi	Onagawa	Izushima	-	26	37	43
Sendai	Miyagi	Higashi-Matsushima	Omagari	62	42	50	53
Sendai	Miyagi	Higashi-Matsushima	Miyato	132	48	55	58
Sendai	Miyagi	Higashi-Matsushima	Shintona	135	59	60	60
Sendai	Miyagi	Shiogama	Uratosabusawa	-	57	59	59
Sendai	Miyagi	Shiogama	Shinhama	-	62	63	63
Sendai	Miyagi	Shinichigahama	Hanabushihama	-	50	50	57
Sendai	Miyagi	Natori	Yuriage	67	59	61	61
Sendai	Miyagi	Watari	Arahama	-	54	57	61
Sendai	Miyagi	Watari	Yamamoto	-	49	50	56
Sendai	Fukushima	Soma	Iwanoko	-	48	49	55
Sendai	Fukushima	Soma	Obama	-	41	42	53
Sendai	Fukushima	Shinchi	Yachigoya	-	47	52	54
Sendai	Ibaraki	Kita-Ibaraki	Hirakata	-	36	36	41
Sendai	Ibaraki	Kita-Ibaraki	Otsucho	-	42	42	42

Table 1 shows tsunami arrival time from surveyed data investigated by means of stopped clocked eyewitnesses after the 2011 Great East Japan tsunami [10]. Since only few surveyed data were observed, tsunami arrival time was further collected from numerical tsunami simulation. Validation of the simulation was performed using the Root Mean Square Error (RMSE), and the parameters K and $\varkappa$ proposed by Aida (1978) for reproduction of tsunami arrival time in this study. Tsunami arrival time from numerical simulation was validated using the surveyed data as shown in figure 4.

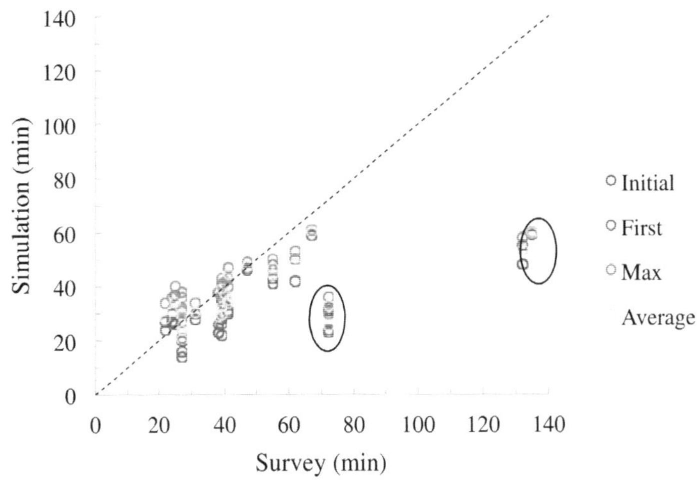

Figure 4. The comparison of tsunami arrival time between simulation and survey.

The values of RMSE, K, and $\varkappa$ were used for validation of tsunami arrival time. The Japan Society of Civil Engineering (JSCE) suggests that $0.95 < K < 1.05$ and $\varkappa < 1.45$ is recommended as "Good agreement" in the evaluation of tsunami source and propagation model. Table 2 shows the results of validation from 26 towns and 22 towns, neglecting Omoe, Koganji, Miyato, and Shintona town, which provide very long arrival time in the surveyed data as shown in figure 4. The $\varkappa$ value can satisfy the standard for all cases. After removing the outliers (4 towns; *see* two circles in figure 4), the values of RMSE and K become much better but still cannot satisfy the standard. Therefore, this numerical simulation is somewhat valid and reliable for reproduction of tsunami arrival time.

Table 2. The results of validation from 26 towns and 22 towns

	26 towns		
	RMSE	K	κ
Initial	27 min	1.40	1.38
First	25 min	1.27	1.42
Max	24 min	1.04	1.44
Average	25 min	1.21	1.40

	22 towns		
	RMSE	K	κ
Initial	10 min	1.25	1.22
First	8 min	1.12	1.24
Max	7 min	0.91	1.20
Average	7 min	1.07	1.19

2.3. Comparison of fatality ratio

The comparison of fatality ratio aimed to check a trend of the insurance data relating to the surveyed data. The surveyed data in this study were collected by the reconstruction support survey archive [6] and prefectural office [11]. Since the insurance data cannot be disclosed, the surveyed data was used instead to develop the relationship between fatality ratio and tsunami arrival time. Figure 5 shows the comparison of fatality ratio between the insurance data and the surveyed data. The result of correlation analysis is shown in figure 5.

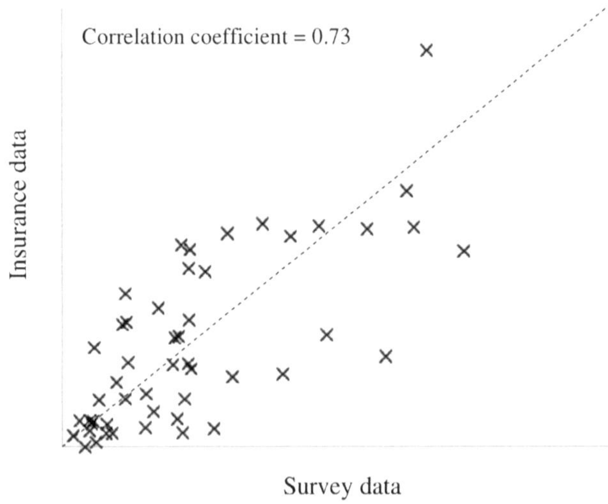

Figure 5. The comparison of fatality ratio between insurance data and surveyed data.

2.4. Regression analysis

2.4.1. Linear regression analysis. In order to develop a linear relationship between fatality ratio and tsunami arrival time represented by estimating equations for future tsunami loss assessment. The method of linear regression analysis is performed by IBM SPSS Statistics (Version 23). Linear regression equation:

$$FR = aT + b \tag{2.1}$$

In which FR = fatality ratio; T = tsunami arrival time; a = slope value of function, which should be a negative value; b = intercept value.

The satisfaction of proposed estimating equations can be checked by the value of R^2, which is obtained from the results of linear regression analysis (R^2= 1.00 satisfy perfectly). The independent variable (tsunami arrival time) and the intercept value are significant, if p-value is less than 0.05.

2.4.2. Nonlinear regression analysis. Due to the scatterplot in the preliminary analysis (*see* Section 4), there is possibility to have the nonlinear relationship. In order to develop a nonlinear estimating equation assuming that fatality ratio increases dramatically with shorter period of tsunami arrival time, the method of nonlinear regression analysis is also performed by IBM SPSS Statistics (Version 23). Logarithmic regression equation:

$$FR = a*\ln(T) + b \tag{2.2}$$

In which FR = fatality ratio; T = tsunami arrival time; a = constant value, which should be a negative value; b = constant value, which should be a positive value.

Inverse regression equation:

$$FR = a*(1/T) + b \tag{2.3}$$

In which a = constant value, which should be a positive value; b = constant value, which can be a positive or negative value.

3. Fatality ratio vs tsunami arrival time

Table 3 shows fatality data arranged by each town, which was separated to Sanriku ria-coast and Sendai plain. The fatality data of each town based on the criteria of small affected area near the coastline was classified to population in inundated area (*N*), number of fatality (D), and fatality ratio (FR). The fatality ratio (FR) was defined as D divided by *N* which is a dependent variable in linear regression analysis. For an independent variable, tsunami arrival time was classified to three cases of tsunami arrival time when tsunami inundation depth is equal or greater than 0.5 m (Initial), tsunami arrival time when the first peak of tsunami inundation depth occurs (First), and tsunami arrival time when maximum inundation depth occurs (Max). Table 3 shows the surveyed data of 63 towns in Iwate, Miyagi, Fukushima, and Ibaraki prefectures, separating to 46 towns in Sanriku ria-coast and 17 towns in Sendai plain.

Table 3. Fatality data for 45 towns in Sanriku ria-coast and 17 towns in Sendai plain

Coastal Type	Town	Fatality data		
		N	D	FR
Sanriku	Mikawame	139	2	1.44%
Sanriku	Osanai	4925	0	0.00%
Sanriku	Noda	4632	27	0.58%
Sanriku	Fudai	3088	8	0.26%
Sanriku	Shimanokoshi	1277	0	0.00%
Sanriku	Wano	3843	24	0.62%
Sanriku	Iwaizumi	1324	0	0.00%
Sanriku	Omoe	1562	17	1.09%
Sanriku	Taro	3062	120	3.92%
Sanriku	Koganji	205	11	5.37%
Sanriku	Takahama	841	3	0.36%
Sanriku	Yamada	410	37	9.02%
Sanriku	Sakaida	286	3	1.05%
Sanriku	Dai13Chiwari	97	2	2.06%
Sanriku	Otsuchi	635	37	5.83%
Sanriku	Heita	3751	24	0.64%
Sanriku	Hakozaki	1274	79	6.20%
Sanriku	Massaki	4718	41	0.87%
Sanriku	Yunagomae	568	3	0.53%
Sanriku	Takonoura	267	6	2.25%
Sanriku	Yamaguchi	251	2	0.80%
Sanriku	Okirai	288	8	2.78%
Sanriku	Ryori	166	1	0.60%
Sanriku	Kesen	566	26	4.59%
Sanriku	Otomo	1911	52	2.72%
Sanriku	Hirota	457	5	1.09%
Sanriku	Yonesaki	472	12	2.54%
Sanriku	Karakuwa	378	0	0.00%
Sanriku	Asane	198	1	0.51%
Sanriku	Nagaisofunahara	2498	21	0.84%
Sanriku	Motoyoshi	321	1	0.31%
Sanriku	Minato	115	4	3.48%
Sanriku	Akaiwaminato	247	12	4.86%
Sanriku	Natari	5068	74	1.46%
Sanriku	Shizugawa	8180	326	3.99%

Sanriku	Kitakami	3896	175	4.49%
Sanriku	Ogatsu	4300	122	2.84%
Sanriku	Ojima	3994	162	4.06%
Sanriku	Ayukawahama	1462	18	1.23%
Sanriku	Futawatashihama	335	0	0.00%
Sanriku	Yagawahama	245	12	4.90%
Sanriku	Yoriisohama	411	10	2.43%
Sanriku	Kyubunhama	243	1	0.41%
Sanriku	Watanoha	7157	226	3.16%
Sanriku	Kitsunezakihama	156	3	1.92%
Sendai	Sakana	130	3	2.31%
Sendai	Shiotomi	864	5	0.58%
Sendai	Onagawahama	1778	175	9.84%
Sendai	Izushima	161	4	2.48%
Sendai	Omagari	7019	313	4.46%
Sendai	Miyato	950	11	1.16%
Sendai	Shintona	243	7	2.88%
Sendai	Uratosabusawa	161	3	1.86%
Sendai	Shinhama	1472	0	0.00%
Sendai	Hanabushihama	416	5	1.20%
Sendai	Arahama	479	12	2.51%
Sendai	Yamamoto	3976	200	5.03%
Sendai	Iwanoko	583	1	0.17%
Sendai	Obama	2460	54	2.20%
Sendai	Yachigoya	1997	33	1.65%
Sendai	Hirakata	2232	2	0.09%
Sendai	Otsucho	1924	2	0.10%

The relationships between fatality ratio in table 3 and tsunami arrival time for the case of Initial, First, and Max in table 2 are shown in figure 6. For Sanriku ria-coast, it was divided into two zones in order to develop a trend that fatality ratio decreases with longer arrival time as well as Sendai plain. As can be seen in figure 6, fatality ratio of Sendai plain is much higher than that of Zone 1 in Sanriku ria-coast and little higher than that of Zone 2.

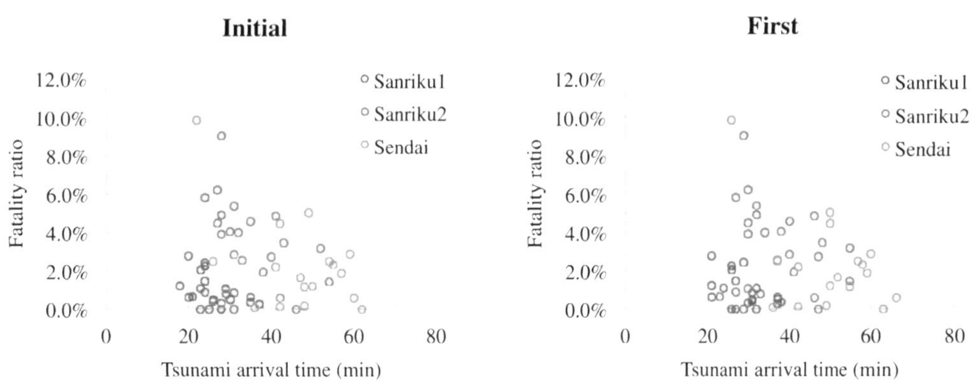

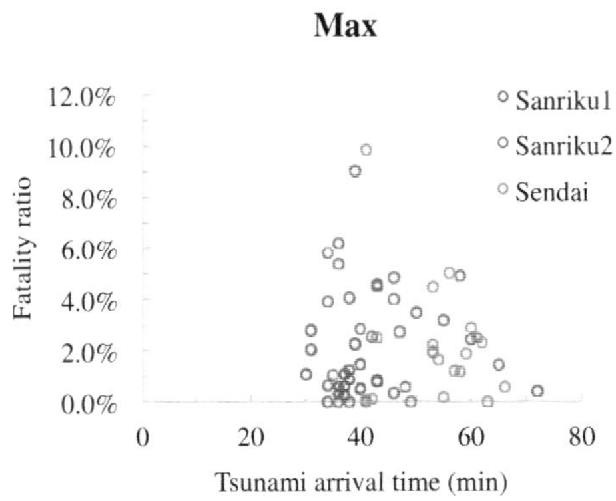

Figure 6. The scatterplots showing relationship between fatality ratio and tsunami arrival time for Initial, First, and Max.

As it can be seen in figure 6, all available data from the 2011 Great East Japan tsunami contains few data for the first 20 minutes of the tsunami arrival time. In order to develop the relationship that covers not only long tsunami arrival time but also short arrival time, which in turn, giving more practical contribution for tsunami loss assessment; more data from previous event were included in the analysis. Table 4 shows the preliminary information of the additional data from the 1993 Okushiri tsunami, collected by Okushiri town office. Tsunami arrival time was calculated by TUNAMI modelling as the same approach as the 2011 tsunami.

Table 4. Fatality data and tsunami arrival time in the 1993 Okushiri tsunami

Town	Fatality data			Tsunami arrival time		
	N	D	FR	First	Initial	Max
Okushiri	242	31	12.81%	10	11	13
Tamaura	50	3	6.00%	9	10	14
Aonae	907	87	9.59%	6	7	7
Inaho	76	13	17.11%	6	6	7

4. Results of statistical analysis

4.1. Linear regression analysis

Linear regression analysis was performed to develop estimating equations of fatality ratio from tsunami arrival time. The results of linear regression analysis provide number of sample (*N*), the value of R^2, and the values of p_1 and p_2 for independent variable and intercept value. The relationships between fatality ratio and tsunami arrival time are shown in figures 7-9 for the case of Initial, First, and Max, respectively. Figure 7 shows fatality ratio of 24 towns in Sanriku ria-coast (Zone 1) and 2 towns in Okushiri which decreases with longer arrival time.

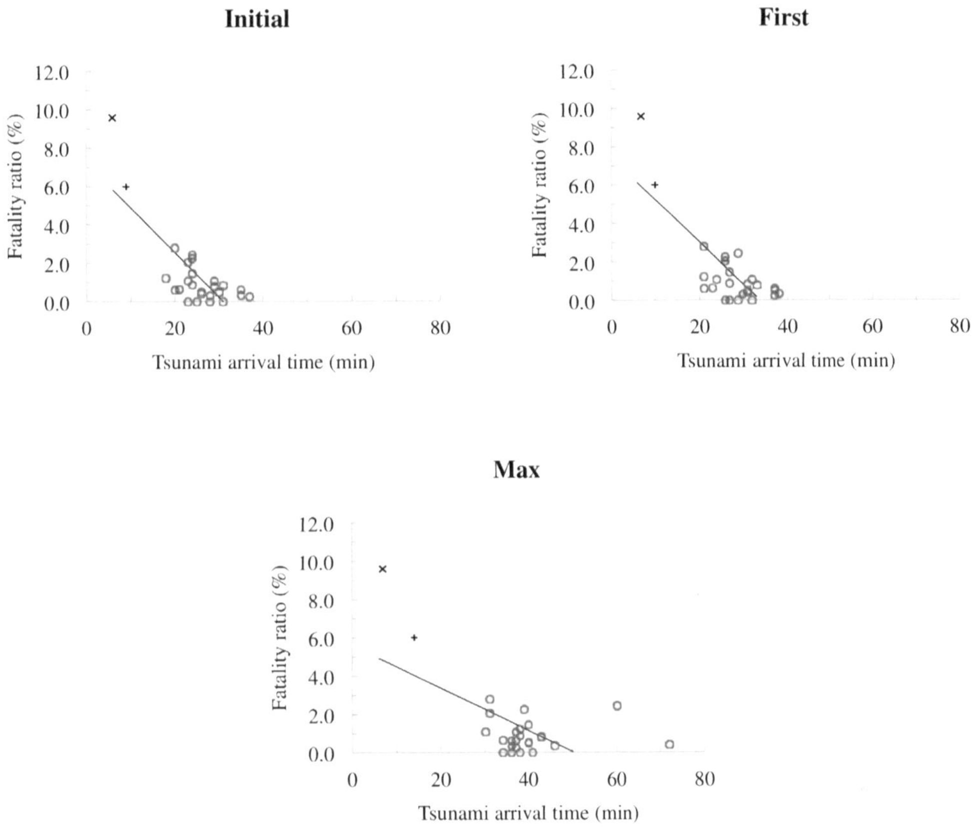

Figure 7. The results of linear regression analysis in Sanriku ria-coast (Zone 1)

For the case of Initial, First, and Max, three estimating equations of Sanriku ria-coast (Zone 1) were developed from the results of linear regression analysis as in equations (4.1) – (4.3). Based on ANOVA analysis, all three model significantly predict FR ($F_{(4.1)}$ = 37.95, $p_{(4.1)} < 0.001$; $F_{(4.2)}$ = 38.39, $p_{(4.2)} < 0.001$; $F_{(4.3)}$ = 15.75, $p_{(4.3)} < 0.001$).

$$FR = -0.23T + 7.21 \quad (N = 26, R^2 = 0.61, p_1 < 0.001, p_2 < 0.001) \quad (4.1)$$

$$FR = -0.22T + 7.44 \quad (N = 26, R^2 = 0.62, p_1 < 0.001, p_2 < 0.001) \quad (4.2)$$

$$FR = -0.11T + 5.59 \quad (N = 26, R^2 = 0.40, p_1 < 0.001, p_2 < 0.001) \quad (4.3)$$

The case of First provides the best fit among three estimating equations in which equals to 0.61 and 0.62, respectively. In addition, all of the predictors (i.e., tsunami arrival time) are significantly contributing to the equations based on the p-value.

Figure 8 shows fatality ratio of 21 towns in Sanriku ria-coast (Zone 2) and 2 towns in Okushiri which decreases with longer arrival time. As can be seen in figure 6, fatality ratio of Zone 1 is higher than that of Zone 2 in Sanriku ria-coast.

IOP Conf. Series: Earth and Environmental Science **56** (2017) 012011 doi:10.1088/1755-1315/56/1/012011

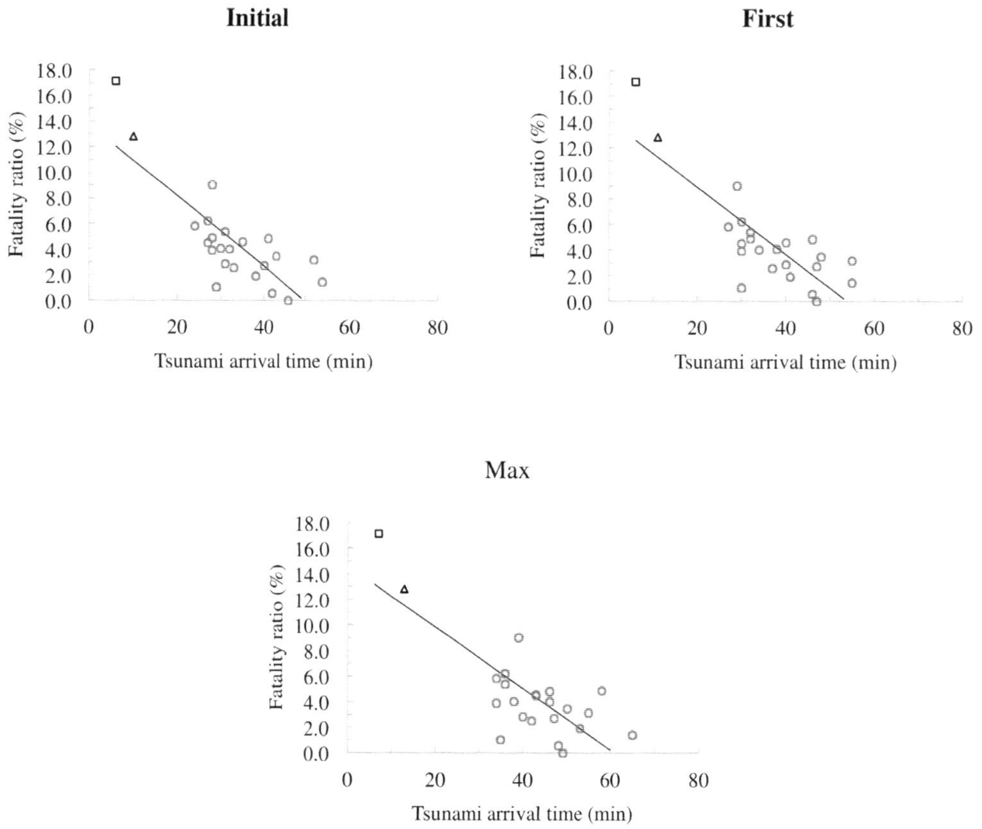

Figure 8. The results of linear regression analysis in Sanriku ria-coast (Zone 2)

For the case of Initial, First, and Max, three estimating equations of Sanriku ria-coast (Zone 2) were developed from the results of linear regression analysis as in equations (4.4) – (4.6). All three models significantly predict FR ($F_{(4.4)} = 37.88$, $p_{(4.4)} < 0.001$; $F_{(4.5)} = 42.66$, $p_{(4.5)} < 0.001$; $F_{(4.6)} = 35.83$, $p_{(4.6)} < 0.001$).

$$\mathrm{FR} = -0.27\mathrm{T} + 13.67 \quad (N = 23, R^2 = 0.64, p_1 < 0.001, p_2 < 0.001) \tag{4.4}$$

$$\mathrm{FR} = -0.26\mathrm{T} + 14.15 \quad (N = 23, R^2 = 0.67, p_1 < 0.001, p_2 < 0.001) \tag{4.5}$$

$$\mathrm{FR} = -0.24\mathrm{T} + 14.63 \quad (N = 23, R^2 = 0.63, p_1 < 0.001, p_2 < 0.001) \tag{4.6}$$

The case of First provides the best fit among three estimating equations in which equals to 0.67. In addition, all of the predictors are significantly contributing to the equations based on the *p*-value.
Figure 9 shows fatality ratio of 17 towns in Sendai plain and 2 towns in Okushiri which decreases with longer arrival time.

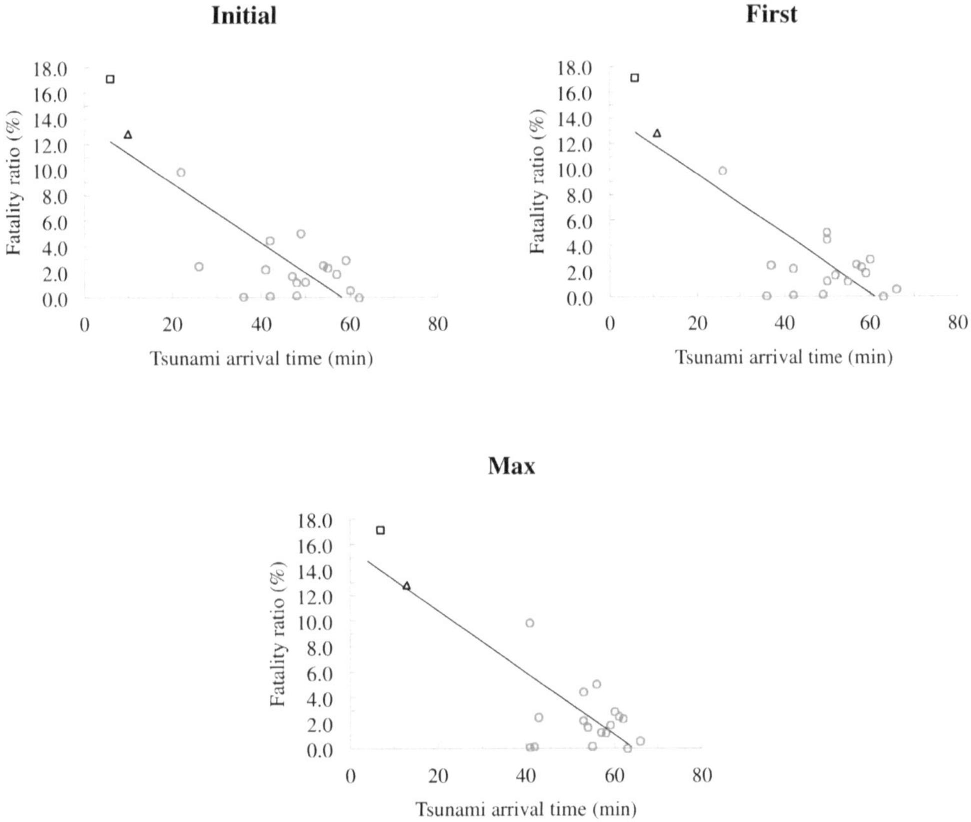

Figure 9. The results of linear regression analysis in Sendai plain

For the case of Initial, First, and Max, three estimating equations of Sendai plain were developed from the results of linear regression analysis as in equations (4.7) – (4.9). All three models significantly predict FR ($F_{(4.7)} = 33.64$, $p_{(4.7)} < 0.001$; $F_{(4.8)} = 36.48$, $p_{(4.8)} < 0.001$; $F_{(4.9)} = 37.32$, $p_{(4.9)} < 0.001$).

$$FR = -0.23T + 13.63 \quad (N = 19, R^2 = 0.66, p_1 < 0.001, p_2 < 0.001) \tag{4.7}$$

$$FR = -0.24T + 14.26 \quad (N = 19, R^2 = 0.68, p_1 < 0.001, p_2 < 0.001) \tag{4.8}$$

$$FR = -0.24T + 15.67 \quad (N = 19, R^2 = 0.69, p_1 < 0.001, p_2 < 0.001) \tag{4.9}$$

The case of Max provides the best fit among three estimating equations in which equals to 0.69. In addition, all of the predictors are significantly contributing to the equations based on the *p*-value.

In comparison of estimating equations between Sanriku ria-coast (Zone 2) and Sendai plain, it was found that fatality ratio of Sendai plain is obviously higher than that of Zone 2 only for the case of Max. For both Sanriku ria-coast and Sendai plain, the different definitions of tsunami arrival time provide the significant difference in estimating fatality ratio.

4.2. Nonlinear regression analysis

With the same data as previous section, nonlinear regression analysis was performed to develop estimating equations of fatality ratio from tsunami arrival time. The results of nonlinear regression analysis provide number of sample (N), the value of R^2, and the values of p_1 and p_2 for independent variable and intercept value as same as linear regression analysis. The relationships between fatality ratio and tsunami arrival time are shown in figures 10-12 for the case of Initial, First, and Max, respectively. Figure 10 shows fatality ratio of 24 towns in Sanriku ria-coast (Zone 1) and 2 towns in Okushiri which decreases with longer arrival time.

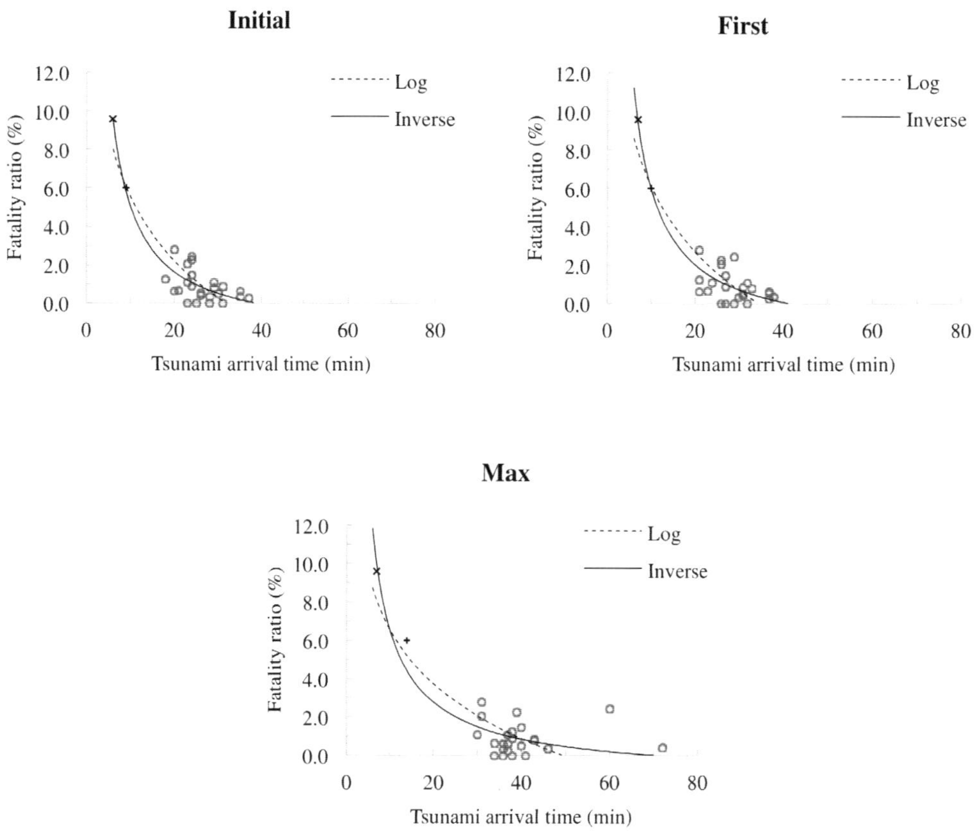

Figure 10. The results of nonlinear regression analysis in Sanriku ria-coast (Zone 1)

For the case of Initial, First, and Max, six estimating equations of Sanriku ria-coast (Zone 1) were developed from the results of logarithmic regression analysis as in equations (4.10) – (4.12) and inverse regression analysis as in equations (4.13) – (4.15). Based on the ANOVA analysis, both the logarithmic regression models ($F_{(4.10)} = 101.66$, $p_{(4.10)} < 0.001$; $F_{(4.11)} = 93.63$, $p_{(4.11)} < 0.001$; $F_{(4.12)} = 61.34$, $p_{(4.12)} < 0.001$) and the inverse regression models ($F_{(4.13)} = 193.57$, $p_{(4.13)} < 0.001$; $F_{(4.14)} = 179.17$, $p_{(4.14)} < 0.001$; $F_{(4.15)} = 116.70$, $p_{(4.15)} < 0.001$) significantly predict FR.

$$FR = -4.80\ln(T) + 16.59 \quad (N = 26,\ R^2 = 0.81,\ p_1 < 0.001,\ p_2 < 0.001) \tag{4.10}$$

$$FR = -4.92\ln(T) + 17.44 \quad (N = 26,\ R^2 = 0.80,\ p_1 < 0.001,\ p_2 < 0.001) \tag{4.11}$$

$$FR = -4.14\ln(T) + 16.16 \quad (N = 26,\ R^2 = 0.72,\ p_1 < 0.001,\ p_2 < 0.001) \tag{4.12}$$

$$FR = 67.83/T - 1.76 \quad (N = 26,\ R^2 = 0.89,\ p_1 < 0.001,\ p_2 < 0.001) \tag{4.13}$$

$$FR = 78.85/T - 1.91 \quad (N = 26,\ R^2 = 0.88,\ p_1 < 0.001,\ p_2 < 0.001) \tag{4.14}$$

$$FR = 77.36/T - 1.08 \quad (N = 26, R^2 = 0.83, p_1 < 0.001, p_2 < 0.001) \tag{4.15}$$

The logarithmic regression and the inverse regression analyses were conducted to determine the best nonlinear relationship model for predicting FR. By comparing between the logarithm regression models and the inverse regression models, the values of R^2 of the inverse regression models are higher than the logarithmic regression models for all Initial, First, and Max cases of Zone 1. In addition, the inverse regression models are more practical because the solid lines can cover longer tsunami arrival time.

Figure 11 shows fatality ratio of 21 towns in Sanriku ria-coast (Zone 2) and 2 towns in Okushiri which decreases with longer arrival time.

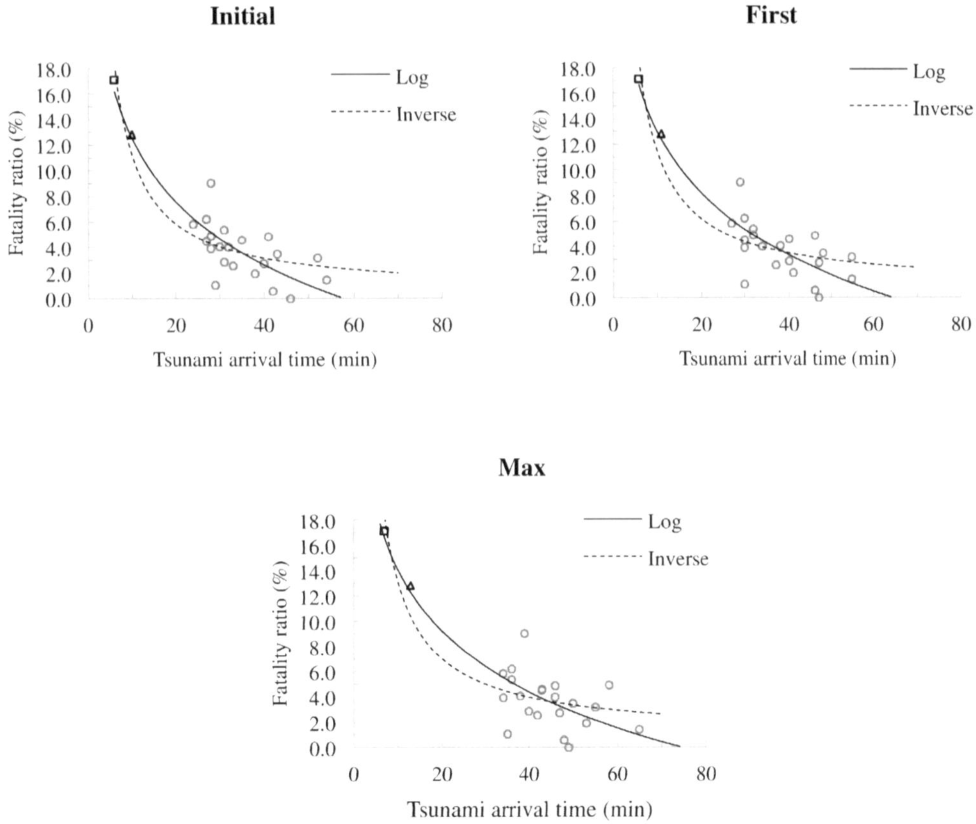

Figure 11. The results of nonlinear regression analysis in Sanriku ria-coast (Zone 2)

For the case of Initial, First, and Max, six estimating equations of Sanriku ria-coast (Zone 2) were developed from the results of logarithmic regression analysis as in equations (4.16) – (4.18) and inverse regression analysis as in equations (4.19) – (4.21). Both the logarithmic regression models ($F_{(4.16)} = 90.65$, $p_{(4.16)} < 0.001$; $F_{(4.17)} = 92.95$, $p_{(4.17)} < 0.001$; $F_{(4.18)} = 67.92$, $p_{(4.18)} < 0.001$) and the inverse regression models ($F_{(4.19)} = 82.58$, $p_{(4.19)} < 0.001$; $F_{(4.20)} = 74.91$, $p_{(4.20)} < 0.001$; $F_{(4.21)} = 61.47$, $p_{(4.21)} < 0.001$) significantly predict FR.

$$FR = -7.16\ln(T) + 29.04 \quad (N = 23, R^2 = 0.81, p_1 < 0.001, p_2 < 0.001) \tag{4.16}$$

$$FR = -7.01\ln(T) + 29.17 \quad (N = 23, R^2 = 0.82, p_1 < 0.001, p_2 < 0.001) \tag{4.17}$$

$$FR = -7.03\ln(T) + 30.28 \quad (N = 23, R^2 = 0.76, p_1 < 0.001, p_2 < 0.001) \tag{4.18}$$

$$FR = 107.98/T + 0.45 \quad (N = 23, R^2 = 0.80, p_1 < 0.001, p_2 > 0.1) \tag{4.19}$$

$$FR = 107.22/T + 0.80 \quad (N = 23, R^2 = 0.78, p_1 < 0.001, p_2 > 0.1) \tag{4.20}$$

$$FR = 123.13/T + 0.87 \quad (N = 23, R^2 = 0.75, p_1 < 0.001, p_2 > 0.1) \tag{4.21}$$

The same nonlinear regression analyses were conducted for Zone 2. The value of R^2 of the logarithmic regression models are higher than the inverse regression models' R^2 for all Initial, First, and Max cases. In addition, the inverse regression models are more practical because FR of the solid lines reaches to zero at the definite tsunami arrival time as well as figure 12.

Figure 12 shows fatality ratio of 17 towns in Sendai plain and 2 towns in Okushiri which decreases with longer arrival time.

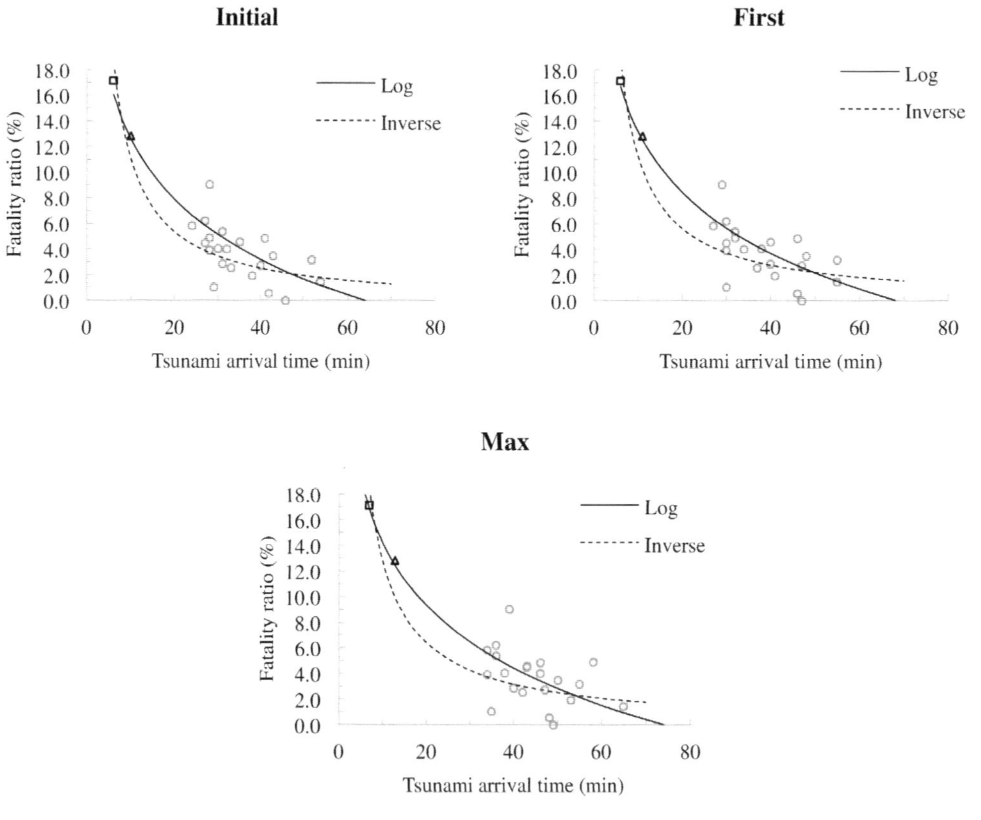

Figure 12. The results of nonlinear regression analysis in Sendai plain

For the case of Initial, First, and Max, six estimating equations of Sendai plain were developed from the results of logarithmic regression analysis as in equations (4.22) – (4.24) and inverse regression analysis as in equations (4.25) – (4.27). Both the logarithmic regression models ($F_{(4.22)} = 76.82$, $p_{(4.22)} < 0.001$; $F_{(4.23)} = 76.77$, $p_{(4.23)} < 0.001$; $F_{(4.24)} = 56.69$, $p_{(4.24)} < 0.001$) and the inverse regression models

($F_{(4.25)} = 78.39$, $p_{(4.25)} < 0.001$; $F_{(4.26)} = 65.15$, $p_{(4.26)} < 0.001$; $F_{(4.27)} = 50.26$, $p_{(4.27)} < 0.001$) significantly predict FR.

$$FR = -6.77\ln(T) + 28.16 \quad (N = 19, R^2 = 0.82, p_1 < 0.001, p_2 < 0.001) \quad (4.22)$$

$$FR = -6.84\ln(T) + 28.92 \quad (N = 19, R^2 = 0.82, p_1 < 0.001, p_2 < 0.001) \quad (4.23)$$

$$FR = -7.13\ln(T) + 30.73 \quad (N = 19, R^2 = 0.77, p_1 < 0.001, p_2 < 0.001) \quad (4.24)$$

$$FR = 113.60/T - 0.33 \quad (N = 19, R^2 = 0.82, p_1 < 0.001, p_2 > 0.1) \quad (4.25)$$

$$FR = 113.49/T - 0.08 \quad (N = 19, R^2 = 0.79, p_1 < 0.001, p_2 > 0.1) \quad (4.26)$$

$$FR = 130.72/T - 0.11 \quad (N = 19, R^2 = 0.75, p_1 < 0.001, p_2 > 0.1) \quad (4.27)$$

Lastly, the same nonlinear analysis techniques were applied with the Sendai plain. The value of R^2 of the inverse regression model is higher than the logarithmic regression model for Initial case. Noted that R^2 of (4.22) was rounded up from 0.819 and R^2 of (4.23) was rounded up from 0.822. On the other hand, the values R^2 of the logarithmic regression models are higher than the inverse regression models' R^2 for First and Max cases. The p-values of the constant in the inverse models (4.25, 4.26, 4.27) show not significant.

5. Conclusions

This study developed estimating equations of fatality ratio based on fatality data of the 2011 Great East Japan tsunami and the 1993 Okushiri tsunami. The different definitions of tsunami arrival time provide the significant difference in estimating fatality ratio from both linear and nonlinear regression analysis. Considering the effect of coastal topography types, a strong correlation between fatality ratio and tsunami arrival time in Sendai plain, whereas Sanriku ria-coast was divided into two zones in order to obtain the strong correlation due different experience and awareness of people.

Based on all results of the linear regression models, all relationship models are good fit. The tsunami arrival time, as a predictor, in all models is significantly contributing to the equations. Since the nature of the data is possible to be nonlinear relationship, the analysis using the nonlinear techniques were done after the linear regression analysis. The results found that it is possible to fit the logarithmic function and the inverse function. Similar to the results of the linear regression analysis, all relationship models are good fit, and the predictor is significantly contributing to the equations. Then, by comparing the linear regression models and the nonlinear models using mean of R^2, there is significant difference between linear regression models' and logarithmic regression models' ($t = -8.726$, $p < 0.001$) and between linear regression models' and inverse regression models' ($t = -4.872$, $p < 0.005$). Thus, the logarithmic and inverse models seem to be more suitable to be used to predict the fatality ratio.

The difference in mean for the R^2 between the logarithmic regression model and the inverse regression model found no significance ($t = -0.828$, $p > 0.1$). Therefore, it is likely that both proposed models can be interchangeable.

However, is it important to consider the limitations of this study. First, the data used in the analyses model is multiple source data (i.e., 2011 Great East Japan tsunami and 1993 Okushiri tsunami) with the intention to provide the prediction in all tsunami arrival time length. While main amount of data came from the 2011 Great East Japan tsunami, only few data were added to the short arrival time length. Second, the statistical models were developed based on the event in specific area. In order to generalize the result, more data from different areas might be necessary to take into account.

6. References

[1] Suppasri A, Koshimura S, Imai K, Mas E, Gokon H, Muhari A and Imamura F 2012 *Coastal Engineering Journal* 54(1) 1250008
[2] Yun N Y and Hamada M 2015 *Earthquake Spectra* 31(3) 1237-1265
[3] Leelawat N, Suppasri A, Charvet I and Imamura F 2014 *Nat. Hazards* 73 449-71
[4] Ueda R 2012 *Proceedings of Institute of Social Safety Science* 18 443-450
[5] Aldrich D P and Sawada Y 2015 *Social Science and Medicine* 124 66-75
[6] Reconstruction Support Survey Archive 2016 *Victims situation* http://fukkou.csis.u-tokyo.ac.jp/
[7] Okada, Y 1985 *Bulletin Seismological Society America* 75 1135-1154
[8] Satake K, Fuji Y, Harada T and Namegaya Y 2013 *Bulletin Seismological Society America*
[9] Endoh K and Takahashi S 1995 *Coastal Engineering 1994* 1016-1029
[10] Muhari A, Imamura F, Suppasri A and Mas E 2012 *Journal of Natural Disaster Science* 34(1) 91-104
[11] Tani K 2012 *Geography research report* (Faculty of Education, Saitama University)

Acknowledgments

The fatality data belong to the damage field survey of the 2011 Great East Japan tsunami provided by Reconstruction Support Survey Archive, University of Tokyo. This research was funded by the Willis Research Network (WRN) under the Pan-Asian/Oceanian tsunami risk modeling through the International Research Institute of Disaster Science (IRIDeS) at Tohoku University.

A decade process of coastal land use changes in Peukan Bada-Aceh after the 2004 Indian Ocean Tsunami

M Fahmi [1,2,4], Syamsidik[1,3,*], E Fatimah [1,2] and M Al'ala [1,2]

[1]Tsunami Computation and Visualization Laboratory, Tsunami and Disaster Mitigation Research Center (TDMRC), Syiah Kuala University, Jl. Prof. Dr. Ibrahim Hasan, Gampong Pie, Banda Aceh, 23233, Indonesia
[2]Civil Engineering Master Program, Syiah Kuala University, Jl. Syech Abdurrauf, Banda Aceh, 23111, Indonesia
[3]Civil Engineering Department, Syiah Kuala University, Jl. Syech Abdurrauf, Banda Aceh, 23111, Indonesia.
[4]Civil Engineering Department, Almuslim University, Jl. Almuslim, Matangglumpangdua, Bireuen-Aceh, 24261, Indonesia

*Email: syamsidik@tdmrc.org

Abstract. Sudden environmental changes due to the impacts of the 2004 tsunami demolished a vast area of land around Peukan Bada of Aceh Besar. There were paddy fields and ponds as major livelihoods supports for coastal communities around this area. However, the tsunami waves demolished all the paddy fields and ponds around this sub-district. Since the 2004, a number of interventions were made to recovery the paddy fields and ponds. This study was aimed at monitoring the land use changes, in term of paddy fields and ponds, during 12-year of recovery process. Geographic Information System tools were used to map every 100 meter of interval in this area to observe the recovery process of the two land use types. After 12-year of the recovery process, total areas successfully being recovered in Peukan Bada were about 80% and only 19% for paddy fields and ponds, respectively. Some reasons were behind these changes. Two of the most frequent reasons are the changes of job types of the residents and the low productivity of the tsunami affected ponds/paddy fields.

1. Introduction

Most of coastal communities in Indonesia are situated in difficult situation. One of them is the stark evidence showing that the coastal areas are prone to a number of coastal hazards. About 80% of coastal areas in Indonesia are prone to tsunami. The 2004 Indian Ocean tsunami has proven the power of the disaster to eliminate several decades' development process made in the affected area. Some of the most destroyed land use types were agricultural land and brackish water ponds. In some area in Aceh coast, paddy fields were situated closely to the sea and were planted at low-lying area. Furthermore, ponds were major economic driven for coastal communities for years. These were also the situation in Peukan Bada, a sub-district of Aceh Besar, located at the western coast of Banda Aceh.

In Aceh and many parts in Indonesia, brackish water ponds are used as aquaculture activities that have significant contributions to people economic and livelihoods of coastal communities [1]. Most of the pond's farmers were small-scale farmers where their ponds area are less than 2 ha. The 2004 tsunami caused a serious disruption to their livelihood. In total, there were about 20,000 ha of ponds and 70,000 of agricultural land destroyed by the 2004 tsunami. In the first second year after the tsunami, BRR Aceh Nias managed to recover about 6,800 ha of ponds and 50,000 of agricultural land in Aceh and Nias [5].

The 2004 Indian Ocean Tsunami has caused a severe erosion at coastal areas in Aceh and Nias Island of Indonesia. One of the most stricken area was Peukan Bada Sub-District of Aceh Besar. The coastal area of Peukan Bada receded to about 200 m from its initial coastline position [2]. One study proven that large shear stress generated by the waves transported the sediment to areas around foothill of the location and to the northern offshore part of Peukan Bada, which is estimated about 2.5 km from the coastline. The impacts of the 2004 tsunami on the land use area was severe. Interestingly, the return of the land use types in this area is seen to gain some momentum following the coastal morphology changes.

Notwithstanding with the rehabilitation and reconstruction process after the tsunami, studies on the recovery of the coastal areas are still limited. This study was aimed at observing the land use recovery after the 2004 Indian Ocean tsunami at the affected area that was accompanied by the coastal morphological changes followed after the 2004 tsunami. Northern coast of Peukan Bada of Aceh Besar District was selected as the study area. The coastal area was totally demolished by the 2004 tsunami. The area is situated between a headland named Ujong Pancu at the western edge of the area and a hill side of Bukit Barisan Mountain at the southern limit of the area. Before the 2004 tsunami this area was largely known for its vast mangrove forest area and aquaculture land use, such as ponds

2. Study area

Northern Peukan Bada sub-district is situated between a steep hilly side of Bukit Barisan Mountains at the west, The Ulee Lheue Bay at the north, and Banda Aceh at the eastern part of the sub-district. **Figure 1** shows the study area. The sub-district is divided into 26 villages, with total population 17,000 people in 2014 [3]. Before the 2004 Indian Ocean tsunami, the study area was sparsely used for settlement around the Ulee Lheue Bay, but it was known for massive mangrove areas and aquaculture such as ponds [4]. The maximum tsunami wave height around this area was about 9.0 m [2].

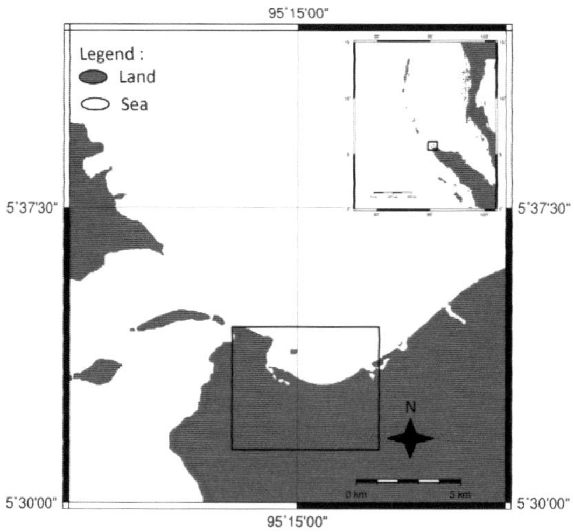

Figure 1. The Peukan Bada Coast as the study area.

3. Methods

We employed Geographical Information System (GIS) method by digitizing types of land use, i.e. ponds and paddy fields. The images were captured in 2003 (before the tsunami, 2005 (just two months after the 2004 tsunami), 2007, 2009, 2011, 2013, 2014, and 2015. The source of the images were obtained from various sources, i.e. IKONOS (for the condition in 2003 and in 2005), Quick Bird (for condition in 2007), and the rest of the data were retrieved from Google Earth. Each of the landuse were measured the distance to the coastline. Accumulation of the digitized land use types were plotted against the distance to the coastline. We also conducted a preliminary interview to local community to identify the factors that drove the changes of the land use during the recovery process. Similar methods employed in this research were applied by other researchers previously [7, 8].

4. Results

There were many interventions made to the recovery process around Peukan Bada. Although the recovery process was not performed through the introduction of hard structures, the development of new land barriers helped some land-use types around Peukan Bada to be recovered. For a simplicity, in this paper we only show four digitized images taken in 2003, 2005, 2007, and 2015. Figure 2(a) shows the land-use types condition in 2003, Figure 2(b) shows the condition in 2005. Condition in 2007 and 2015 are shown in Figures 3(a) and 3(b), respectively.

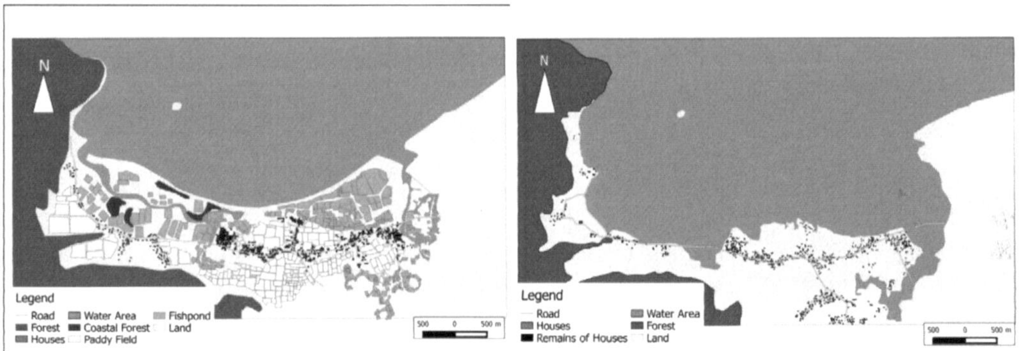

(a) (b)

Figure 2. (a) Land-use types before tsunami digitized based on 2003 condition and (b) after tsunami, digitized based on 2005 condition.

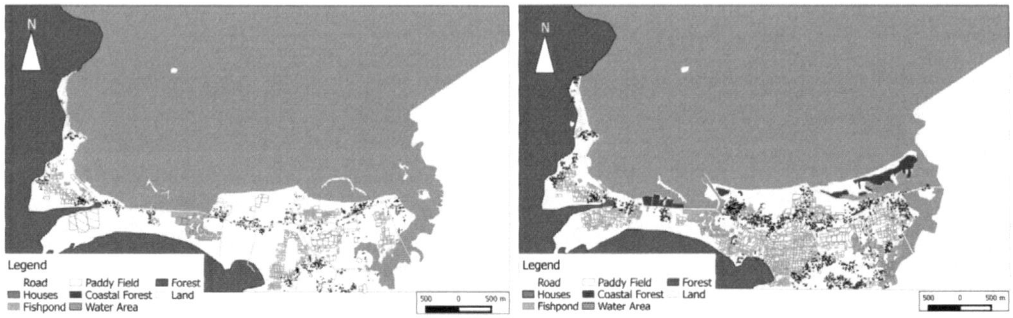

(a) (b)

Figure 3. (a) Land-use types based on 2007 condition and (b) recently land use changes digitized based on 2015 condition.

4.1. Paddy Fields

Before tsunami, paddy fields in this area were about 189 ha. Mainly the paddy fields used dryland farming system. Due to the 2004 tsunami, the paddy fields were totally destroyed. The ill function paddy fields after the 2004 tsunami were caused by the length of the time the agricultural land inundated by salt water, the thickness of tsunami deposit in the paddy fields, damaged irrigation system that could dilute the salt contamination, and low precipitation rates after the tsunami [9]. In 2015, the recovery of the paddy fields reached about 151 ha or about 80% compared to the situation before the tsunami. Recovery of the paddy fields was done by several measures depending on the level of the damages. For thin tsunami deposit, lower than 20 cm, run-off process generated by precipitation is sufficient to deplete the salt contamination and neutralize the sea-sand into original soil. However, tsunami deposit thicker than 20 cm required mechanical removal of the sea-sand to make the paddy field cultivable again. In another area, a bunding technology has proven effective to reduce the salinity level and was also successful to recover the agronomic productivity from the tsunami-affected paddy field [10]. However, the mechanical methods were not performed in this study area according to our interview process. Considering the thicknesses of tsunami deposit at several places were thick, therefore some of paddy fields failed to be recovered. In 2007, a small retention basin (*embung*) construction named *Embung Lambadeuk* was started. The basin is expected to irrigate the paddy field around this area [11]. **Figure 4** shows the recovery of the paddy field in the study area from spatio-temporal analysis.

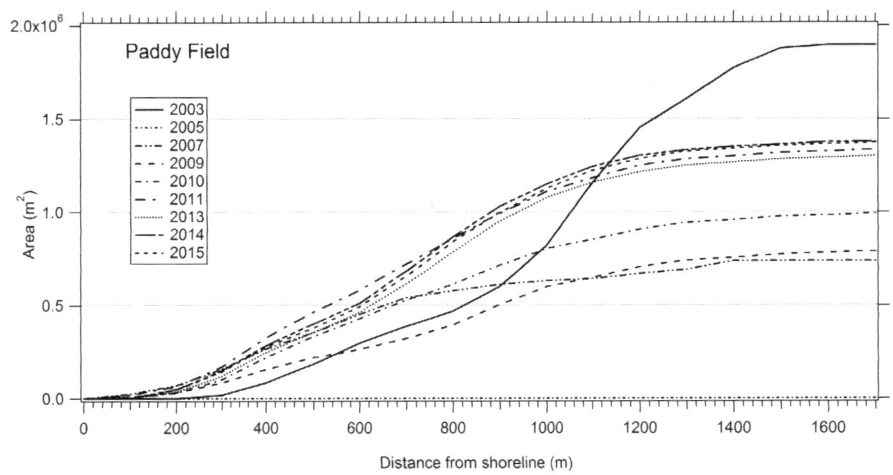

Figure 4. Recovery process of paddy field area in Peukan Bada from 2005 until 2015.

4.2. Ponds

This area had about 149 Ha of ponds before the 2004 tsunami. The tsunami waves damaged all of the ponds. The damages at ponds were mainly caused by the collapse of pond's dykes, sedimentation, and severe erosion. Rehabilitation of ponds were performed in the first four years after tsunami by International Donors and later continued by Marine and Fishery Department/DKP [5]. Until 2015, the recovery of ponds reached only about 28 ha or about 19% of total ponds area before the 2004 tsunami. According to the interviews made at former ponds farmers, they revealed that during the first two years after the tsunami, they had to look for another type of job. Some of them became construction workers or fishermen with their own boat or employed at some one's fishing boat. Another reasons for the low rate of the ponds recovery was the unsuccessful recovery of the ponds' area with minimum facilities provided. There was a conflicting report regarding the ownership of the lands near the coastline. This also hampered process of constructing facilities to rebuild the ponds. Furthermore, most of non-productive ponds are located near the coastline where the fresh water inputs is limited and no canal to discharge the fresh water to maintain the brackish water environment suitable for ponds. One study revealed that the salinity level around this area was above 25ppt [6]. This is beyond of the brackish water salinity environment, which is between 0.5ppt-17.0ppt. **Figure 5** shows the return of the ponds area until 2015. **Figure 6** shows an aerial image of *Gampong* (village) Lambadeuk coast in Peukan Bada. In the figure, a large area of non-productive ponds can be seen, close to the settlement in the village.

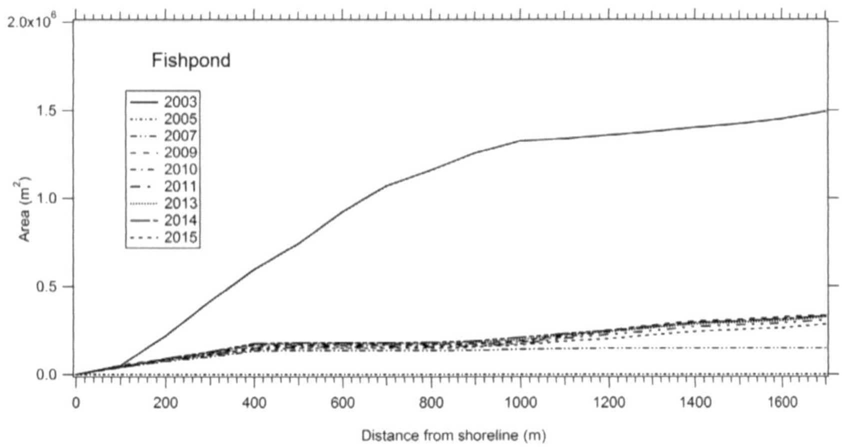

Figure 5. Recovery of ponds in Peukan Bada compared to ponds area before the tsunami (black-solid line).

Figure 6. An aerial image of non-productive ponds around Gampong Lambadeuk in Peukan Bada, captured by drone in November 2016.

The non-productive ponds, as shown in **Figure 6**, are now being left by local communities. There is no significant economic activities done to reactivate the ponds. Some Non Governmental Organisations are trying to replant mangroves (Rhizopora Sp.) in the area. However, due to limited knowledge on the ecological situation post the tsunami, many of the replanted mangroves did not grow properly.

5. Conclusions

This research concludes that the recovery of paddy field areas in Peukan Bada and Ponds, both did not reach a complete recovery process. About 80% of the paddy fields were recovered and productive. However, only about 19% of ponds were successfully being recovered. A low percentage of the ponds recovery was driven by several reasons, i.e. the change of the job of the pond's farmers, the unavailability facility to cater fresh water to recreate brackish water environment, and the severe level of damages due to the 2004 Indian Ocean Tsunami. A promising return of the paddy field productivity in this area was due to a new irrigation system from the *Lambadeuk* basin, introduced to maximize water supply to the paddy fields. A further recovery process of ponds in this area requires more attention to construct canal to cater fresh water to the ponds area. This will enable to re-create a brackish water environment to rerun the ponds productively.

6. References

[1] Philips M and Budhiman, A 2005 Fisheries Tsunami Emergency Programme: An assessment of the impacts of the 26th December 2004 earthquake and tsunami on aquaculture in the Provinces of Aceh and North Sumatra, Indonesia (Food and Agriculture Organization of the United Nations, Indonesia)
[2] Syamsidik, Tursina, Meutia M, Al'ala M, Fahmi M, and Meilianda E, 2016 Numerical Simulations of Impacts of the 2004 Indian Ocean Tsunami on Coastal Morphological Changes Around the Ulee Lheue Bay of Aceh, Indonesia *J. Earthquake and Tsunami*, http://dx.doi.org/10.1142/S179343111740005X
[3] BPS Aceh Besar 2015 Aceh Kecamatan Peukan Bada Dalam Angka 2015 (BPS-Aceh Besar)
[4] Syamsidik, Fahmi M, Al'ala M, and Tursina 2016 Tsunami wave impacts on Coastal Morphological changes and one decade process of Coastal Line Recovery after the 2004 Indian Ocean Tsunami around Banda Aceh, Indonesia. Proceedings of the Twenty Sixth International Ocean and Polar Engineering (Rhodes, Greece June 26-July 1, 2016) p 796
[5] Zainun I, Budidarsono S, Rinaldi Y, and Adek MC 2007 Socio-economic aspects of Brackish Water Aquaculture (Tambak) Production in Nangroe Aceh Darussalam: Integrated Natural Resource Management and Livelihood Paradigms in Recovery from Tsunami in Aceh (ICRAF Southeast Asia 2007-Jakarta)
[6] Syamsidik, Iskandar A, and Rasyif TM 2014 Progress of Coastal Line Rehabilitation after the Indian Ocean Tsunami around Banda Aceh Coasts *Recovery from the Indian Ocean Tsunami* (Springer-Tokyo) p 175
[7] Griffin C, Ellis D, Beavis S, and Zoleta-Nantes D 2013 Coastal resources, livelihoods and the 2004 Indian Ocean tsunami in Aceh, Indonesia *Ocean & Coastal Management* **71**, p176
[8] Moore LJ, Ruggiero P, and List JH 2006 Comparing mean high water and high water line shorelines: should proxy-datum offsets be incorporated into shoreline change analysis? *J. Coastal Research* **22**(4) p894
[9] Australian Center for International Agricultural Research (ACIAR) 2014 Restoring agriculture after a tsunami: the experience from Aceh, Indonesia (New South Wales Department of Primary Industries, Australia)
[10] Velmurugan A, Swarnam TP, and Lal R 2015 Effect of land shaping on soil properties and crop yield in tsunami inundated coastal soils of Southern Andaman Island *Agriculture, Ecosystem and Environment* **206** p 1
[11] BWSS I (River Section for Sumatra Region I) Lambadeuk Retention Basin Developmen *in Bahasa Indonesia* bwssum1.net/index.php/kegiatan-strategis/akan-berfungsi/item/4600pembangunan-embung-lambadeuk (accessed on November 11, 2016)

Acknowledgments
Authors are grateful with the research grant through Partnership Enhanced Engagement in Research (PEER) Cycle 3 sponsored by USAID and National Academy of Sciences (NAS), through Sponsor Grant Award Number: AID-OAA-A-11-00012 and Sub Grant Number PGA-2000004893 with research project title: *Tsunami Waves Impacts on Coastal Morphological Changes Based on Sediment Transport Simulations*.

The behaviour of reinforced concrete structure due to earthquake load using Time History analysis Method

M Afifuddin[1,2], M A R Panjaitan[1] and D Ayuna[3]

[1]Civil Engineering Departement, Engineering Faculty, Syiah Kuala University, Banda Aceh 23111, INDONESIA
[2]Tsunami Disaster and Mitigation Research Centre member, BandaAceh, INDONESIA
[3]Former Civil Engineering Departement Student, Engineering Faculty, Syiah Kuala University, Banda Aceh 23111, INDONESIA.

Abstract. Earthquakes are one of the most dangerous, destructive and unpredictable natural hazards, which can leave everything up to a few hundred kilometres in complete destruction in seconds. Indonesia has a unique position as an earthquake prone country. It is the place of the interaction for three tectonic plates, namely the Indo-Australian, Eurasian and Pacific plates. Banda Aceh is one of the cities that located in earthquake-prone areas. Due to the vulnerable conditions of Banda Aceh some efforts have been exerted to reduce these unfavourable conditions. Many aspects have been addressed, starting from community awareness up to engineering solutions. One of them is all buildings that build in the city should be designed as an earthquake resistant building. The objectives of this research are to observe the response of a reinforced concrete structure due to several types of earthquake load, and to see the performance of the structure after earthquake loads applied. After Tsunami in 2004 many building has been build, one of them is a hotel building located at *simpang lima*. The hotel is made of reinforced concrete with a height of 34.95 meters with a total area of 8872.5 m2 building. So far this building was the tallest building in Banda Aceh.

Introduction

In more than 300 natural disasters in year 2011, over 30,000 people lost their lives and 206 million people were affected and $366 Billion were the economic losses, which made it the costliest year in the history of the catastrophes [1]. Indonesia has a unique position as an earthquake prone country. It is the place of interaction for three tectonic plates, namely the Indo-Australian, Eurasian and Pacific plates. Banda Aceh is one of the cities that located in earthquake-prone areas. Due to the vulnerable conditions of Banda Aceh some efforts have been exerted to reduce these unfavorable conditions. Many aspects have been addressed, starting from community awareness up to engineering solutions. One of them is all buildings that build in the city should be designed as an earthquake resistant building.

The objectives of this research are to observe the response of a reinforced concrete structure due to several types of earthquake load, and to see the performance of the structure after earthquake loads applied. Observation will be done on the story displacement, story drift, and base shear. Hotel building in Banda Aceh will be used as an object of the research. The building will be loaded by 4 types of big earthquake data that have occurred in the world.

Response of Hotel Building will be observed by using Time History Analysis Method. Time-history analysis provides for linear or nonlinear evaluation of dynamic structural response under loading which may vary according to the specified time function. STERA (Structural Earthquake Response Analysis) 3D ver.7.1 developed by Prof. Taiki Saito from Toyohashi University of Technology, Japan, will be used in this research.

2. Literature review

2.1. Stera 3D

STERA_3D (Structural Earthquake Response Analysis 3D) is integrated computer software for the seismic analysis of steel and reinforced concrete buildings in three dimensional spaces developed by T. Saito and distributed for free for research and educational purposes [2]. The software was developed since 2007. STERA_3D has a visual interface to create a building model and show the results easily and rapidly. Figure 1 shows the element models used in STERA_3D. A beam is modelled as a line element with nonlinear flexural springs at both ends. The degrading tri-linear slip model is used for the hysteresis. A column is modelled in a similar manner, and nonlinear interaction between axial force and moment is expressed using axial springs of concrete and steel arranged in sections at both ends [3]

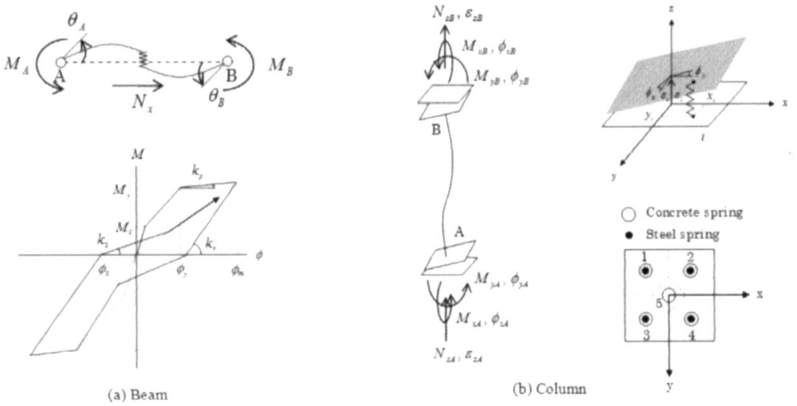

Figure 1. Nonlinear model used in STERA_3D

2.2. Time history analysis

There were four types of earthquake history recorded from PEER Ground Motion Database Application [4]. Those are El-Centro earthquake that occurred in California in 1940, Irpinia earthquake that occurred in Italy in 1980, Kobe earthquake that occurred in Kobe Japan in 1995, and the Chi-Chi earthquake that occurred in Taiwan in 1999. El-Centro earthquake has 6.4 SR magnitudes with 9 km depth from epicenter, and 15 second period of time. Irpinia earthquake has 6.9 SR magnitudes with 22.5 km depth from epicenter, and 39.3 second period of time. Kobe earthquake has 6.9 SR magnitudes with 7.1 km depth from epicenter, and 20 second period of time. Chi-Chi earthquake has 7.3 SR magnitudes with 9.6 km depth from epicenter, and 90 second period of time. Figure 2 shows all the acceleogram of each earthquake.

IOP Conf. Series: Earth and Environmental Science **56** (2017) 012013 doi:10.1088/1755-1315/56/1/012013

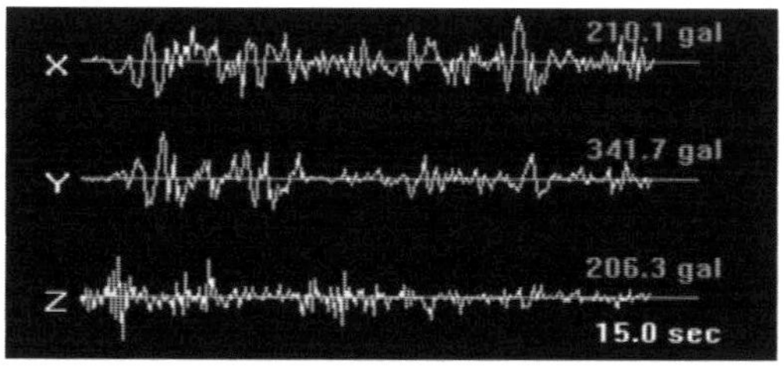

(a) El-Centro Earthquake

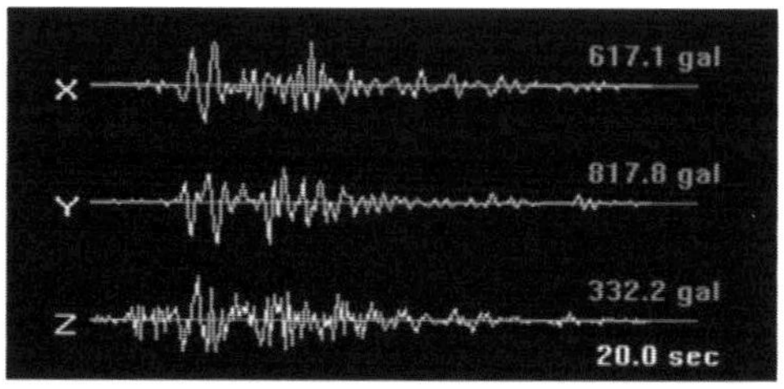

(b) Kobe Earthquake

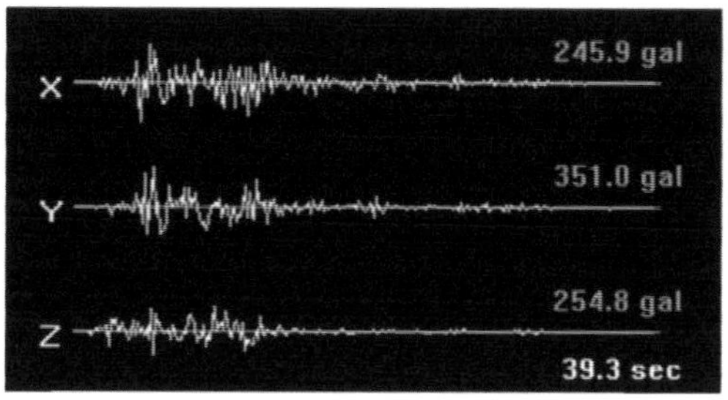

(c) Irpinia Earthquake

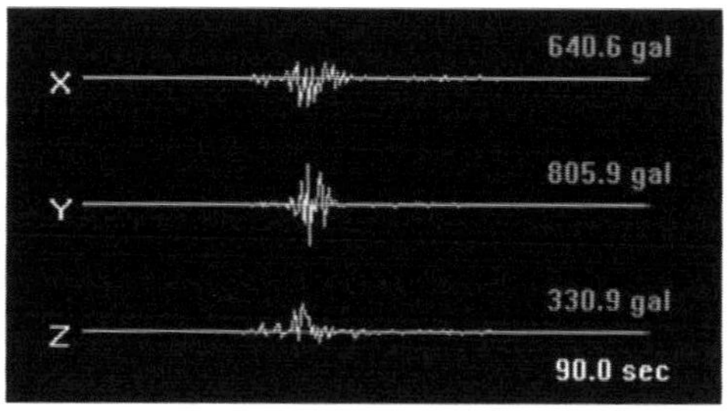

(d) Chi-Chi Earthquake

Figure 2. Acceleogram for each earthquake

3. Methodology

3.1. Object Introduction

The object is a hotel located in *simpang lima* Banda Aceh city, Aceh Province. The hotel is made of reinforced concrete with a height of 34.95 meters with a total area of 8872.5 m^2 building. So far this building was the tallest building in Banda Aceh. Compressive strength of concrete is 30 MPa, tensile yield strength for main reinforcement is 400 MPa, on the other hand for shear reinforcement is 240 MPa.

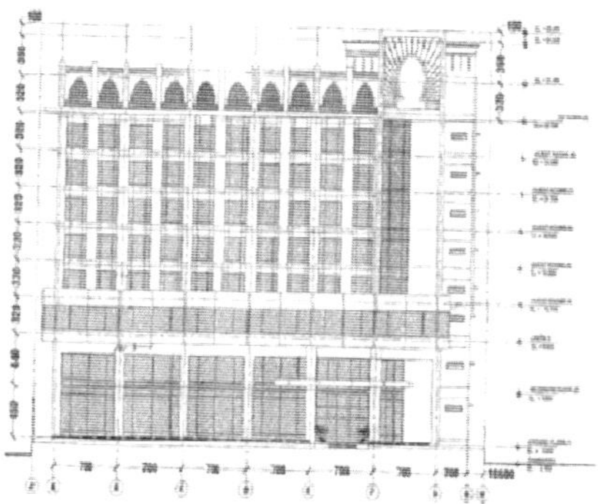

FRONT OF VIEW

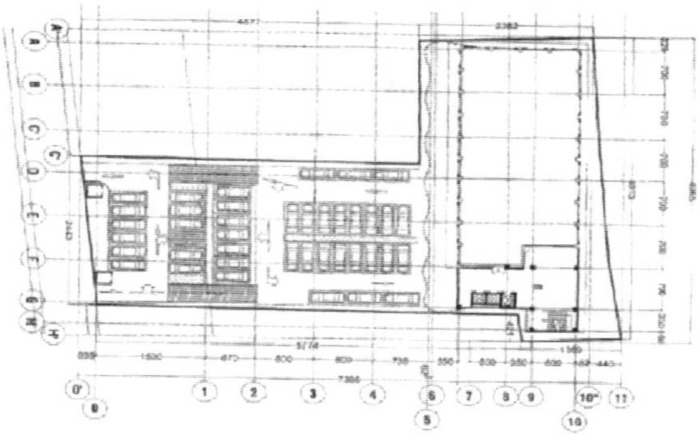

PLAN OF VIEW

Figure 3. Front of View and Plan of View of the hotel

3.2. Model Calculation

The object is modelled by using STERA_3D software. All procedure to create the model and how to input the earthquake loading is shown in Figure 4. Starting with giving the length and depth of building, then following with entering the dimension of beams and columns including the number of reinforcement, concrete cover, and material properties. Dimension for beams and colomn just following dimension in as built drawing.

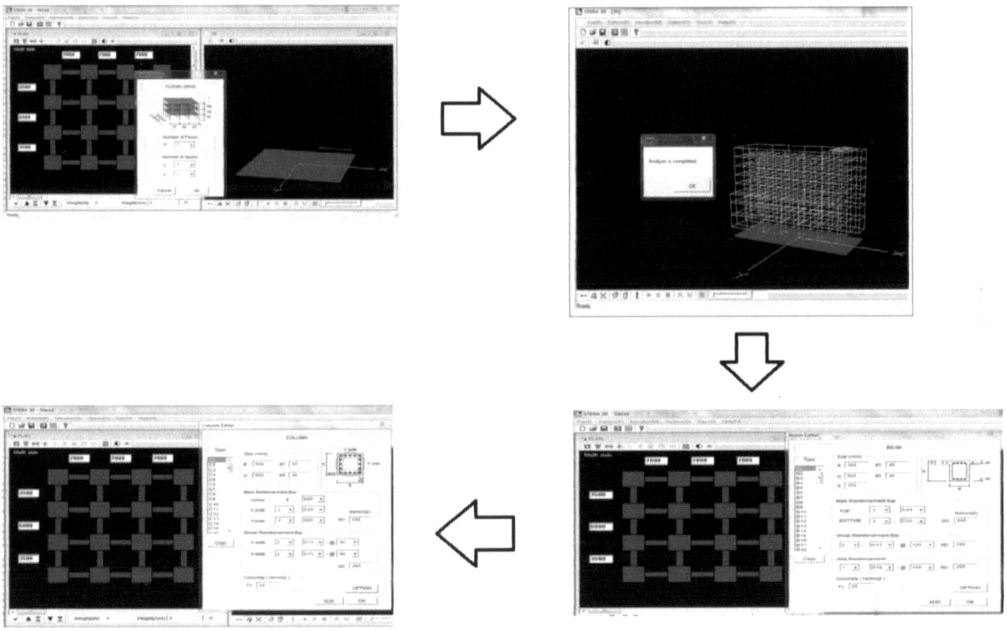

Figure 4. Procedure how to create the model

Dead load, live load, and wind load will be applied manually. Earthquake load will be applied by using four types of earthquake history recorded from PEER Ground Motion Database Application as shown in Figure 2. Procedure to input earthquake load in STERA_3D is shown in Figure 5.

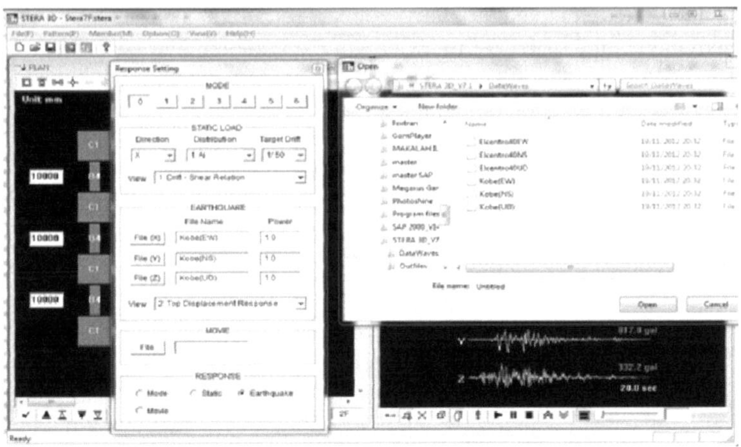

Figure 5. Procedure how to input earthquake load

4. Results and Discussion

4.1.Story Displacement

Story displacement result for X and Y direction is shown in Figure 6. It can be seen from the figure that the largest story displacement in X direction occurred was 56.64 cm due to El-Centro earthquake, followed by Chi-Chi earthquake, Kobe earthquake, and Irpinia earthquake. In Y direction the largest story displacement was 62.24 cm due to Chi-Chi earthquake, followed by Irpinia earthquake, El-Centro earthquake, and Kobe earthquake. Based on Indonesian standard SNI-1726-2012, it can be seen that the hotel building is in safe condition, because all the largest displacement was less than 87.4 cm.

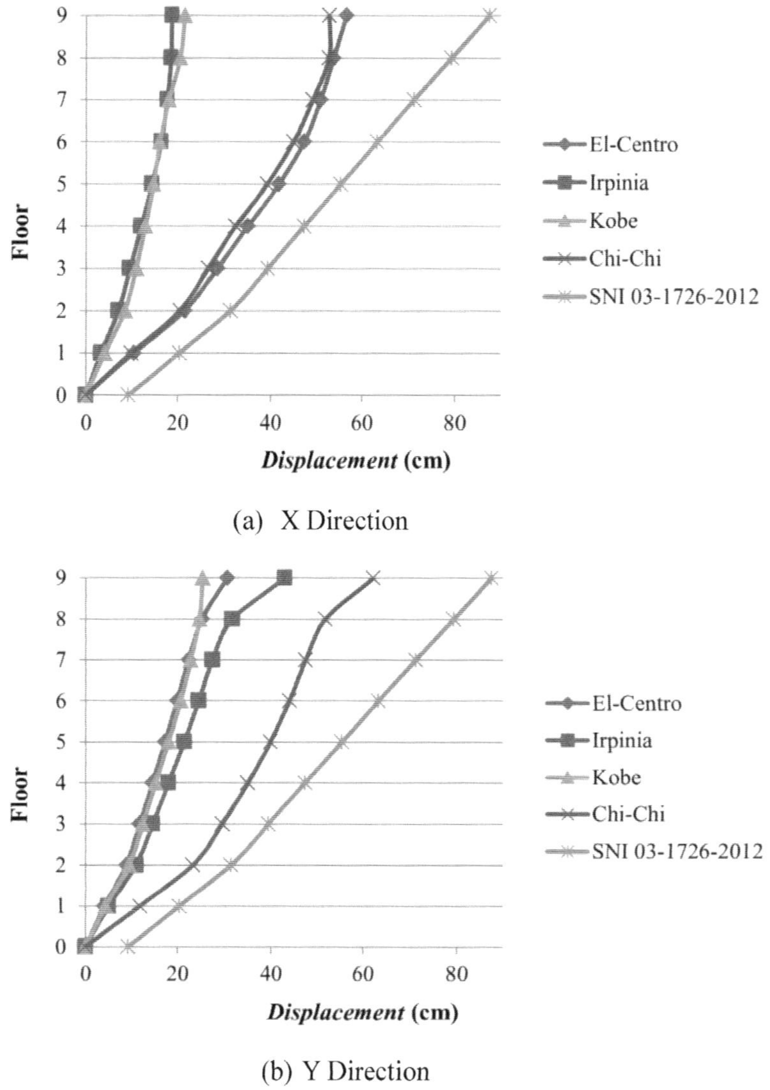

Figure 6. Story Displacement for X and Y Direction

4.2.Story Drift

Story drift result for X and Y direction is shown in Figure 7. It can be seen from the Figure that the largest drift in X direction occurred was 9.74 cm on between base floor and first floor due to Chi-Chi earthquake. Based on Indonesian standard SNI-1726-2012, the maximum allowable story drift was 11.30 cm, so the hotel building is in safe condition.

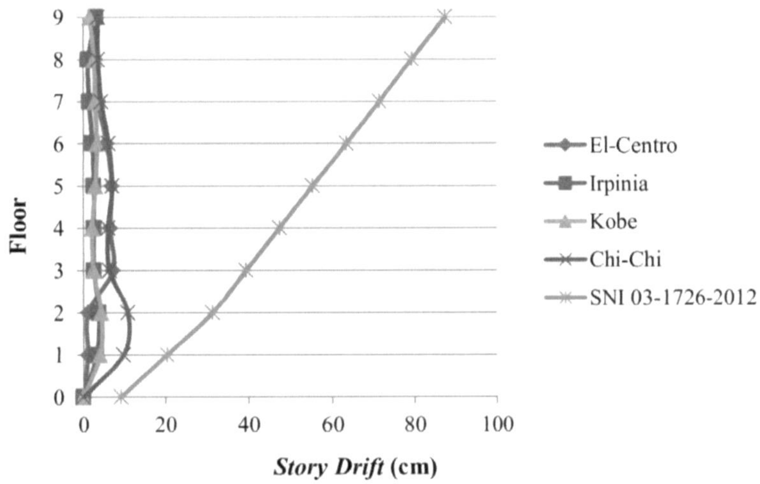

(a) X Direction

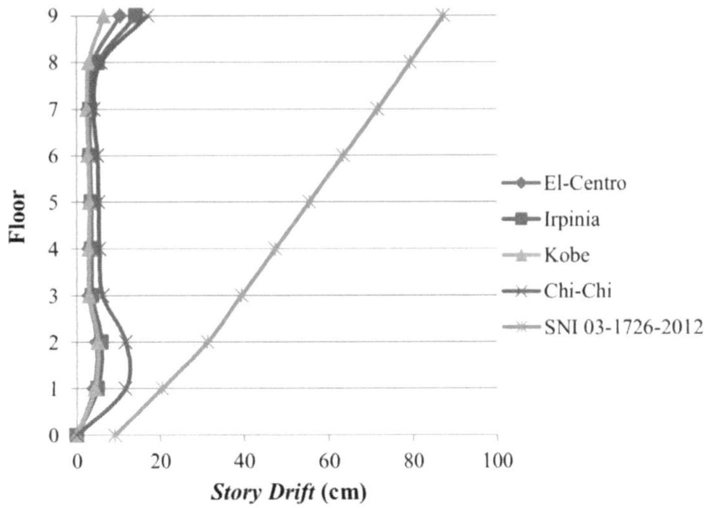

(b) Y Direction

Figure 7. Story Drift for X and Y Direction

4.3. Base Shear

Base shear result for X and Y direction is shown in Figure 8. It can be concluded from the Figure that the largest base shear in X direction occurred was 15,830 kN due to Chi-Chi earthquake, followed by El-Centro earthquake, Kobe earthquake, and Irpinia earthquake. In Y direction the largest story displacement was 14,160 kN due to Chi-Chi earthquake, followed by Irpinia earthquake, El-Centro earthquake, and Kobe earthquake. Based on Indonesian standard SNI-1726-2012, it can be seen that the hotel building is in safe condition, because all the base shear is larger than 4,280.02 kN.

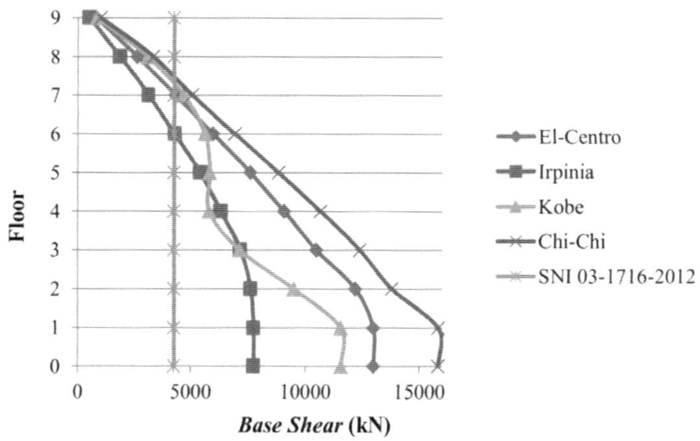

(a) X Direction

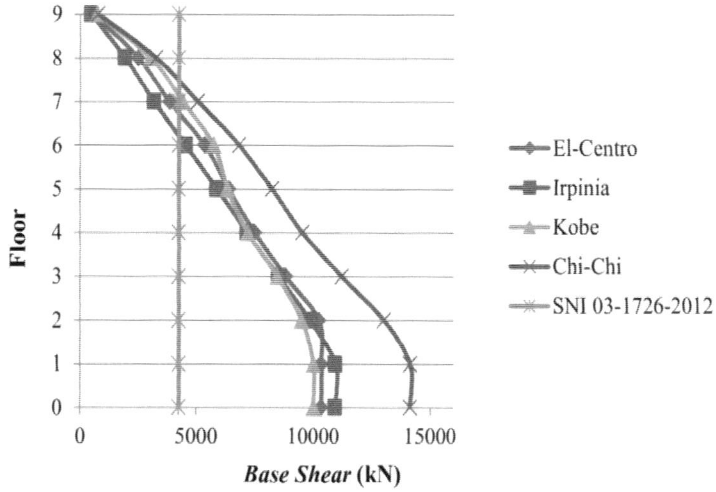

(b) Y Direction

Figure 8. Base Shear for X and Y Direction

5. Conclusions

1. Based on the results, it can be concluded that all response of the hotel structure are still comply with the requirement in Indonesian Standard SNI-1726-2012.

2. The largest story displacement in X direction was 56.64 cm, and in Y direction was 62.24 cm, those value still less than 87.4 cm as required by SNI.

3. The largest drift in X direction was 9.74 cm on between base floor and first floor, this value was less than 11.30 cm as required by SNI.

4. The largest base shear in X direction was 15,830 kN, and in Y direction was 14,160 kN, those values larger than 4,280.02 kN as required by SNI.

5. Finally, it was found that hotel building is in safe condition after loaded by 4 types of earthquake loads.

References

1. EM-DAT, 2001, Internasional Disaster Database, The OFDA/CRED Belgium.
2. T. Saito. STERA_3D Software. [Online]. Available: http://www.rc.ace.tut.ac.jp/saito/software-e.html
3. T. Saito., 2016, Response of High-Rise Buildings under Long Period Earthquake Ground Motions., *International Journal of Structural and Civil Engineering Research*, **Vol. 5, No. 4**, pp. 308-314.
4. Anonymous, 2011, PEER Ground Motion Data Base web Application.

IOP Conf. Series: Earth and Environmental Science **56** (2017) 012014 doi:10.1088/1755-1315/56/1/012014

The pattern of spatial flood disaster region in DKI Jakarta

M P Tambunan[1*]

[1]Department of Geography Faculty of Mathematics and Natural Sciences
University of Indonesia
*email: mangapul.parlindungan@ui.ac.id and mptgeoui@gmail.com

Abstract. The study of disaster flood area was conducted in DKI Jakarta Province, Indonesia. The aim of this research is: to study the spatial distribution of potential and actual of flood area The flood was studied from the geographic point of view using spatial approach, while the study of the location, the distribution, the depth and the duration of flooding was conducted using geomorphologic approach and emphasize on the detailed landform unit as analysis unit. In this study the landforms in DKI Jakarta have been a diversity, as well as spatial and temporal pattern of the actual and potential flood area. Landform at DKI Jakarta has been largely used as built up area for settlement and it facilities, thus affecting the distribution pattern of flooding area. The collection of the physical condition of landform in DKI Jakarta data prone were conducted through interpretation of the topographic map / RBI map and geological map. The flood data were obtained by survey and secondary data from *Kimpraswil* (Public Work) of DKI Jakarta Province for 3 years (1996, 2002, and 2007). Data of rainfall were obtained from BMKG and land use data were obtained from BPN DKI Jakarta. The analysis of the causal factors and distribution of flooding was made spatially and temporally using geographic information system. This study used survey method with a pragmatic approach. In this study landform as result from the analytical survey was settlement land use as result the synthetic survey. The primary data consist of landform, and the flood characteristic obtained by survey. The samples were using purposive sampling. Landform map was composed by relief, structure and material stone, and process data Landform map was overlay with flood map the flood prone area in DKI Jakarta Province in scale 1:50,000 to show. Descriptive analysis was used the spatial distribute of the flood prone area.The result of the study show that actual of flood prone area in the north, west and east of Jakarta lowland both in beach ridge, coastal alluvial plain, and alluvial plain; while the flood potential area on the slope is found flat and steep at alluvial fan, alluvial plain, beach ridge, and coastal alluvial plain in DKI Jakarta. Based on the result can be concluded that actual flood prone is not distributed on potential flood prone

1. Introduction

Floods as natural disasters that could potentially damage and harm life even human sacrifice [9] states that the flood become a problem when it interferes with the activity of human lives and livelihoods even threaten her safety. Gupta [5] noted the incidence of flooding and its geographical locations such as major flood events in the north and northeast of Australia [16]. Bemmelen [3] and Verstappen [14] and [15] wrote about classification of Jakarta landform based on the origin of its formation namely landform from the process of fluvial, fluvial- coastal, and coastal. A characteristic of landform's origin was identified as landform that was layered by sedimentary rocks of volcanic-fluvial and coastal-

fluvial. The topography of landform was gently sloping to flat with low altitude and even on the coastal alluvial plain landform whose altitude lower than sea level. Aside of this landform, there was another landform in the north side of Jakarta, e.g. beach ridge that was parallel to its coastline.
The location of beach ridge landform prevented surface flow derived from central and southern Jakarta during heavy rainfall which headed to northern part of Jakarta Bay. Beach ridge usually had hollow between shelves that could serve as water reservoir, but this area had been built as well. If heavy rains continuously flushed in the basin between shelves, this then would be the Prone area of flood as well. Bappenas [2] noted that built up land use in Jakarta reached 610,752 km^2 (61,075.2 hectares) or 93.44% of total area of alluvial landforms in Jakarta. Even, the land use for settlement reached 456,626 km^2 (45,662.6 Ha) or 69.86% of total area of built land use which was equivalent to 65.27% from its total area. Agricultural land, parks, and urban forest were the green open space and container land or water park which was left only in amount of 42.94 km^2 (4,294.5 hectares) or 6.57%.
Besides, damage on agricultural land use and wet land e.g. reservoir/lake/*situ* which acted as water retention and water reservoir (retarding basin) to settlements and their facilities, were assumed as cause of the increasing and spreading of flood intensity in lowland of Jakarta, both spatially and temporally patterned as prone area of flood. The Prone area of actual and potential floods on the alluvial plain in DKI Jakarta province was always changing, both spatially and temporally (e.g. in 1980 flood = 769,603 Ha., and in 1996 amounted = 2,258.53 Ha). On February 1st, 2002 and 2007 there was a flood that had been in large categorized in Jakarta Province [6] .
Thus, the landform can be used as a basis for determining flood prone areas of flooding. Flooding that occurs on landform including potential flooding due to natural landform have any vulnerability to flooding. According to the reality on the field was not all landform are in Jakarta or elsewhere are not entirely flooded at the same time. Flooding that occurred at some point it is called the actual flood that a real flood occurred. Prone areas of potential flooding have not necessarily flooded. Based on the background and problems of flooding in Jakarta, the formulation of research problems are as follows. Where is actual and potential flood prone area in DKI Jakarta province?

2. Methods

Methods This study contains study approach, selection of research area, and time, preparation of materials and research materials, determination of research variables, data collection, and data processing and data analysis.

2.1 Study Approach

Study approach in the study of approach spatial flood area in the landform as geomorphological survey approach [14] which survey is divided into three analytic, synthetic survey, and survey pragmatic. This study uses a survey approach pragmatic, because the object studied is associated with flooding. In a survey pragmatic directed to solve problems, in this case is the problem of flooding. The data used in the survey is pragmatic geomorphological survey data analytic and synthetic survey. Analytic survey data is required in relation to the magnitude of floods is morphology, and morphogenesis, whereas the synthetic data for the survey is landform, climate, and land use.
Based on a variety of materials in the literature review can be understood that the study of geomorphology in the perspective of geography more emphasis on the similarities and differences of space and time. From the standpoint of fluvial geomorphology in the lower mainland of Jakarta is interesting to study the flooding distributed spatially and temporally. Distribution of flood prone area can be approximated by a unit of flood landform. Pragmatic geomorphological approach, can be used to understand the characteristics of flooding from rivers and fluvial activity fluvio-marine, and can provide answers to problems where the original landform unit as flood prone area flooding.
Low land in Jakarta has the magnitude and distribution of rainfall variability. Variations in local rainfall have high rainfall distribution both intensity and thickness. Rainfall in February is the maximum rain that falls on the landform and diverse land use. There are differences in diversity and intensity of heavy rainfall in the north to the south nor the west and east. This difference is due to the location and height landform. When heavy rain with high intensity falls to a variety of landform due to

IOP Conf. Series: Earth and Environmental Science **56** (2017) 012014 doi:10.1088/1755-1315/56/1/012014

the hardening of water-proof buildings and facilities house the rain water runoff into water that flooded / flooding in the lower mainland Jakarta different. Genesis there is diversity in lowland landform in Jakarta such as fluvial landform (alluvial fan and alluvial plain), fluvial coastal (beach ridge), and fluvial-marine (coastal alluvial plain). Differences due to differences in topography unit landform (slope and altitude), geology (rocks), and geomorphic processes in the lower mainland Jakarta. In the northern part of Jakarta landmass is dominated by alluvial plain of coastal landform. In addition, on the coast of landform formed beach ridge extending parallel to the coast which blocks the flow of water into the Bay of Jakarta. The origin of fluvial landform, fluvial -marine, and coastal as a potential Prone areas in the flood plain of Jakarta (Figure 1).

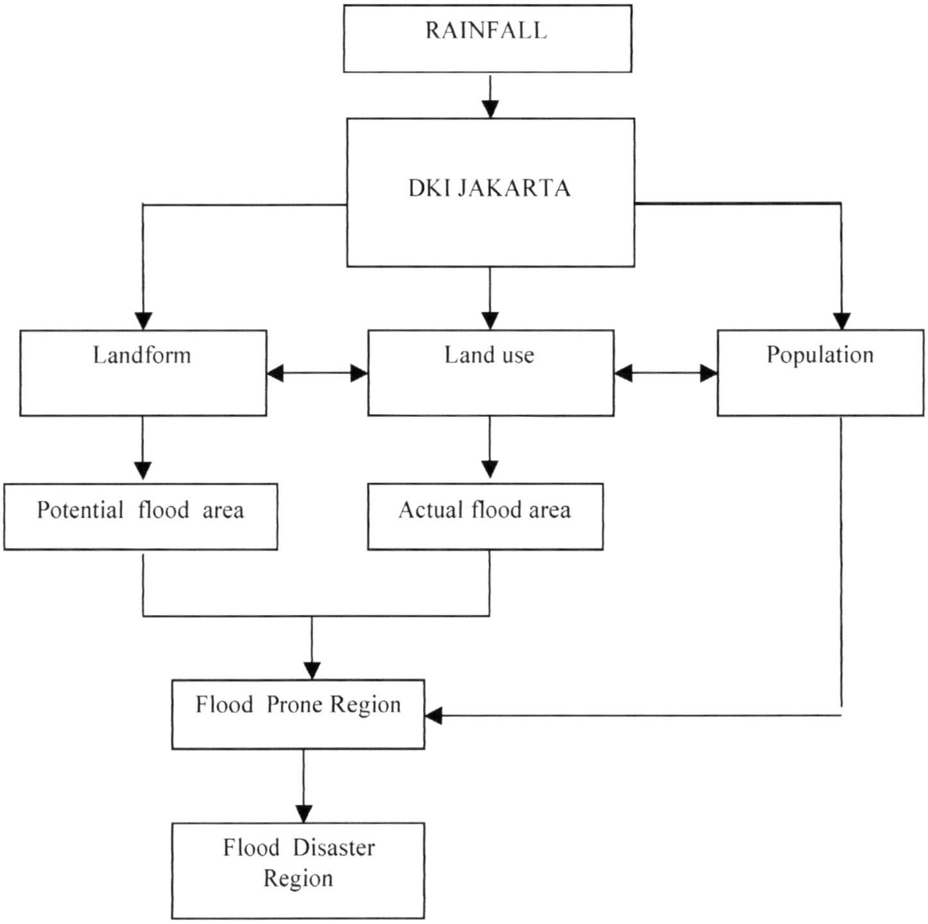

Figure 1. Flow chart of theoretical Framework

2.2 Selection of Research Area and Time

Research areas studied is in the mainland DKI Jakarta 65,493 hectares. The administration areas of study consists of five areas of the city administration, 42 districts, and 261 sub districts. Assumptions of the real (empirical) used in selecting areas that flooded sub districts are as follows: 1)flood due to high daily rainfall. 2)frequently flooded area of 261 sub districts in 107 sub districts. 3)there are differences and similarities in the characteristics of the floodplain areas subject to flooding. 4) population (density and distribution) is relatively large. 5)changing in agricultural land use and wetlands that serve as a rain water to settlement areas.

Flood events discussed in this study is limited to the major flooding that occurred in February of 1996, 2002 and 2007. By reason of the spatial distribution of flood events is the most extensive and diverse and also shows the temporal aspect.

2.3 Materials and Research Equipments

Materials research includes all data related directly and indirectly with the study landform, rainfall, land use, population, and the floods in DKI Jakarta. Research tools used include geological surveying equipment such as geological compass, GPS, helling, altimeter, current meter, flood depth ruler, digital cameras, and motorcycles. Research tools that come from government agencies. measuring equipment in the form of rain is rain gauge instrument manual for daily rainfall data from the agency BMKG, equipment for river water discharge data, and high river water from the Irrigation Department of the Provincial Government of DKI Jakarta.

2.4 Research Variables

1)Variable of climate parameters analyzed were annual rainfall (mm), ten days rainfall (mm), daily rainfall (mm), rainfall duration (hours), and distribution of rainfall (stations). 2)Variable of landform parameters analyzed were relief (altitude and slope), b.material, and c.geomorphic processes. 3)Variable of floods analyzed were flood sites (sub districts), .flood area (hectares), depth of flooding (cm), and d. long flood (hour). 4)Variable of land use analyzed were agricultural land area (hectares), wetland area (ha), and settlement land area (hectares).

2.5 Data Collection

Primary data collection was done at the time of field survey in DKI Jakarta are as follows. 1)survey of flood location with a visit to the field. 2)survey flood depth is noticed of marks on the walls of homes flooded as water limits the depth flood. 3)survey of the flood area is obtained from measurements in the sub district flooded by using GPS devices, and Kementerian PU. 4)the survey asked when the old flood flooding and how long homeowners flood events to residents who experienced flooding in their respective units of landform.

Flooding the sample locations in a 37 sub districts was conducted purposively in land-use settlement. This sample is to answer the problem to the correlation flood (area and depth) by land settlement area (Figure 2).

IOP Conf. Series: Earth and Environmental Science **56** (2017) 012014 doi:10.1088/1755-1315/56/1/012014

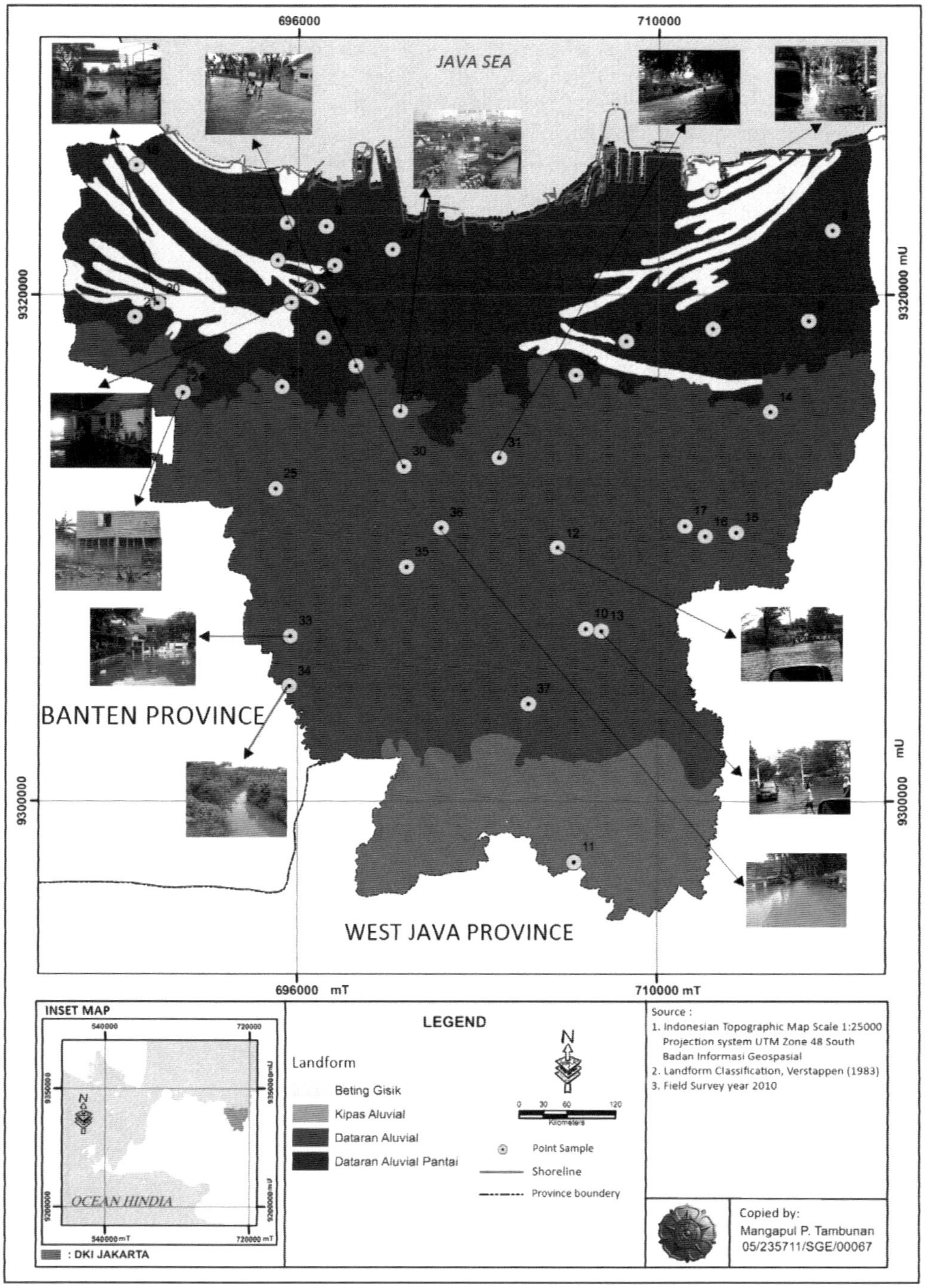

5.22 Photo area sample target actual flood in DKI Jakarta on February 2007 180

Figure 2. Map Distribution of Sampling Flood location in DKI Jakarta province

3. Research Result and Discussion

3.1 Rainfall

The results of the analysis of rainfall within 12 months (January to December) in 1996, 2002 and 2007 in Jakarta has a thickness variation and the rain are very diverse in their respective stations owned by BMKG Jakarta [1]. There are 8 months rainfall >100 mm, while the 4 months rainfall <100 mm / month. The highest monthly rainfall in January and February respectively more than 300 mm, while the other six months average of 150 mm / month. The graph of rainfall and Map Isohyet on February in Jakarta region in 1996, 2002 and 2007 are presented in Figure 3, and 4.

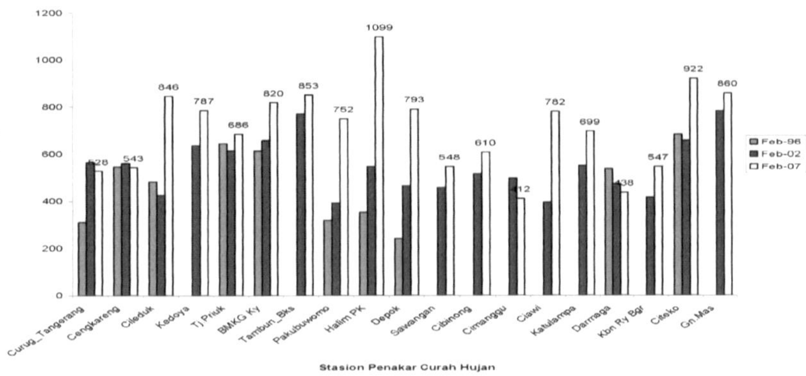

Figure 3. Bar chart rainfall in DKI Jakarta on February

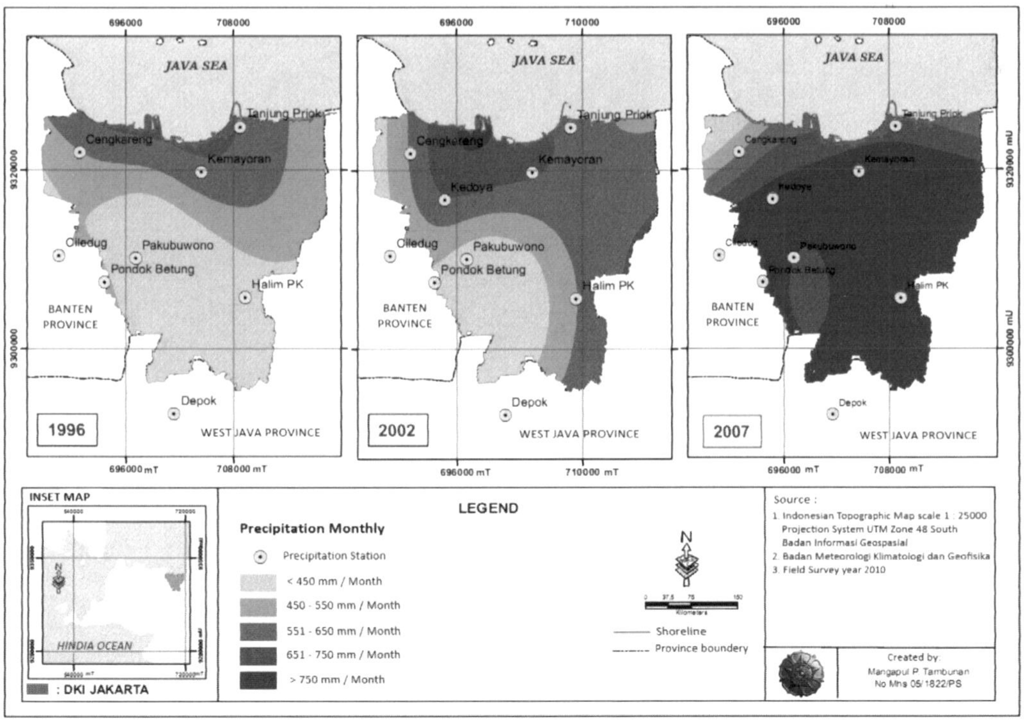

IOP Conf. Series: Earth and Environmental Science **56** (2017) 012014 doi:10.1088/1755-1315/56/1/012014

Figure 4. Map of Isohyet on February in DKI Jakarta

3.2 Distribution of Spatial Potential and Actual in Prone Area

Potential flood prone area in DKI Jakarta land is described as follows (Figure 5). 1)Very high potential flood is located in northern DKI Jakarta. 2)High potential flood is located in central and southern DKI Jakarta. 3)Rather high potential flood is located in eastern DKI Jakarta. 4)Low potential flood is located in the eastern part to south DKI Jakarta. 5)Very low potential flood in DKI Jakarta is located in the southeast on the steep slopes

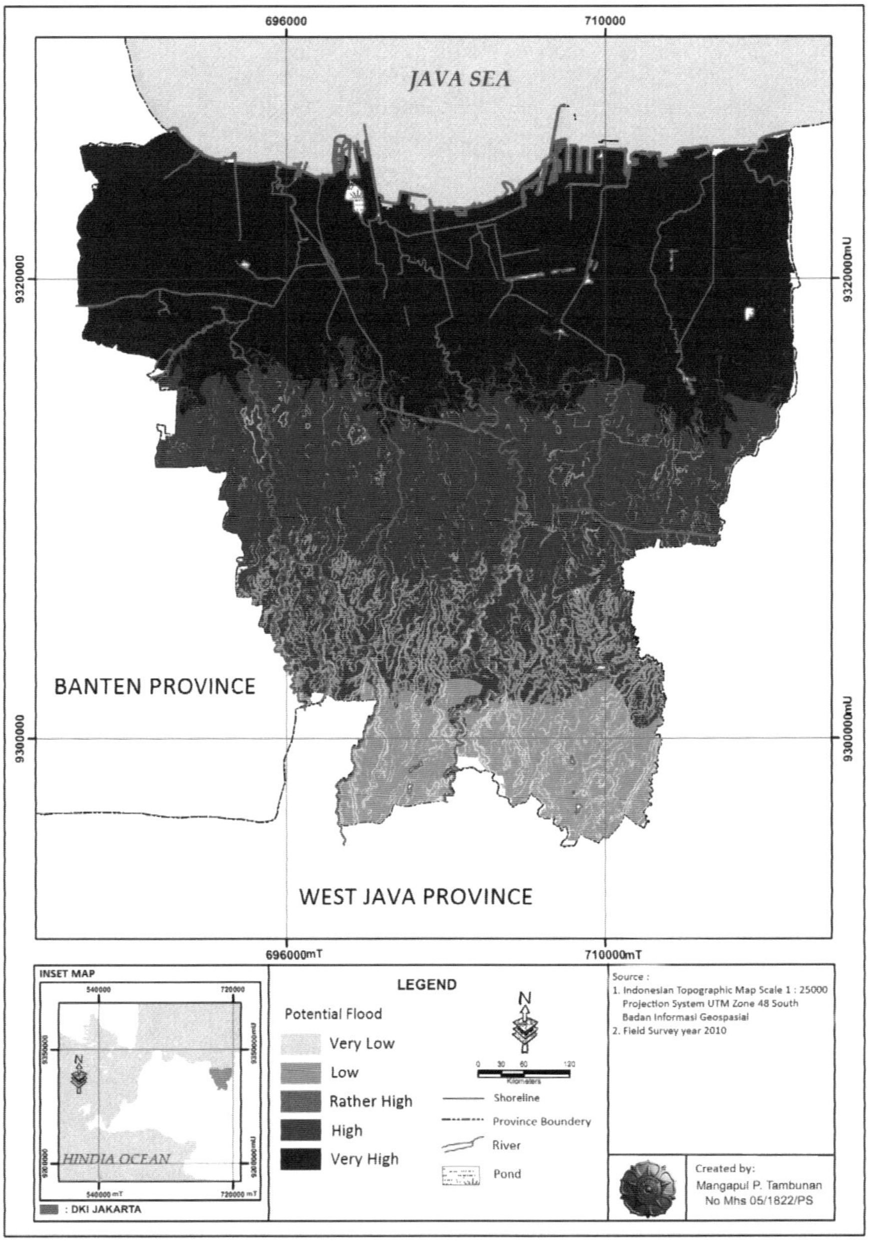

Figure 5. Map of Potential flood prone area in DKI Jakarta

The actual Prone area flooding in Jakarta in February 1996 accured in 181 sub districts (69.4%). Actual flood prone area on February 2002 has in common with the floods in February 1996. There is a good addition of flooded areas in Jakarta, West Jakarta, Central Jakarta, North Jakarta and South Jakarta. Actual flood prone area in DKI Jakarta province on February 2007 has in common with the floods in February 1996 and 2002. There are additional areas of flooding in East Jakarta, West Jakarta, North Jakarta and South Jakarta except Central Jakarta, and North Jakarta (Figure 6).

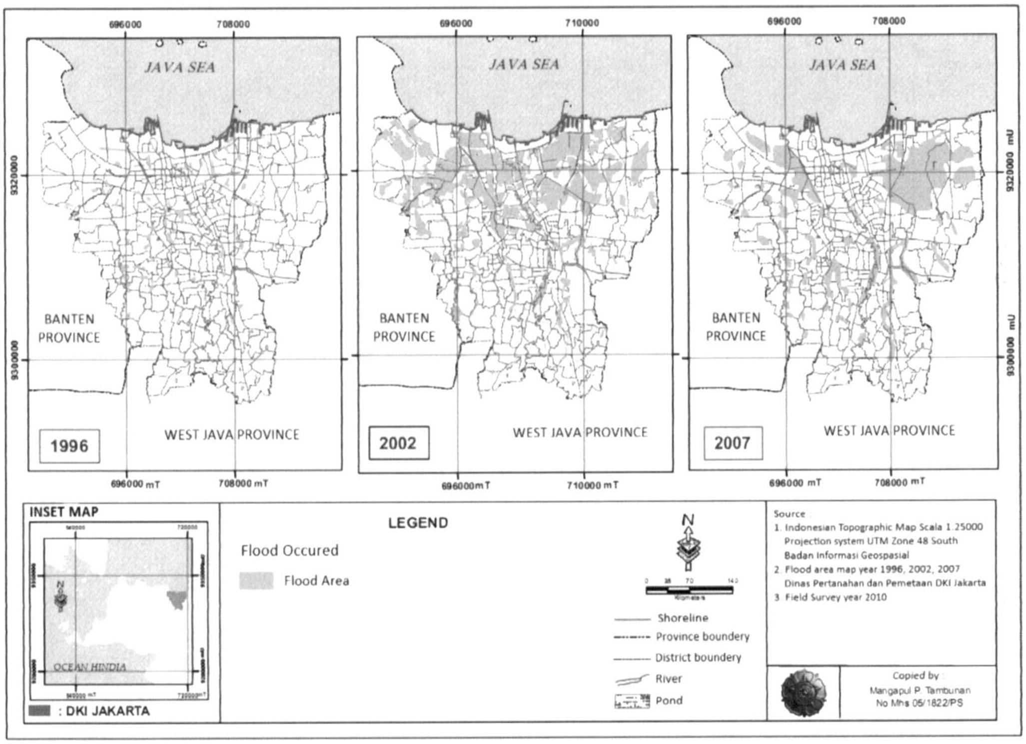

Figure 6. Map Actual flood prone area in DKI Jakarta province

3.3 Comparative Potential and Actual in Flood Prone Area

Geomorphological approaches that have been used by Thornbury [11], and Spark [10] and Verstappen [14] and Sutikno [11] in his research. While the synthesis of geomorphological approach used by Oya M., and Shigeko H [7], and Oya M [8] in examining landform lowland flooding in urban Tokyo, Japan. Neither DKI Jakarta Province almost every year there was a flood of magnitude too varied and diverse spatially and temporally in fluvial landform, fluvial-coastal, and coastal in the lower mainland Jakarta. The geomorphological approach to the synthesis approach is a form of fluvial landform. This approach can be used in the synthesis or answer any spatial-temporal problems of flooding in Jakarta and in other areas or as Dakka, Kuala Lumpur, and Tokyo in which these cities have in common morphologically fluvial landform are flat and low altitude. In this respect the findings of this study can be used as an approach to the classification of spatial and temporal variability of floods in flood prone area landform. Prone areas affected by flooding residential land use (area and dispersion) in fluvial landform.

Prone areas of flooding in beach ridge, coastal alluvial plain, and alluvial plain more breadth, depth and length than the alluvial fan. The spread of flood prone area flooding in the northwest, and northeast of Jakarta. Topography in flood prone area is located on the slope of the flood until basin flat with the surface altitude of less than 3 m above sea level. Beach ridge located in the northwest, and northeast of Jakarta, where the tendency is landform direction perpendicular to the coastline. Distribution of flooded areas in beach ridge increasingly widespread, high and long on the slopes of the flat to very flat compared to the rather steep slope.Prone areas of alluvial flood plain are located on the north coast of Jakarta includes all the sub districts in North Jakarta area and a small sub district in the City of west Jakarta and east Jakarta bordering the sub district administration in north Jakarta. Topography in flood prone area is located on the slope of the flood flat with the surface altitude of less than 3 m above sea level. Distribution of flooded areas in coastal alluvial plain of the widespread, high and long on the slopes are flat to very flat compared to the rather steep slope (Figure 7).

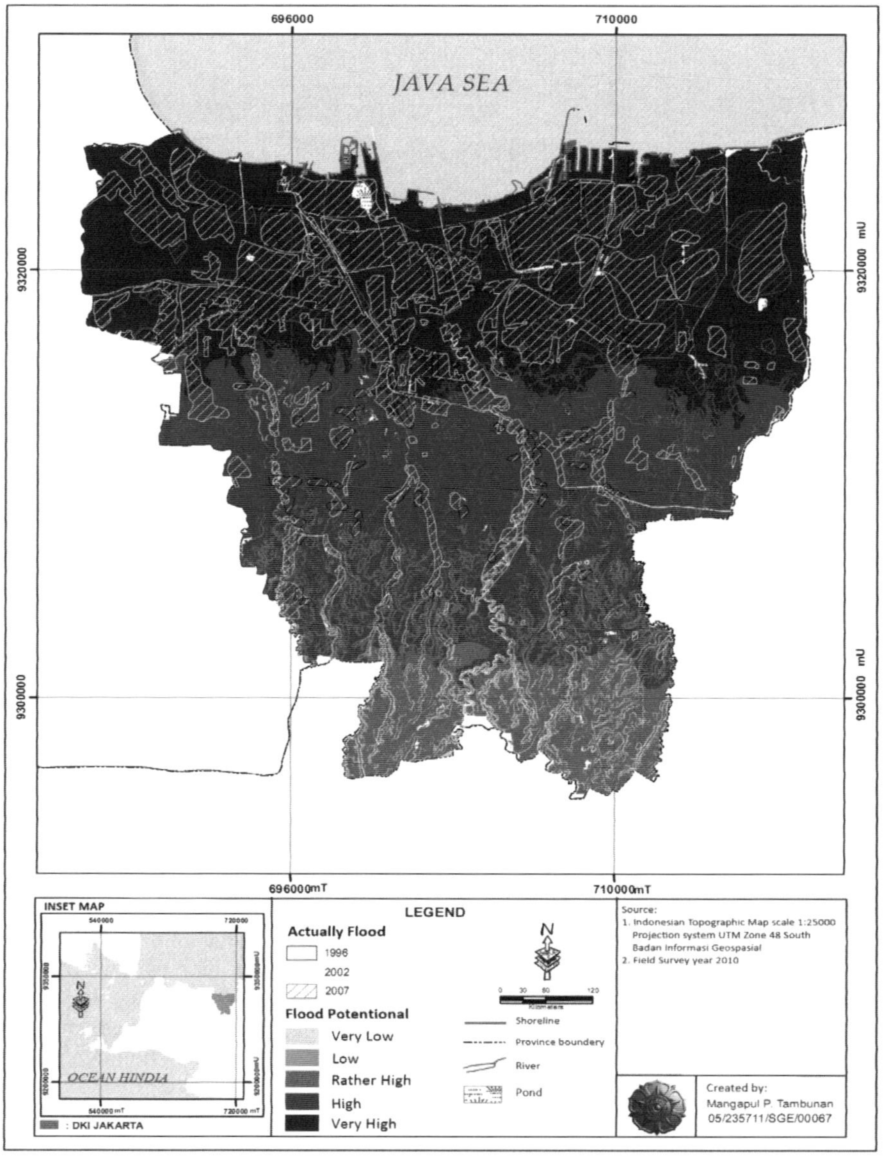

Figure 7. Map potential and actual flood prone area in DKI Jakarta

The flood disaster in 1996, 2002 and 2007 in Jakarta is the major natural disasters that includes all the characteristics of a good flood extents, long, depth, and human victims. The flood disaster in Jakarta February 4 in 2007 resulted in damage to public facilities, social services, and the impact on economic activities and services with total losses amounting to 8.78 trillion rupiah, and the displaced population of 450 thousand people in government offices and schools [2]. According to the UNDP Indonesia [12], the last time the massive flooding in the lowlands Jakarta occurs in late January to February 4, 2007. The floods have caused 57 deaths, displaced 422 300 people to safety, building 1500 houses were damaged, and the loss of the city Jakarta from floods reaches USD 695 million, like example at central business district Bundaran HI, central Jakarta, and slum settlement area at Kampung Pulo, south Jakarta.

4. Conclusion

Potential prone area is not similar with actual prone area. Actual of flood prone area in the north, west and east of Jakarta lowland both in beach ridge, coastal alluvial plain, and alluvial plain; while the flood potential area on the slope is found flat and steep at alluvial fan, alluvial plain, beach ridge, and coastal alluvial plain in DKI Jakarta.

5. References

[1] Badan Meteorologi, Klimatologi dan Geofisika (BMKG). 2008. *Publikasi Data Curah hujan di Jakarta, Bogor, Depok, Tangerang dan Bekasi selama 40 tahun terakhir.* Jakarta: BMKG Pusat Kemayoran.
[2] Bappenas.2007. *Pengendalian Banjir Jakarta. Informasi Ringkas.* Direktorat Pengairan dan Irigasi-Kedeputian Sarana dan Prasarana Kementerian Negara Perencanaan Pembangunan Nasional.
[3] Bemmelen, R.W, van. 1970. *The Geology of Indonesia.* Vol IA General Geology, Martinus Nijhoff, the Netherland: The Hague.
[4] Flood disaster at Central Business Distrik at Bundaran HI, DKI Jakarta. liputan6.com, 2015 download on 22 Agustus 2016 pkl 4.30 pm.
[5] Gupta, A., 2009. The hazardousness of high-magnitude floods. In Irasema Alcantara Ayala and Andrew Goudie (eds.), *Geomorphological Hazards and Disaster Prevention.* Cambridge University Press.
[6] Japan International Cooperation Agency. Direktorat Jenderal Pengembangan Sumber Daya Air Departemen Pekerjaan Umum Republik Indonesia. 2002. *Studi On Urban Drainage And Wastewater Disposal Project In The City Of Jakarta.*
[7] Oya Masohito, dan Shigeko Haruyama, 1987. Flooding and Urbanization in the Lowlands of Tokyo and Vicinity, *Natural Disaster Science, Volume 9, Number 2,* pp.1-21. Japan.
[8] Oya, Masahiko. 2001. *Applied Geomorphology for Mitigation of Natural Hazards.* The Netherlands: Published by Kluwer Academic Publishers. P.O.Box 17,3300 AA Dordrecht.
[9] Smith, K., 1996. *Environmental Hazards: Assessing Risk and Reducing Disaster*, 2nd edition. London: Routledge.
[10] Spark, B.W., 1960. *Geomorphology.* London: Longman.
[11] Sutikno,. 1995. *Geomorfologi dan Prospeknya di Indonesia.* Pidato Pengukuhan Jabatan Guru Besar pada Fakultas Geografi Universitas Gadjah Mada, Yogyakarta.
[11] Thornbury W.D., 1959. *Principles of Geomorphology.* New York: John Wiley & Sons.
[12] UNDP, 2007. *Sisi lain perubahan iklim: Mengapa Indonesia harus beradaptasi untuk melindungi rakyat miskin.* Jakarta: Penerbit UNDP di Indonesia.
[13] Verstappen, H. Th., 1953. *Djakarta Bay, a Geomorphological Study on Shoreline Development.* Doctoral Thesis, Univ. Utrecht, Drukkerij Trio's Gravenhage, The Netherlands.
[14] Verstappen, H. Th., 1983. *Applied Geomorphology, Geomorphological Surveys for Environmental Development.* Amsterdam: Elsivier.
[15] Verstappen H. Th., 2000. The outline of the Geomorphology of Indonesia: A case study on tropical geomorphology of a tectogene region. *ITC Publ. No.79.* Enschede. The Netherlands.
[16] Wohl, E. E., 1992. Bedrock benches and boulder bars: floods in the Burdekin Gorge of Australia.

IOP Conf. Series: Earth and Environmental Science **56** (2017) 012014 doi:10.1088/1755-1315/56/1/012014

Bulletin, Geological Society of America.

Numerical simulations of tsunami waves impacts on Ulee Lheue Harbour in Banda Aceh-Indonesia

Fachrurrazi[1,*], Syamsidik[1,2], M Al'ala[1,3] and W Mahardi[4]

[1]Tsunami Computation and Visualization Laboratory, Tsunami and Disaster Mitigation Research Center (TDMRC), Syiah Kuala University, Jl. Prof. Ibrahim Hasan, Gampong Pie, Banda Aceh, 23233, Indonesia.
[2]Lecture at Civil Engineering Department, Syiah Kuala University, Jl. Syeh Abdurrauf, No. 7, Banda Aceh, 23111, Indonesia.
[3]Student at Civil Engineering Master Program, Syiah Kuala University, Jl. Syeh Abdurrauf, No. 7, Banda Aceh, 23111, Indonesia.
[4]Directorate of Coastal and Marine Affairs Ministry of Maritime Affairs and Fisheries, Republic of Indonesia, Jl. Medan Merdeka Timur No. 16, Jakarta Pusat, 10110, Indonesia.

*email: fachrurrazi@email.com

Abstract. This paper reports the effects of 2004 Indian Ocean tsunami onto the Ulee Lheue harbour facility, Banda Aceh – Indonesia. The breakwater that had damaged after tsunami were rebuilt into its original design once more due to UNDP funding source. As the existing construction knowing how much the chance it stands againts the tsunami in various terms would be decent for further improvement. This research aim is to measure the capabilities of the breakwater againts the various tsunami scenario. performing the numerical simulation to analyze the hydrodynamics we used both COMCOT and Delft3D-flow for tsunami propagation in line with the hydrodynamics. Several observation points were deployed representing each part of the breakwater. The process revealed that the breakwater only able to hold 8.0 Mw induced wave from overtopping. uniquely when 8.2 Mw tsunami wave strikes the breakwater till overtopped but not giving enough energy to move the boulder aside. Potential movement of the boulder occurred when the 8.4 Mw tsunami wave come through the breakwater produced 77.12 m (5.88 %) damaged structure. The 8.6 Mw single fault highest magnitude gave 209.32 m long (15.97%) destruction upon this Ulee Lheue Harbour breakwater.

1. Introduction

The 2004 Indian Ocean tsunami left many lessons learned. Among the most important lessons were the destructions of harbour area around Banda Aceh where their images were often used to describe the tsunami prowess. One study also revealed that large shear stresses generated by the 2004 tsunami were able to detach a portion of land creating a new island as in the case of Ujong Seudeun of Aceh Jaya District, located about 70 km to the western part of Banda Aceh [1]. Despite the firm evidence, future development of the coastal area in Banda Aceh is questioned regarding its resiliency toward future tsunamis. Smaller magnitude of tsunamis than 2004 have higher possibility to hit the coastal zone. Nonetheless, no study have been conducted to assess the possibility of the impacts of the future tsunami on coastal area in Banda Aceh. The reasons inspired this study, to evaluate the possible effects of the smaller tsunami generated by earthquakes smaller than 8.6 Mw that could happen around the same rupture area, as in the 2004 Indian Ocean Tsunami. The Ulee Lheue Ferry Port is one of the most important complexes in Aceh that connect the mainland of Sumatra Island to some small islands around Aceh Besar and Sabang Districts. During the 2004 Indian Ocean tsunami reconstruction process, this harbor was frequently used to deliver aids and other emergency supports for the affected area. Considering its important status, this study was performed to see the impacts of the future tsunami regarding of wave heights, time of arrivals, and shear stresses.

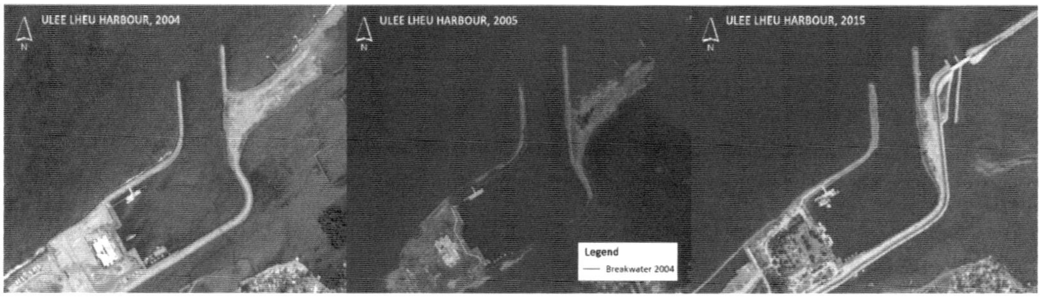

Figure 1. Satellite images of the Ulee Lheue Harbor taken in 2004 before the tsunami (left), after the tsunami taken in 2005 (center), and after reconstruction taken in 2015.

2. Study Area

Ulee lheue harbour became the site where the research took place. Massively destructed, the harbour area wears a 8 hectares, which consists of the passenger terminal as the main building, parking area, ferry port, harbour basin etc. After the 2004 mega disaster, almost all the facility in the harbour area were seriously damaged with almost nothing left except the ruins. This unexpectable damage also occurred in the breakwater. It had destructed along 699 m from 1311 m in total, representing 53.32% of damage (according to breakwater line comparations on Google Earth images taken on June 23th, 2003 and January 28th, 2004). As the people will always try to learn and do better, the facilities were rebuilt with the UNDP's fund with no modification in the former design. The breakwater rebuilt with 464 m long on the left side, and 847 m on the other side. The construction built of the big stones that assembled with 1 m diameter in average. The study area is shown in figure 2.

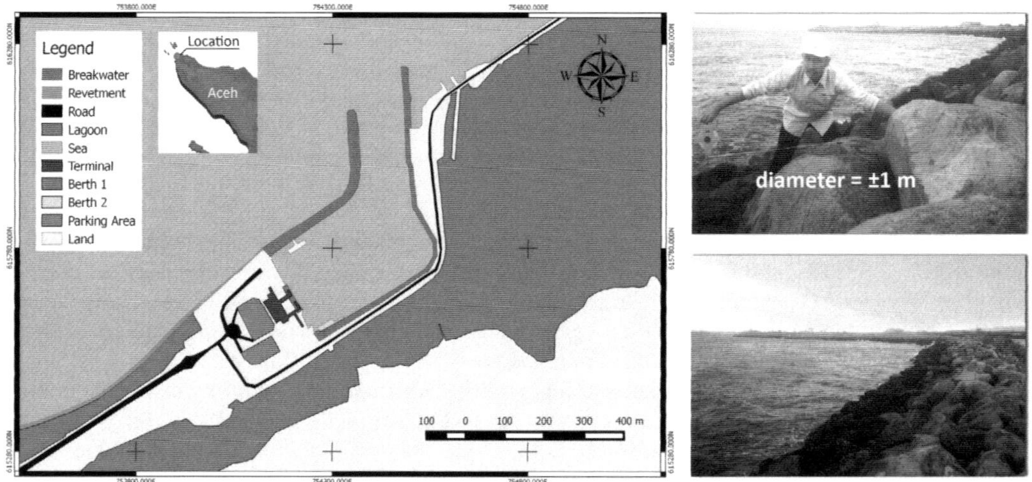

Figure 2. The map of Ulee Lheue Harbour.

3. Methods

Two numerical models were used in this research, namely COMCOT (Cornell Multi-grid Coupled Tsunami Model) and Delft3D-FLOW. COMCOT was used to perform tsunami propagation from the sources since the Delft3D-FLOW has a limitation to generate a linear shallow water equation and to include initial waves produced by presumably rupture areas. At the innermost layers of the simulation, we impose Dellft3D-FLOW where its open boundary condition was adopted from the wave heights produced by COMCOT simulations. The Delft3D FLOW is able to simulate non-linear shallow water equation with the small size of the grid. This gives the advantage to create grids that represent breakwaters and another type of facilities in the harbour area.

3.1 Grid setup

Due to this research, the COMCOT model proposed 5 horizontal layer to complete the process (figure 3). The fault rupture will be accommodated in the first outer layer while the 2nd, 3rd and 4th layer were smaller in area and grid size. The GEBCO data were used as the bathymetry which updated first with the available chart from DISHIDROS TNI – AL (Indonesia Navy). All 4th outer layer were using the spherical coordinate system while the finest one using the cartesian. The latest field measurement in 2014 data was applied for the innermost domain which consists of 5x5m grid size. The domain consists the detail measurement of both topography and bathymetry from 2014 field survey. After compilation, the grid was resampled to produce the final modelling grids. Simulating the flow, the 0.013 roughness coefficient were applied in the wet area while 0.02 determined in the dry area. As a focus the grid that represents breakwater had 0.035 value for roughness coefficient. Bathymetry model used in this study is shown in Figure 3.

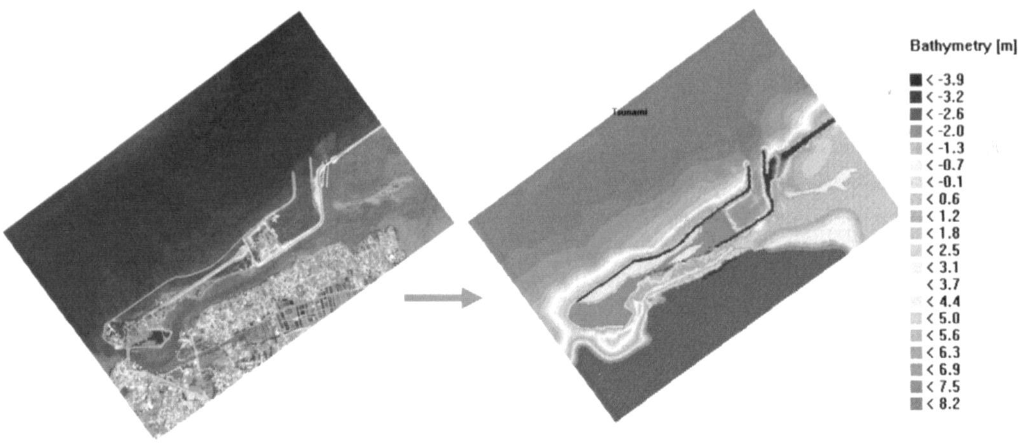

Figure 3. Simulation domain in Delft3D-Flow.

3.2 Fault model

Using the same historical data as based on 2004 megathrust earthquake, some modification on the fault scenario applied to represent the different magnitude. The model varied with single fault scenario from 7.8 Mw, 8.0 Mw, 8.2 Mw, 8.4 Mw, to 8.6 Mw. The adjustment of the fault parameter was calculated in scale based on well Coppersmith equation by follows a finite rectangular rupture area [10]. started at 01.00 UTC (08.00 am local time) on December 26, 2004, the megathrust earthquake which located at 95.982° E and 3.295° N had produced a significant dislocation an amount volume of water moved dynamically. This event caused by the duration of the earthquake within 8 to 10 min which caused the western coast of northern Sumatra Island ruptured as far as 1200–1300 km from the

IOP Conf. Series: Earth and Environmental Science **56** (2017) 012015 doi:10.1088/1755-1315/56/1/012015

Andaman–Sunda Sea. The complex fault mechanism of the rupture process is completed to explain in detail. Okada had purposed a theory which explains tsunamis are generated by the deformation of the sea floor, which is instantly reproduced at the sea surface [5]. This theory is also known as the deformation model. Several researchers later on also have proposed earthquake source parameters to describe the complexity of the multi-fault mechanism, one of them was developed by Romano [6]. This model was also used by Liu et al. [3] to investigate the morphological changes in Aceh due to the 2004 Indian ocean tsunami.

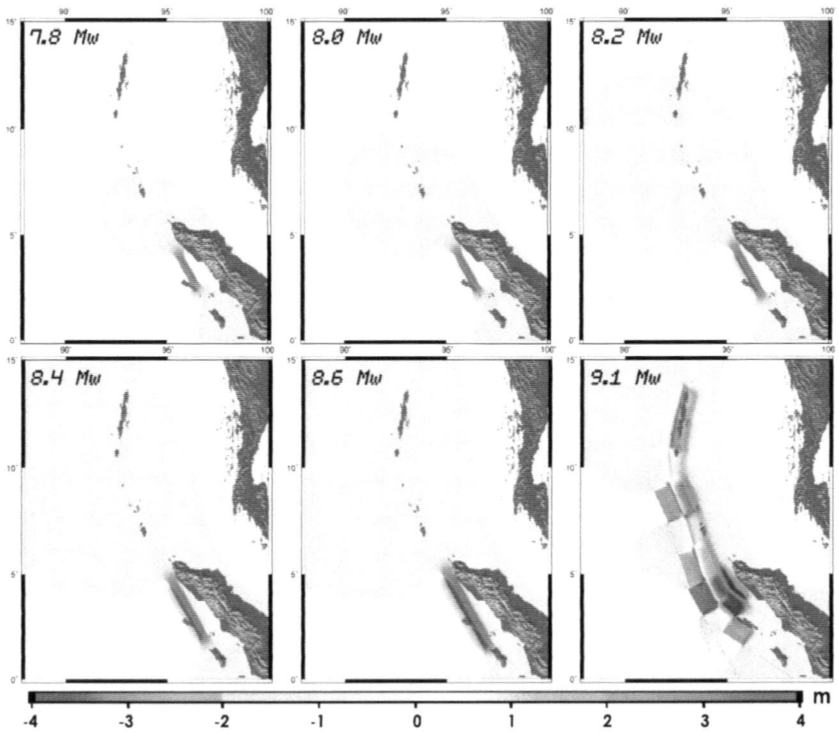

Figure 4. Initial waves produced by various single fault scenario in comparison with 2004 event.

A multi-fault model as proposed by Romano [6] was used to generate the tsunami waves around the rupture area of the 2004 Indian Ocean tsunami. Each of the nine faults was used to calculate the waveforms around the faults using Masinha and Smyle Theories combined with Okada analytical solutions for waveforms produced by a fault [4]. The model also had been validated by Syamsidik et al [7] using a transect line captured by Satellite Jason 1 about two hours after the first earthquake [2]. The fault was divided into nine segments of rupture area. The initial waves around the sources based on this scenario can be seen in figure 4. Negative waves were observed in the eastern part of the source. This multi fault scenario used as a comparison for the variation of magnitude wave that generated the tsunami and affecting the coastal protection (table 1).

Table 1. Fault mechanism variation parameter.

Event (Mw)	Long, Lat (deg)	Depth (Km)	Strike;Dip;Rake (deg)	Mo (N*m)	Width (m)	Length (m)	Dislocation (m)
7.8	95.982° E 3.295° N	30.00	329°; 8°; 110°	6.31 x 10^{20}	30619.63	145211.16	4.05
8.0				1.26 x 10^{21}	35481.34	190546.07	5.32
8.2				2.51 x 10^{21}	41114.97	250034.54	6.98
8.4				5.01 x 10^{21}	47643.1	328095.29	9.16
8.6				1.00 x 10^{22}	55207.74	430526.61	12.02
9.1				[6]	[6]	[6]	[6]

3.3 Numerical simulations

Two model were used to performed the simulation, i.e. Cornell Multi-grid Coupled Tsunami Model (COMCOT) which generated the hydrodynamic condition during the tsunami waves propagation and Delft3D continued the sediment transport process that later produces morphological changes around the domain of the simulations. The Delft3D was applied at the innermost layer of the simulation.

COMCOT uses Shallow Water Equations (SWE), i.e. Linear SWE and Nonlinear SWE. The linear SWEs are as follows.

$$\frac{\partial \eta}{\partial t} + \frac{1}{R\cos\varphi}\left\{\frac{\partial P}{\partial \psi} + \frac{\partial}{\partial \psi}(\cos\varphi Q\right\} = -\frac{\partial h}{\partial t} \tag{1}$$

$$\frac{\partial P}{\partial t} + \frac{gh}{R\cos\varphi}\frac{\partial \eta}{\partial \psi} - fQ = 0 \tag{2}$$

$$\frac{\partial Q}{\partial t} + \frac{gh}{R}\frac{\partial \eta}{\partial \psi} + fP = 0 \tag{3}$$

The nonlinear equations that include friction factors can be seen in Eqs 4-6 as follows.

$$\frac{\partial \eta}{\partial t} + \frac{1}{R\cos\varphi}\left\{\frac{\partial P}{\partial \psi} + \frac{\partial}{\partial \psi}(\cos\varphi Q\right\} = -\frac{\partial h}{\partial t} \tag{4}$$

$$\frac{\partial P}{\partial t} + \frac{1}{R\cos\varphi}\frac{\partial}{\partial \psi}\left\{\frac{P^2}{H}\right\} + \frac{1}{R}\frac{\partial}{\partial \psi}\left\{\frac{PQ}{H}\right\} + \frac{gH}{R\cos\varphi}\frac{\partial \eta}{\partial \psi} - fQ + F_x = 0 \tag{5}$$

$$\frac{\partial Q}{\partial t} + \frac{1}{R\cos\varphi}\frac{\partial}{\partial \psi}\left\{\frac{PQ}{H}\right\} + \frac{1}{R}\frac{\partial}{\partial \psi}\left\{\frac{Q^2}{H}\right\} + \frac{gH}{R}\frac{\partial \eta}{\partial \psi} + fP + F_y = 0 \tag{6}$$

where,

$$H = \eta + h \tag{7}$$

$$f = 2\Omega\sin\varphi \tag{8}$$

$$F_x = \frac{gn^2}{H^{7/3}}P\sqrt{P^2+Q^2} \tag{9}$$

$$F_y = \frac{gn^2}{H^{7/3}}Q\sqrt{P^2+Q^2} \tag{10}$$

$$P = \int_{-h}^{\eta} udz = u(h + \eta) = uH \quad (11)$$

$$Q = \int_{-h}^{\eta} vdz = v(h + \eta) = vH \quad (12)$$

Here, g is gravitational acceleration, f is Coriolis coefficient, P is discharged flux at x-axis (west-east), Q is discharge flux at x-axis (north-south). Meanwhile, φ,ψ are *latitude* and *longitude* respectively, R is the earth radius; h is depth, η is water elevation H is total depth, Ω is the earth rotational speed (7.292 x 10^{-5} rad/s); F_x, $F_y$are friction at x- and y-dimensions respectively, n is manning coefficient, and u,v are velocity at x- and y- directions, respectively. The equations were schematized using Leap-Frog Method of Finite Different Method. Details of the COMCOT model development were released in 2004 and 2009 [8,9].

3.4 Observation Point

In objective to observe the hydrodynamic process due to the tsunami in affecting the coastal protection in this case breakwater, several observation point were applied. The observation points represented the breakwater as the object that receiving the tsunami energy. There was two side of the breakwater which on the south-east and north-west. 119 observation point were deployed to represent each grid on breakwater area as shown in figure 5. Wave height, velocity and shear stress were recorded during the simulation process in time series.

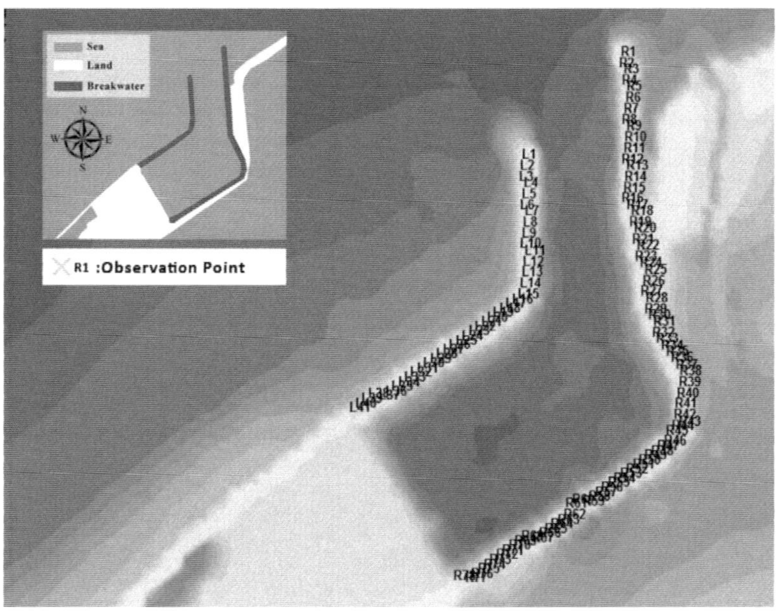

Figure 5. Locations of the numerical observation points and cross sections.

3.4 Critical shear stress

Shear stress is the force that gave by current, wave or both of them combined. As a fuel for sediment transport, shearstress gave the force which hold by critical shear stress as a reaction. Critical shear stress represents the ability of a particle to react and holds the stand againts the external forces. Particle start moving (initial movement) when the bed shear stress (τb) produced by current or waves exceeds it's critical shear stress (τc), in other words, τb > τc. but mathematically it only explains whether it was potential to move or not. In the harbour case, the particle was substitute with boulder-size (diameter) while the calculation produced the indication of the boulders respond whether it moved or not. Producing the velocity and waves for fueling the shear stress, Delft3D-Flow using tsunami wave input from COMCOT get the job done. The following equation was used to determine critical shear stress for stone motion:

Komar and Miller 1974,

$$\tau_{cr} = \theta^*(\rho_s - \rho)gd_{50} \tag{12}$$

$$\theta^* = \frac{0.30}{1 + 1.2D_*} + 0.055[1 - \exp(-0.020D_*)] \tag{13}$$

With,

$$D_* = \left[\frac{g(s-1)}{v^2}\right]^{\frac{1}{3}} d_{50} \tag{14}$$

Here, τ_c is the critical bed shear stress (N/m^2), θ^* is parameter for the given particle size (dimensionless), s is the specific gravity of the particles and is calculated as the ratio of specific weight of stone ρ_s, to specific weight of water ρ (kg/m^3), g is the constant for acceleration of gravity (m/s^2), D_* is grain size (dimensionless), d_{50} is median boulder size (m).

4. Results

As the construction that had a submerged part while other part didn't, breakwater as potential protection due to the tsunami, but how far it stands again tsunami was the question about to answer. Creating the tsunami wave COMCOT had produced the water level in the outer finest layer as an input to Delft3D-flow boundaries as shown in figure 6. Different wave height had produced due to different magnitude waves of the earthquake. The difference of the water volume movement had contributed to the tsunami waves height. The result shows the waves started to peak rapidly exceed 1 meter after 8.0 Mw scenario applied and keep raising until the highest magnitude waves of single fault mechanism which represent by 8.6 Mw. 0.72 m high peak wave released from 7.8 Mw scenario which follows by 2.06 m from 8.0 Mw. The next scenario 8.2 Mw gave further gap with 4.18 m high wave which continues higher, even twice with 8.4 m high in peak. The higher single fault scenario produces the highest wave with 13.28 m.

IOP Conf. Series: Earth and Environmental Science **56** (2017) 012015 doi:10.1088/1755-1315/56/1/012015

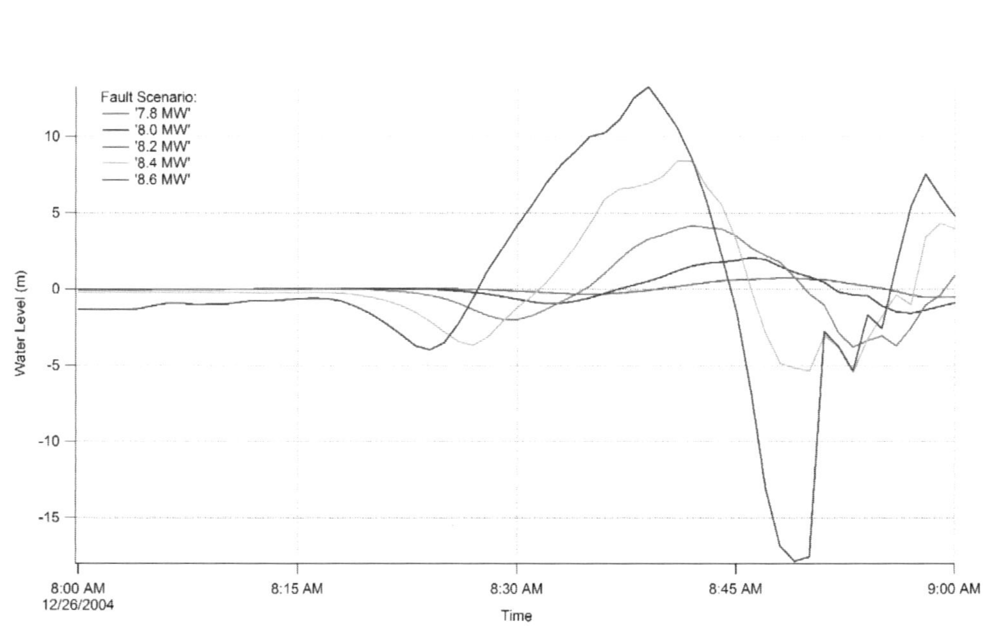

Figure 6. COMCOT tsunami wave height with difference of magnitude wave.

The tsunami waves came and amplified inside the harbour mouth and circling inside the harbour basin. The hydrodynamic condition inside the basin gave the destruction around the southern part of the basin and also the north, which are prependicular with the waves direction.

4.1 Tsunami hydrodynamic

Interacting with a shallower depth, the wave getting higher on the break water. The result revealed that after all scenario waves going through breakwater, only with 8.2 Mw minimum the waves succeed overtopped the breakwater. Three scenarios which 8.2 Mw, 8.4 Mw, and 8.6 Mw each of them produced 8 m, 13.9 m and 19.2 m wave height when interacting with the breakwater. Beside all the differences with the wave height, the arrival time was also different for each of the scenario. The higher magnitude wave gave, the sooner time to arrive for the tsunami waves.

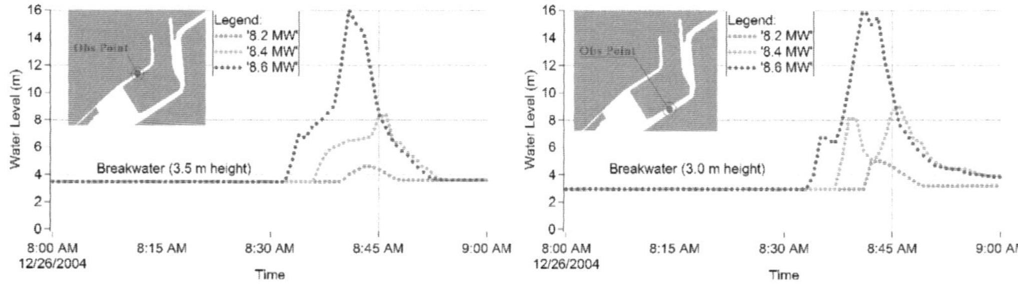

Figure 7. Differences in tsunami water level on breakwater construction.

Tsunami continues waving from the Delft3D boundary and interacting with the bathymetry around the harbour. The first 7.8 Mw scenario didn't give any significant wave run up while the next 8.0 Mw scenario gave the tsunami wave run-up although did not overtop the breakwater construction. Deducing the propagation and the run up the 8.0 scenario become the furthest limit for the breakwater to be able in facing the tsunami wave run-up. The breakwater construction couldn't do much for the higher magnitude wave, started from 8.2 Mw the breakwater were not able to hold the tsunami wave from amplifying and flooding far inland.

4.2 Bed shear stress

Complementing the former process which was the wave run up, the bed shear stress explained how the breakwater interacted with massive tsunami energy that varies due to the proposed scenario. The respond, the ability of the breakwater to stand agains the tsunami wave will be described with the process of forces which worked around the harbour. As the calculation went the critical shear stress were done by using the boulder size (diameter) regarding 25 degree which gave 0.89×10^{-3} kg/ms of viscosity value. While 2600 kg/m^3 were the value of the boulder density and 1 m of diameter the calculation induce 873.814 N/m^2 for the internal force that holds the particle alias critical shear stress as shown in figure 8.

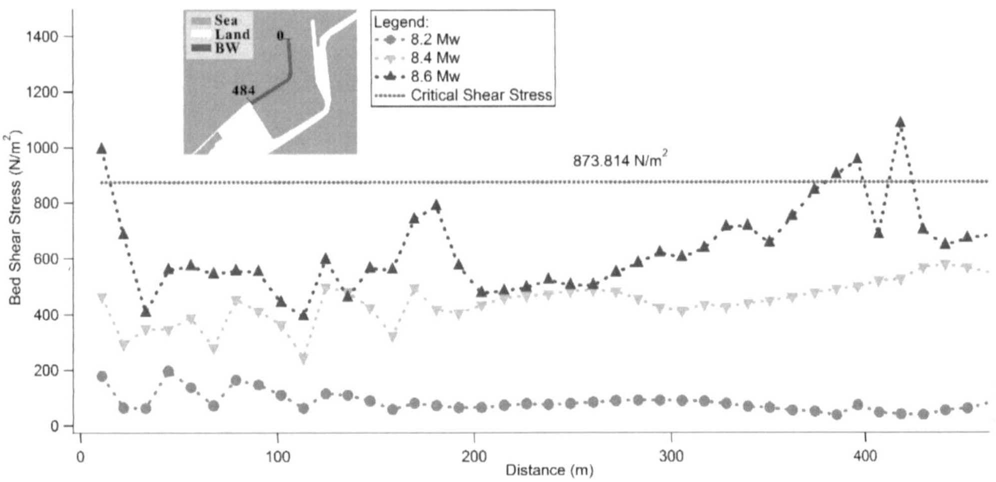

Figure 8. Bed shear stress maximum on left breakwater.

The similar process with wave run-up happened during the shear stress analysis. But the overtopping didn't always mean enough energy to move the boulder. This conclusion came from the 8.2 scenario which had overtopped the construction, but in the shear stress that induced didn't have enough energy to exceed the critical shear stress. Its appealed that 8.2 Mw scenario became the limit for the 1 m diameter of the boulder to hold the tsunami energy from 2004 historical fault parameter. The exceeded value in the next scenario confirmed that in 8.4 Mw the maximum shear stress that induced, 1530 N/m^2 had overtaken 873.814 N/m^2 of boulder critical shear stress.

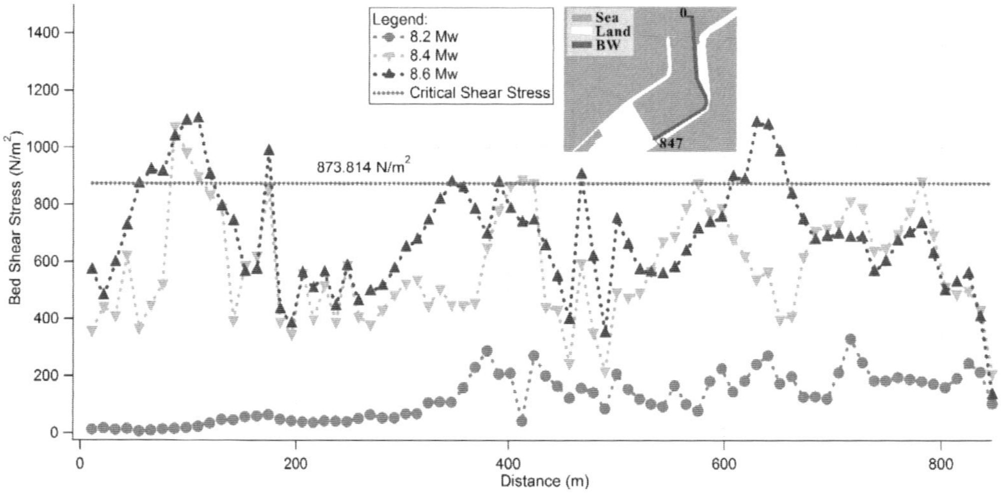

Figure 9. Bed shear stress maximum on right breakwater.

Uniquely the shear stress only exceeds on the south-east part of breakwater while leaving the southwestern part remain save from the energy. It happened to be the current that produced the shear stress amplified around the corner of the south-east part while the wave and current were directed into that spot and had nowhere to go, then turn back. When the wave that turns and circling back met the wave that keeps coming from the mouth of the breakwater in harbour basin, it amplified the waves and the current, produced the massive energy of shear stress that exceeds the limit of boulder critical shear stress. The spot that located in the corner of the harbour basin were accommodating the amplification through the propagation and run-up the process. The spot where the exceeding value of shear stress against its critical limit was getting wider in 8.6 scenario. The breakwater couldn't hold the wave and the energy any longer in the southwest part. The spot is wider and it amplified both southeast and northwest part of the breakwater. This simulation shows that the placement and the design of the break water gave an effect for the tsunami wave to interact with the construction and how it will be after getting through the construction.

Table 2. Breakwater damage of several tsunami scenario

Description	Event				
	7.80 Mw	8.00 Mw	8.20 Mw	8.40 Mw	8.60 Mw
Total Observation Points (Obs)	No Overtopping	No Overtopping	119	119	119
Exceed Critical Shear Stress (Obs)			0	7	16
Percentage of Fail (%)			0	5.88	15.97
Breakwater Failed (m)			0	77.12	209.32

5. Conclusion

COMCOT had produced the tsunami waves along with 1-hour numerical simulation. A 4.18 m high tsunami wave continued and overtopped through the breakwater. In the boulder case, the overtop not always can be transferred to movement (transport) of the boulder with 1 m diameter which produced 873.814 N/m^2 critical shear stress. The shear stress started exceeding its critical shear stress in 8.4 Mw scenario which brought 8.4 m high wave into the boundary which later on produced 1530 N/m^2 which exceed its critical shear stress and so on with 8.6 scenario. With the limitation, design whom still old style harbour, the breakwater could sustain a tsunami generated by an earthquake smaller than 8.2 Mw but started in 8.4 Mw the damage began to happen 5.88 % and followed by 8.6 Mw that gave 13.45 % damage (table 2). Presumably, the source is the same with the recent 2004 tsunami. However, with bigger earthquakes, the breakwater could be detached and failed as in the 2004 Indian Ocean tsunami case.

6. References

[1] Al'ala M, Syamsidik, Rasyif T M and Fahmi M 2015 Numerical Simulation of Ujong Seudeun Land Separation Caused by 2004 Indian Ocean Tsunami, Aceh-Indonesia *J. of Tsunami Society* **34** No.3 159-172

[2] Gower J 2005 Jason-1 detects the December 26, 2004 tsunami *EOS Transaction AGU* **86** 37–38

[3] Li L, Qiu Q and Huang Z 2012 Numerical Modeling of the Morphological Change in LhokNga, west Banda Aceh, during the 2004 Indian Ocean Tsunami : Understanding Tsunami Deposits Using a Forward Modeling Method *Nat Hazard* **64** 1549-1574

[4] Masinha L and Smylie D E 1971 The displacement fields of incline faults *Bull. Seismol. Soc. Am.* **61** (5) 1433–1440

[5] Okada Y 1985 Surface deformation due to shear and tensile faults in a half-space *Bull. Seismol. Soc. Am.* **75** (4) 1135–1154

[6] Romano F 2009 The Rupture Process of Recent Tsunamigenic Earthquake by Geo- physical Data Inversion *Doctoral Thesis* University of Degli Bologna Italy

[7] Syamsidik, Rasyif T M and Kato S 2015 Development of Accurate Tsunami Estimated Times of Arrival for tsunami-prone cities in Aceh Indonesia *Int. J. of Disaster Risk Reduction* **14(4)** 403-410

[8] Wang X and Liu P L F 2004 An Analysis of 2004 Sumatra Earthquake Fault Plane Mechanisms and Indian Ocean Tsunami *Journal of Hydraulic Research* **00** No. 0 (2006) pp 1–8

[9] Wang X 2009 User Manual for Comcot Version 1.7 (First Draft) http://ceeserver.cee.cornell.edu//pll-group/comcot_down.htm, accessed on 12 January 2013

[10] Wells D L and Coppersmith K J 1994 New empirical relationship among magnitude, rupture length, rupture width, rupture area, and surface displacement *Bull. Seismol. Soc* **84** (4) 974–1002

Acknowledgments

Authors would like to thank the Ministry of Research, Technology and Higher Education (RISTEKDIKTI) of Indonesia through its grant named International Collaboration and Publication (*Kerjasama Luar Negeri dan Publikasi International*) Year 2017 under research title; "Mitigating Impacts of Tsunami Waves on Coastal Structures and Harbor Facilities". Authors are also grateful to the member of Coastal Research Group and TDMRC Syiah Kuala University for support and help.

Development of Web GIS for information of renewable energy in Aceh Province after rehabilitation and reconstruction process

Nizamuddin[1,*], Hizir[2], Ardiansyah[3], D Pertiwi[1] and P Handayani[1]

[1]Department of Informatics, Faculty of Mathematics and Natural Sciences, Syiah Kuala University, Darussalam, Banda Aceh 23111, Indonesia
[2]Department of Mathematics, Faculty of Mathematics and Natural Sciences, Syiah Kuala University, Darussalam, Banda Aceh 23111, Indonesia
[3]Department of Electrical, Faculty of Engineering, Syiah Kuala University, Darussalam, Banda Aceh 23111, Indonesia

*email: niz4muddin@unsyiah.ac.id

Abstract. There are a lot of renewable energy potentials that is spread in numerous locations in Aceh Province. These potentials can be developed into energy source that can be utilized to fulfill the need of energy in the area. It is unfortunate that the information about the potentials is still hard to find, regardless the abundance of spatial data that were produced during the rehabilitation and reconstruction process, including data related to the renewable energy. Therefore, this research was conducted to develop an application of information technology, especially Web Geographic Information System (GIS) for renewable energy potential. The objective of this research was to fulfill the needs of relevant agencies for the Web GIS-based information technology applications to manage information on renewable energy potential in an interactive and integrated way. This research used components of Free Open Source System (FOSS) to develop a Web-based geospatial information technology application. The component consists of GeoServer as a GIS server, PostgreSQL as a geographic databases management system, QGIS as Desktop GIS for spatial data preparation and GeoExt as the framework in developing user interfaces for the Web GIS. The result of combining those components was a Web GIS that displayed areas with its renewable energy potential which was organized for each area along with legends for the map information. The Web GIS is equipped with search feature and tooltips function to select any areas and bring up detailed information of renewable energy potential of the area.

1. Introduction

Aceh is one of the provinces in Indonesia that is rich in oil, gas, and also other minerals. Oil and natural gas, along with other minerals give the largest contribution to Aceh's local revenue. However, the management and utilization of mineral resources and renewable energy in Aceh is facing problems and challenges that increase every year. Exploitation of natural resources that ignore spatial aspects in a continuous manner will cause environmental carrying capacity to be exceeded. This is because the geographical information on mineral and renewable energy potential itself is not integrated with other geographical information such as utilization of land, including forest protected areas, data infrastructure, and settlements.

This article discusses a solution of the problem of managing information on mineral and renewable energy potential that has been described above. The development of information technology applications, interactive Web-based GIS is one of the solutions for the integration of information on mineral and renewable energy potential in Aceh.

2. Literature

2.1. Geographic Information System (GIS)
Web-based Geographic Information System or WebGIS is an application that can be accessed via the Internet to manipulate, analyze, store, and use GIS features without having to use GIS software.

2.2. Geoserver
GeoServer is open source web server software. GeoServer provides features for reforming, abolishing, inputting, and publishing geospatial data using a Web Feature Service (WFS), which can be done via online and offline [1].

2.3. GeoExt
GeoExt is an Open Source JavaScript Framework and used to build GIS-desktop application that provides a web service to make the basics of web based mapping. GeoEXT combines GIS functions of OpenLayers and user interface of Ext JS Library. GeoExt also provides services for accessing information on the services of Open Geospatial Consortium (OCG), the object OpenLayers and other remote sensing data [2].

2.4. PostgreSQL
PostgreSQL is one of database (Object-Relational Database Management System) that offers great flexibility, stability and high performance. PosgreSQL database can be used on multiple platforms and supported by many programming languages such as Java, Perl and PHP. PostgreSQL provides flexibility for users to add data types, functions, operators, aggregates function, methods of indexing and procedural languages. One capability of PostgreSQL is to store geospatial objects as data in a database. To accommodate spatial data, PostgreSQL required one spatial extension, which is named PostGIS. This extension also allows users to perform analysis and processing of geospatial data [3].

3. Research Methods
Web GIS development process flow to integrate the information of mineral potential and renewable energy in Aceh can be seen in Figure 1. The method used in this study was established based on the study of the literature on the potential of mining and renewable energy in the area. The data were obtained by collecting data directly from the relevant departments or agencies in the city or district and also from the annual reports of the relevant subject in the province.

Identification, verification and standardization were the next step that must be done during the data processing stage. This stage was conducted to determine the aspects of data and information were correct and to ensure all the data were going to be published were in conformity with the standards of geospatial data. The next stage was to build a geodatabase using PosgtreSQL with the PostGIS extension. All of the data that have been verified was inserted into the database.

Publishing online geospatial data and information services were the most important stage in the development of a Web-based geospatial information technology application. The Web-based GIS utilized this service, which allow geospatial data and information to be accessed from anywhere using a web browser or a desktop GIS. This study used GeoServer software to publish geospatial data into an online service.

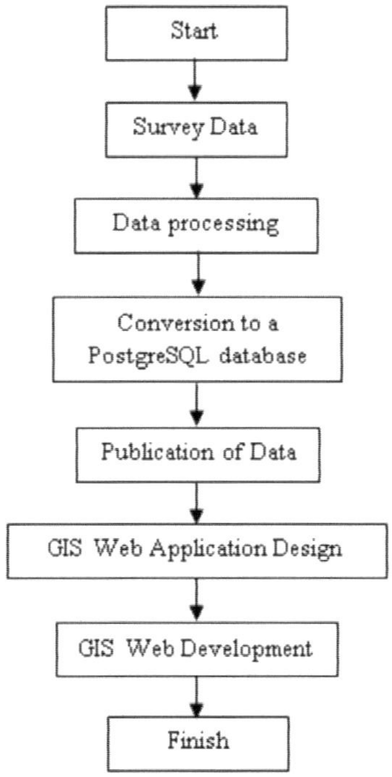

Figure 1. Process Flow Diagram of Web GIS Development

Architectural design for Web GIS to integrate information of renewable energy was implemented using Open Source applications of OpenGeoSuite. WebGIS application system architecture was supported by the GeoServer application, PostgreSQL database, OpenLayers and Framework GeoEXT. The complete list of software used for developing the Web GIS is shown in Table 1.

Table 1. Software used for developing the Web GIS

Application	Purpose
PostgreSQL/PostGIS	Database server with spatial extension
GeoServer	GIS Server for publishing OCG web services (WMS, WFS)
JavaScript	Programming Language
Openlayers/GeoExt	Library for WebGIS GUI
Apache	Server for Web GIS

4. Result and Discuss

After all of the research stages were carried out, the Web GIS for mineral and renewable energy potential of Aceh Province was realized. The GUI of the Web GIS is shown in Figure 2. The main page of the Web GIS shows the whole map of Aceh which is equipped with a list of map layers of mineral and renewable energy potential in Aceh. The important features included in the Web GIS are as the followings:

- **Legend**, which explains the meaning of symbols used on maps displayed in the Web GIS. Legend will appear every time a spatial data layer on the left panel is enabled via check boxes.

IOP Conf. Series: Earth and Environmental Science **56** (2017) 012016 doi:10.1088/1755-1315/56/1/012016

- ***Map layers***, which is a menu list of maps related to mineral and renewable energy sources that appear on the left side. The list is organized as tree layout of folders. The folders are added in order of the registered layer name so they will appear organized. They are labeled with districts or cities in the province. Guests can access the information about the locations of renewable energy in each district or cities. Renewable energy information includes geothermal energy and hydropower.
- ***Tools* Map Navigation**, which is a feature used or navigating the map, such as zoom-in and zoom-out to change the scale of the map.
- ***Popup* Information,** which displays that additional detail information on a feature in the map. The information is shown in the form of material name, area code and minerals information. Popup will appear when user clicks on one area of the map. This process of showing extra details for each feature on the map utilizes WMSGetFeatureInfo from the OpenLayers library.
- **Search Menu,** which serves as a search menu that can be used by users to search information based on the attributes of the map layer. Users can search for mineral deposits that is scattered in several districts or cities. To perform a search, user can choose the name of mineral to be searched, and then press the search button on the right side. Results of the search will be shown as symbolized points in the current map layer.

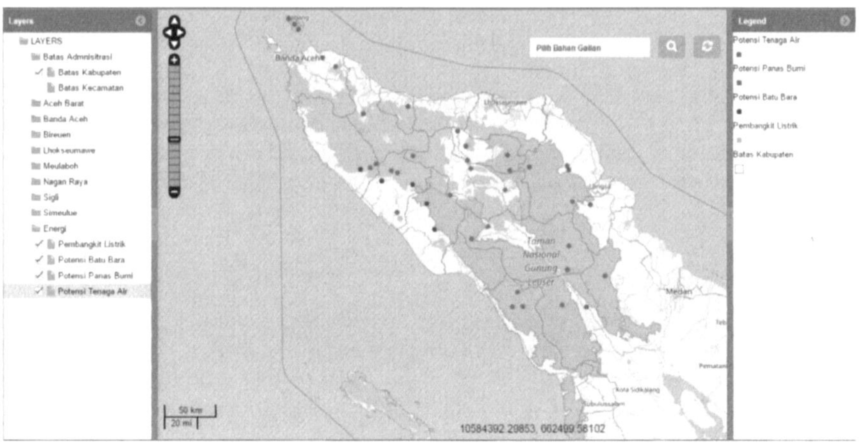

Figure 2. Main Display of the Web GIS

5. Conclusion

The study has developed a WebGIS application that displays spatial and non-spatial information regarding mineral and renewable energy potential in Aceh which organized by the district. The application can facilitate user in finding information on renewable energy potential, such as potential of geothermal and hydroelectric power in the province.

References

[1] Li G, Zhou K, Wang J, Sun L, Wang Q, and Qin Y 2010 Implementation of publication system of geological data using Open Source GIS *Information Science and Engineering (ICISE) 2010 2nd International Conference*.
[2] About GeoExt 2016 http://geoext.github.io/geoext2/ [Online]
[3] Priyanto E, Muludi K and Irawati A R 2013 Perancangan Sistem Informasi Geografis (SIG) Berbasis Web untuk Penyediaan Informasi Fasilitas dan Personalia di Universitas Lampung *Prosiding SEMIRATA 2013*.

Groundmagnetic survey used to identify the weathered zone, in Blang Bintang, Aceh, Indonesia

M Syukri[1,4], Marwan[1], R Safitri[1], Z Fadhli[2], F Andika[3] and R Saad[3]

[1]Department of Physics, Faculty of Mathematics and Natural Siences, Syiah Kuala University, Darussalam, Banda Aceh, 23111, Indonesia
[2]Departmentof Geophysical Engineering, Faculty of Engineering, University of Syiah Kuala, Darussalam, BandaAceh, 2311, Indonesia
[3]Geophsyics Section, School of Physics, Universiti Sains Malaysia, 11800 Penang, Malaysia
[4]Tsunami and Disaster s Mitigation Research Center (TDMRC) , Syiah Kuala University, Ulee Lheue

Abstract. The objective of a geomagnetic survey is to investigate subsurface characteristic of geology, especially the weathered zone on the basis of the anomalies in the earth's magnetic field resulting from the magnetic properties of the underlying rocks. The magnetic properties of rocks called as magnetic intensity is extremely variable depending on the type of rock and the environment of the region. The result of the research shows that the local magnetic intensity varies of -697 nT and 484 nT in the study area 1 and about -697 nT and 103 nT in the study area 2 whichare indicative of weathered zones. The weathered zones were interpreted by highly magnetic contrast. The total magnetic anomaly of Blang Bintang, Aceh Besar (Indonesia) shows higher magnetic anomalies over the eastern part (area 1) and low magnetic anomalies in the western part (area 2) of the study area, which indicate the presence of weathered zonedue to the presence of sediments such as clay, gravel and schist within the zone.The study has shown that magnetic method could be used as an efficient tool for delineating weathered zone in a study area.

1. Introduction

Geophysical techniques are often significant for discovering unknown subsurface conditions. Magnetic methodis the oldest geophysical technique applied in studying subsurface characteristics to measure lateral variation in the magnetic field in igneous and metamorphic rocks. The magnetic survey technique generally involves the measurement of the earth's magnetic field intensity or vertical gradient of the earth's magnetic field. Anomalies in the earth's magnetic field are caused by induced or remanent magnetism.Magnetization of any structure is due to induction in the geomagnetic field and permanent magnetization (remanent) [1]; [2]. The earth's magnetic field is a vector measured by a magnetometer [3]. Geological structures, rocks, minerals, ore deposits and engineering geological structures have magnetic properties that differ by orders of magnitude rather than percentages [15]; [16]. Magnetic rocks contain various combinations of induced and remnant magnetizations that perturb the Earth's primary field [2]; [17]; [23]. Total field and gradient are the two common types of magnetic measurements [4]; [18]; [19]. Total field measurement responds to the total magnetic field of the earth, natural and cultural magnetics, and any changes caused by a target [20]; [21]; [22].

Magnetic survey is a rapidand low-cost technique representing one of the mostwidely used geophysical methods in terms of line length surveyed [5]. One of the main objectives of the magnetic method is to map the changes in the magnetization which are related tothe distribution of magnetic minerals. The detection depends on the amount of the presented magnetic material and its distance to the sensor [6] [23]. This work is a study of geophysical characterization of weathered zone based on the anomalies of magnetic field variation. The location of the study area is shown in Figure 1. This study is important to analyze the landslides prone area, and disaster mitigation.

IOP Conf. Series: Earth and Environmental Science **56** (2017) 012017 doi:10.1088/1755-1315/56/1/012017

Figure 1. The study area of the magnetic survey in Blang Bintang, Aceh Besar (Indonesia).

Geological Setting

The geology of Aceh Besar Quadrangle has been mapped by Bennet [7] (Figure 2). The Blang Bintang valley is bounded by the Aceh Fault in the south-west and by the Seulimeum Fault in the northeast. The lithology of the Blang Bintang area is dominated by Lam Teuba volcanic composed of andesitic to daciticvolcanics, pumiceous breccia, tuffaceous, calcareous sometimes cross-bedded sandstones, conglomerates, agglomerate, minor mudstones and ash flows which intruded of the Seulimum formation. The Seulimeum formation is composed of tuffaceous and calcareous sandstones, conglomerates and minor mudstones [7]. The Seulimeum Formation was deposited on the Pliocene to Pleistocene boundary. The adjacent Sumatra in the west coast constitutes volcanic arc with many Quaternary volcanoes such as Pulau Weh and Seulawah Agam. The volcanic belt is extended along the tectonically weak SFS and this tectonic weakness is supposed to have triggered volcanism [8]. In the northern most Sumatra, the Sumatran Fault System (SFS) splits into two major dextral strike-slip faults which are the Seulimeum and the Aceh Fault [7]. From upstream of the town of Jantho to downstream of the town of Indrapuri, the Pleistocene coarse-grained partly volcanic sands and gravels form a prominent terrace surface on both sides of the Krueng Aceh. Downstream of Indrapuri, the alluvial deposits can be subdivided into a shallow aquifer system and a deep aquifer system. Upstream of Indrapuri, the alluvial sandy-gravelly deposits in the vicinity of the river courses, the older terrace sand-gravel deposits and the semi-consolidated sandstones are assumed to constitute the main aquifers of the upper part of the Krueng Aceh valley [9].

To the northeast of the Aceh Fault, extensive Plio-Pleistocene volcanism is formed the area of the Seulawah Agam volcano.To the southwest the Aceh Fault builds a direct border to the highly structurally deformed Pre-Tertiary formations of the Barisan Mountains.The volcano is located within the Barisan Mountain Range which expands along the whole Sumatran west coast [10]. The survey area is close to the Krueng Raya mount and the Ie Se'uem hotspring. The tectonic setting of the area forms a topographic depression, occupied by alluvial flat and low, flat-topped hills within the Barisan Range, a rugged mountain range that runs along the entire western edge of the island of Sumatra [11]; [12]. The crest of the Barisan Range is a continuous system of axial valleys, including the Kr. Tangse valleys. This is essentially a right lateral fracture system [13]; [14]. The topographical morphology of the study area (Blang Bintang) is weak (soft) because the rocks are strongly fractured and altered.

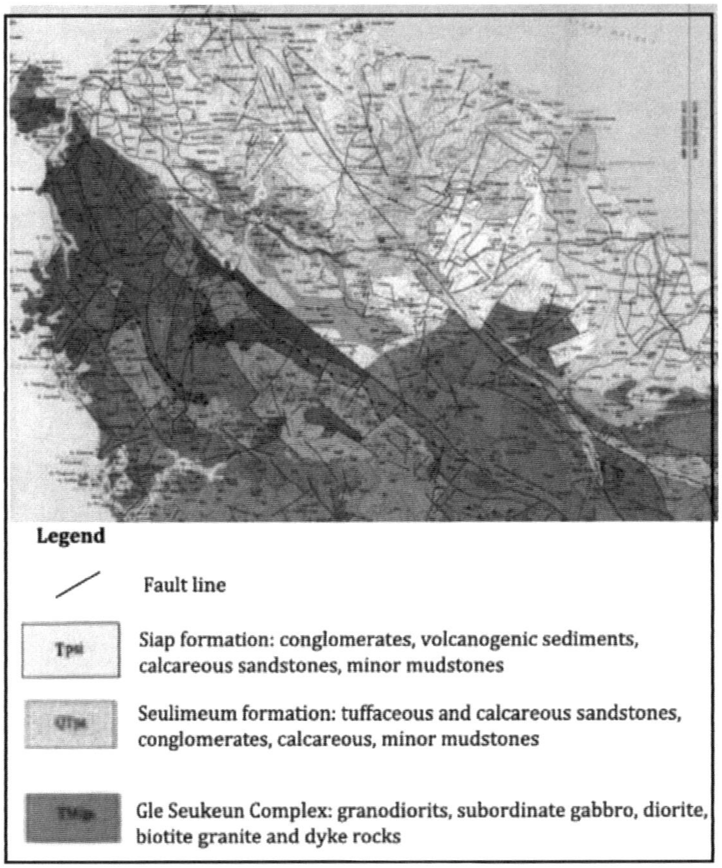

Figure 2. Geology map of study area [7]; [11].

2. Methods

The magnetic survey was carried outto understandthe subsurface structureswhich could be the targets to investigate the potentiality of weathered areas. The magnetic survey of two study areas were conducted with scattered moving station ranging from 10 to 20 m in order to detect the subsurface structure. Measurements were taken using two units of portable G-856 proton precession magnetometer providing a standard resolution of 0.1 nT and 0.5 nT absolute accuracy. One unit was used as a base station to record automatically, every 60 seconds to correct the diurnal variations of the earth geomagnetic field from the measurements.Base station data was used to correct the moving data, and finally, a total intensity magnetic anomaly map was produced, reflecting the subsurface structure. The other unit was used to measure the total intensity of the magnetic field at each point of observations along the profiles. To obtain a representative reading, the sensor should be operated well above the ground. The proton magnetometer measures a radio-frequency voltage induced in a coil by the reorientation (precession) of magnetically polarized protons in soil containing water. No magnetic storms were recorded during the periods of measurements.

The data were presented as magnetic map by plotting the magnetic values against station separations for each traverse. Magnetic surface map (2D plot) were also constructed for a more qualitative interpretation using Surfer 10 software. Processing the magnetic data enhances and sharpens the anomalies and the trends of the data and helps in the interpretation. In this survey, two techniques were applied in order to estimate the weathered area.First step, the process was the inspection of raw data for spikes, gaps, instrument noise or any irregularities in the data. The second step involved diurnal variation correction and IGRF correction to produce magnetic residual. Data are usually displayed in the form of a contour map of the magnetic field, but interpretation is often made on profiles.

3. Results and Discussion

Magnetic results show lateral view of the weathered zone in the study area (Figure 3). Magnetic survey data was plotted as a contour map using Surfer10 software and was displayed in nanoteslas (nT). The local magnetic intensity contrast observed between -697 nT and 484 nT at study area 1 and between -697 nT and about 103 nT at study area 2 which are indicative of probable weathered zones.

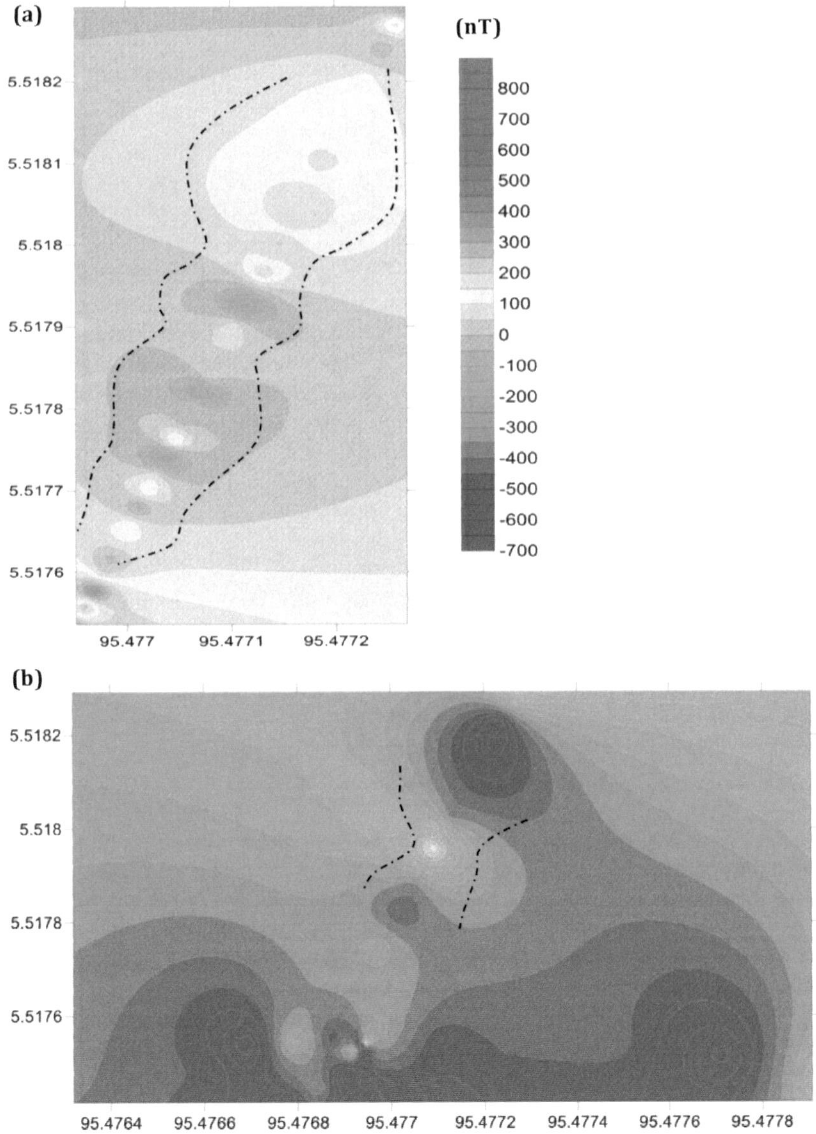

Figure 3. Magnetic Survey contour map in Blang Bintang (a). Survey are 1 (b) Survey area 2.

The weathered zone was interpreted if there was a high magnetic contrast.Generally, the total intensity magnetic anomaly map of Blang Bintang, Aceh Besar (Indonesia) shows higher magnetic anomalies over eastern part (area 1) and low magnetic anomalies over western part (area 2) of the study area, which indicate the presence of weathered zone covering these parts. As can be seen from the images, high magnetic signatures were noted in most of the area in the north and some sections in the south. This weathered zone is directly related to the disaster-prone zones, and need to take the mitigation action, before the disaster.

Hence, the areas with lower magnetic values than the surroundings could be due to the presence of weathered zone beneath the surface (black dashed lines).Such values of magnetic intensity respons could be due to the characteristic of geological sediments such as clay, gravel and schist within the zone. Beside that, the iimportant lateral variations and contrasts in magnetic properties of the shallow formations are brought singularly or by some combination of faulting, deposition and mineralization associated with volcanic rock. These volcanic rocks may produce high amplitude of magnetic variations in a very local extent. From the magnetic maps, several of the anomalies can be clearly correlated with geological map.

4. Conclusions

A simple pattern recognition technique can be used to locate weathered zones that have a varied magnetic response. The study has shown that the ground magnetic method can be used as an efficient method for delineating weathered zone in study area. There was a good relationship found among the geological characteristics and yield data in the study area.The maximum variation of magnetic intensity observed along area 1 might be an indication of intrusion of the overlying layer by the weathered basement. The low magnetic intensity of the weathered basement recorded along this profile (area 1 and 2) suggests that weathered basement is composed of consolidated earth materials. The magnetic responses of the weathered layer along the study area are characterized by clay, gravel and schist. The appearance of this zone is more dominant in area 1 than area2. The variations in the magnetic intensity responses of the weathered basement suggest the heterogeneous nature of the layer. This is related to the location of landslides, so it is necessary to mitigate disaster.

Acknowledgements

The authors express their gratitude to the Ministry of Education and Culture, Indonesia for financial support in the scheme of Decentralized Research Competitive Grants Program 2016. We thank the member of Geophysics group, Universiti Sains Malaysia (USM), Penang, Malaysia, students and staffs of Faculty of Sciences and Faculty of Engineering, Syiah Kuala University, Banda Aceh (Indonesia) who helped acquire the geophysical field data.

References

[1] Parasnis, D.S., 1972, Principles of Applied Geophysics. Second Edition. Champman and Hall Ltd. London. pp. 180-214.

[2] Reynolds, R.L, Rosebaum, J.G., Hudson, M.R and Fisherman, N.S, 1990, Rock magnetism, the distribution of magnetic minerals in the Earth's crust and aeromagnetic anomalies in Hana, W.F., ed., Geologic Applications of Modern Aeromagnetic Surveys: U.S.Geological Survey Bulletin 1924 pp.24-45.

[3] Lowrie, W., 2007, Fundamentals of Geophysics, second edition, Cambridge University Press, New York, pp. 281- 334.

[4] Benson, R.C., 2006. Remote Sensing and Geophysical Methods for Evaluation of Subsurface Conditions. In: David M. Nielsen (editor), Practical Handbook of Ground Water Monitoring. Lewis Publishers, Inc. Chelsea, Michigan. pp.249-296.

[5] Kearey P., Brooks M., Ian Hill, 2002, An Introduction to Geophysical Exploration, Third edition, Blackwell Science Ltd, London, pp. 155-183.

[6] Mariita, N.O., 2013, Strengths And Weaknesses Of Gravity And Magnetics As Exploration Tools For Geothermal Energy, *Short Course VIII on Exploration for Geothermal Resources*,

organized by UNU-GTP, GDC and KenGen, at Lake Bogoria and Lake Naivasha, Kenya, Oct. 31 – Nov. 22, 2013.

[7] Bennet, J. D., Bridge, D. Mc., Cameron, N. R., Djunuddin, A., Ghazali, S. A., Jeffery, D. H., Kartawa,W., Keats,W., Rock, N. M. S., Thomson, S. J. and Whandoyo, R. 1981, Peta Geologi Lembar Banda Aceh, Sumatra (1:250.000). Geologic Map of the Banda Aceh Quadrangle, Sumatra, Lembar (Quadrangle) 0421 (Banda Aceh), pp. 19, Bandung.
[8] Gasparon, M. 2005, Quaternary volcanicity. (In: Barber, A. J. et al. (Eds.): Sumatra, Geology, Resources and Tectonic Evolution), pp. 120 – 130, London.
[9] Ploethner D.,and Siemon B., 2006, Hydrogeological reconnaissance survey in the province Nanggroe Aceh Darussalam, Northern Sumatra, Indonesia – Survey area Banda Aceh / Aceh Besar 2005, Helicopter Project Aceh (HELP ACEH). *BGR Report Vol. C-1*, Archives No. 0126196 C-1, Hanover.
[10] Milsom, J. S., 2005, Seimology and neotectonics. (In: Barber, A. J. et al. (Eds.): Sumatra,Geology, Resources and Tectonic Evolution), pp. 7 - 15; London.
[11] Nordiana, M.M., Syukri, M., Saad, R., Marwan, and Kamaruddin, N.A., 2014, The Identification of Fault Zones in Krueng Raya, Aceh Besar (Indonesia) Using Magnetic Method, *The Electronic Journal of Geotechnical Engineering,* Vol. 19., Bundle A. pp.1-9.
[12] Syukri, M., Marwan, Safitri, R., and Saad, R., 2015, Delineation of Potentially Contaminated Zones by Electrical Resistivity Method in Aceh Besar, Indonesia, *The Electronic Journal of Geotechnical Engineering,* Vol. 20, Bundle 22, pp.12195 – 122012.
[13] Katili, J., andHehuwet, F., 1967, On the occurrence of large transcurrent faults on Sumatra Indonesia: Osaka, Japan, *Osaka City University Journal of Geoscience*, Vol. 10, pp 1-17.
[14] Page, B. G. N., Bennett, J. D., Cameron, N.R., Bridge, D., McC Jeffery D. H., Keats, W. and Thaib, J., 1979, A review of the main structural and magmatic features of northern Sumatra," *Journal of Geological Society of London*, Vol. *1*, 36, pp. 569-579.
[15] Begonha A. and Sequeira Braga M.A. 2002.Weathering of the Oporto granite: Geotechnical and physical properties.*Catena* **49**, pp. 57–76.
[16] Dewandel B., Lachassagne P., Wyns R., Maréchal J.C. and Krishnamurthy N.S. 2006.A generalized 3-D geological and hydrogeological conceptual model of granite aquifers controlled by single or multiphase weathering.*Journal of Hydrology* **330**, pp. 260–284.
[17] Hall, K. 1993: Enhanced bedrock weathering in association with late-lying snowpatches – evidence from Livingston Island, Antarctica. Earth Surface Processes and Landforms,vol. 18, pp. 121–129.
[18] Hall, K. 1997: Rock temperatures and implications for cold region weathering. I: New data from Viking Valley, Alexander Island, Antarctica. Permafrost and Periglacial Processes,vol. 8, pp. 69–90.
[19] Jones, J., 2002, Managing small teams, Penguin, Sydney.
[20] Lan H.X., Hu R.L., Yue Z.Q., Lee C.F. and Wang S.J. 2003. Engineering and geological characteristics of granite weathering profiles in South China.Journal of Asian Earth Science,vol. 21, pp. 353–364
[21] Goudie, A.S., Migon, P., Allison, R.J. and Rosser, N. 2002: Sandstone geomorphology of the Al-Quwayra area of south Jordan. *ZeitschriftfürGeomorphologie* 46.
[22] Olona, J., Pulgar, J.A., Viejo, G.F., Fernández, C.L., and Cortina. J.M.G., 2010, Weathering variations in a granitic massif and related geotechnical properties through seismic and electrical resistivity methods, *Near Surface Geophysics, vol.*8, pp.585-599.
[23] Sharma, V.P.,1997, Environmental and Enginering Geophysics, published by Cambridge University Press, United Kingdom. pp. 40-45.

Indigenous knowledge management to enhance community resilience to tsunami risk: lessons learned from *Smong* traditions in Simeulue island, Indonesia

A Rahman[1,3], A Sakurai[2] and K Munadi[3]

[1]Department of International Environment and Resources Policy, International Post Graduate Program in Human Security and Society, Graduate School of International and Cultural Studies, Tohoku University, 41 Kawauchi, Aoba-ku, Sendai, Miyagi, 980-8579, Japan, alfi@tdmrc.org
[2]Division of Disaster Information, Management and Public Collaboration, International Research Institute of Disaster Science (IRIDeS), Tohoku University, Sendai, Miyagi, 980-0845, Japan
[3]Tsunami and Disaster Mitigation Research Center (TDMRC), Syiah Kuala University, Banda Aceh, 23233, Indonesia

Abstract: Knowledge accumulation and production embedded in communities through social interactions meant that the Smong tradition of indigenous knowledge of tsunami risk successfully alerted people to the 2004 tsunami, on the island of Simeulue, in Aceh, Indonesia. Based on this practical example, an indigenous management model was developed for Smong information. This knowledge management method involves the transformation of indigenous knowledge into applicable ways to increase community resilience, including making appropriate decisions and taking action in three disaster phases. First, in the pre-disaster stage, the community needs to be willing to mainstream and integrate indigenous knowledge of disaster risk reduction issues into related activities. Second, during disasters, the Smong tradition should make the community able to think clearly, act based on informed decisions, and protect themselves and others by using their indigenous knowledge. Last, in the post-disaster phase, the community needs to be strong enough to face challenges and support each other and "building back better" efforts, using local resources. The findings for the Smong tradition provide valuable knowledge about community resilience. Primary community resilience to disasters is strongly related to existing knowledge that triggers appropriate decisions and actions during pre-disaster, disaster, and post-disaster phases.

Keywords: Tsunami risk, indigenous knowledge, knowledge management, community resilience, disaster risk reduction, Smong

1. Introduction

The increasing number of disasters in recent years has already exceeded the disastrous events predicted, and this upward trend has become even more apparent in the over 200 million people affected each year since 1994 [1]. Understanding disasters is important to evaluating correctly the challenges that such cataclysms pose and the growing need to find appropriate ways to reduce disaster risk. The role of indigenous knowledge in managing this risk has been recognized in international frameworks for disaster risk reduction (DRR).

One of the success stories that has emerged—stimulating a new interest in the concept of indigenous knowledge for DRR—is the Simeulueans' knowledge gained from living off the coast of Aceh, Indonesia. When the 2004 tsunami hit Aceh on December 26, the Smong tradition successful alerted people to run away to higher ground, allowing them to survive.

2. Research Purpose and Methods

This study's purpose was to examine a process of preserving effective indigenous knowledge of tsunami risk on the island of Simeulue, in Aceh, Indonesia. This research applied a qualitative method based on an extensive literature review and semi-structured interviews.

3. Knowledge, Knowledge Management, and Indigenous Knowledge Enhancing Community Resilience

3.1 Knowledge and Knowledge Management

The critical role of knowledge in reducing the worst impacts of disasters has been addressed in the Sendai framework for disaster risk reduction (SFDRR) for 2015–2030. The SFDRR's first priority of action relates to the topic of knowledge's important contribution to community resilience, but improved knowledge by itself has failed to reverse the downward trend of community resilience [2]. In other words, disaster awareness has strengthened disaster knowledge but, in the process, communities have lost still more of their resilience. Therefore, the questions remain of why existing knowledge is not always applied or not used appropriately and how this knowledge can be maintained by improving on existing knowledge, especially community knowledge (see Figure 1). Researchers must examine the nature of this knowledge, including understanding knowledge production processes, the existence of different type of knowledge and their sources, and the transfer and use of information that increases knowledge. These processes can be improved, thereby greatly contributing to community resilience in the face of disaster [3].

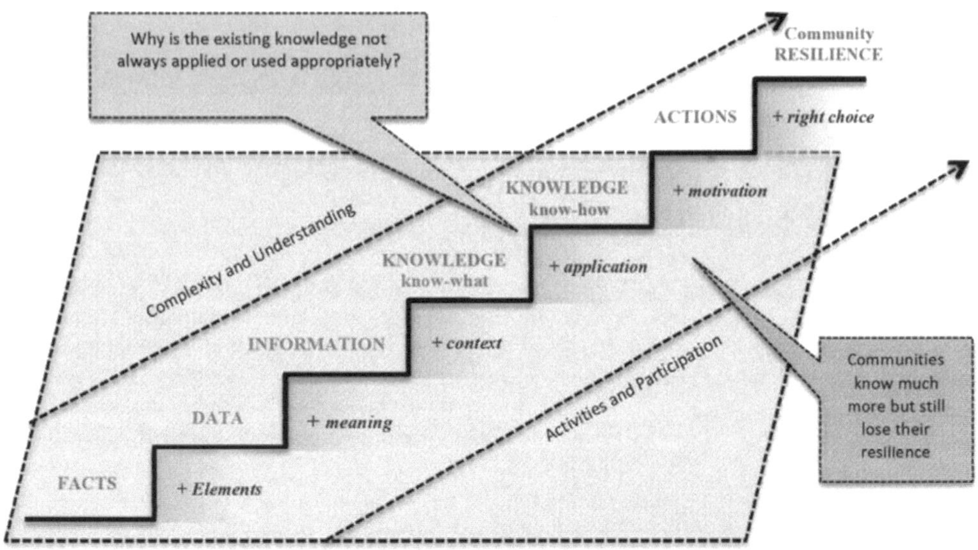

Figure 1: Knowledge ladder through community resilience

Source: Adapted from Klaus North (2014) and Weichselgartner (2015)

Knowledge management is a new vital issue in the DRR strategies involving social constructions [4]. Lessons learned from every disaster have proved that reusing and sharing knowledge can enhance community resilience to manage disaster risk. The challenge of implementing knowledge management in the construction of community resilience is the lack of systematic management strategies for developing and applying community knowledge. Knowledge management models of community resilience have not been widely recognized by academics and practitioners as a tool to support community knowledge development. However, communities' existing knowledge may present challenges, which prevent it from being used efficiently and effectively [5]. The present study sought to provide a knowledge management approach that could overcome such problems and provide a successful and efficient way to manage existing knowledge in the strengthening of community resilience to disaster risk.

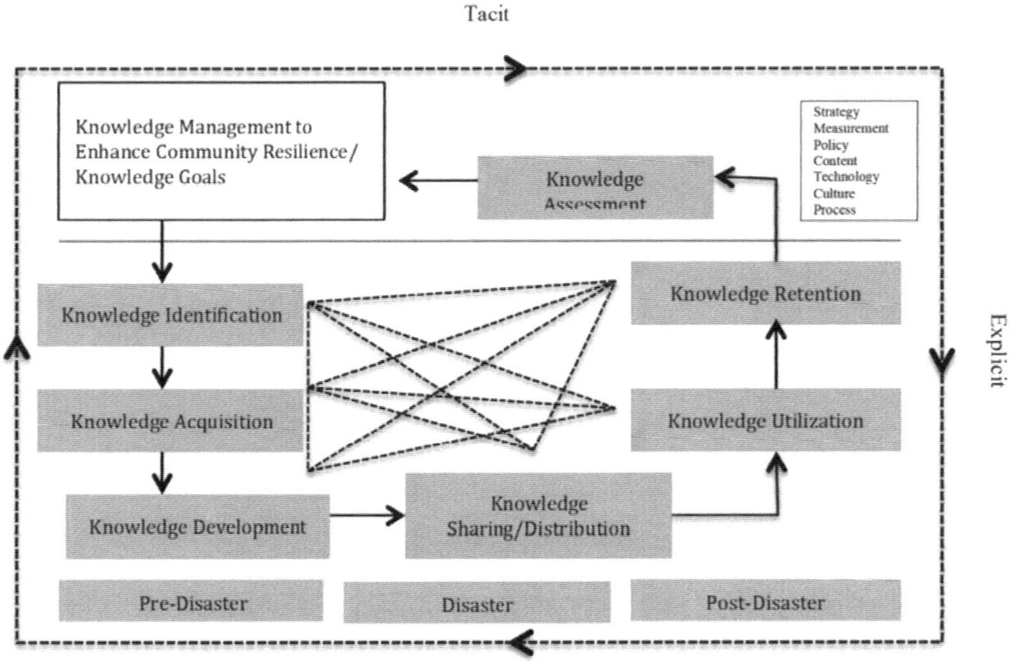

Figure 2: Framework of knowledge management to enhance community resilience

Source: Adapted from Probst, Raub, and Romnardt (2000)

Figure 2 shows the framework of knowledge management to enhance community resilience as a basic approach to understanding the flow of knowledge in communities. Effective local community resilience that manages risks and recovers from crises involves steps in knowledge management: 1) identification, 2) acquisition, 3) development, 4) sharing/distribution, 5) utilization, and 6) retention. Another important thing in knowledge management is assessment, which is responsible for maintaining and ensuring that knowledge can sustainably achieve community goals [6].

3.2 Community Resilience

The first conceptualization of resilience was recognized in the field of systems ecology and introduced in the literature on disasters in the 1970s, but only spread widely in the 1990s [7]. The concept of community resilience is defined in different ways depending on the background and viewpoint of each researcher, and, thus, this concept has many interpretations in academic discourse. The word "resilience" derives from the Latin word *resilio*, which means "to jump back," and definitions emphasize a capacity for successful adaptation in the face of stress, disturbance, or adversity [8].

The scope of resilience here includes the ability of communities to manage, respond to, and recover from disasters by reusing existing knowledge. A focus on resilience means putting greater emphasis on what communities can do for their members and how to strengthen these capabilities [9]. This makes it possible to use local perspectives and initiatives to build resilience that increases communities' capacity to recover effectively [10]

The significant contribution of knowledge to community resilience through managing disaster risks is mentioned by Twigg in his book entitled *Characteristics of a Disaster Resilient Community: A Guidance Note*. Although the concept of community resilience to disaster is complex and based on multiple factors, one of the specific characteristics that builds up community resilience is a focus on knowledge and education. More specifically, this includes the possession of appropriate technical and organizational skills and knowledge of risk reduction and disaster response at a local level. Community resilience, thus, more particularly refers to how local communities develop their capacity to cope with disaster by reusing local resources.

The definition of resilience used in the present study is communities' capacity to adapt to, reduce, manage, and recover from the worst impacts of hazards by utilizing their local resources through appropriate, actionable decisions in pre-disaster, disaster, and post-event phases. The adaptive component of resilience in the context of local communities is their capacity to make suitable and actionable decisions in the face of events. The proposed model argues that communities' existing knowledge is central to resilience when managing risk, including their ability to sustain systems and respond to, and recover from, disasters.

More specifically, good decisions are one of the most important ways to reduce the loss of life during crises. Comfort, Boin, and Demchak divide crisis management systems into four subsets of decisions: 1) detection of risks, 2) recognition and interpretation of risks in the immediate context, 3) communication of risks to the community, and 4) community structure and mobilization of systems to reduce risk and respond to events [11] (see Figure 3). In the present conceptualization of knowledge management to enhance community resilience, utilization of knowledge is one of many processes within communities' knowledge management system. Indigenous knowledge is a source of resilience because when indigenous knowledge is integrated into socio-ecological systems, these have demonstrated the ability to deal with complexity and uncertainty [12].

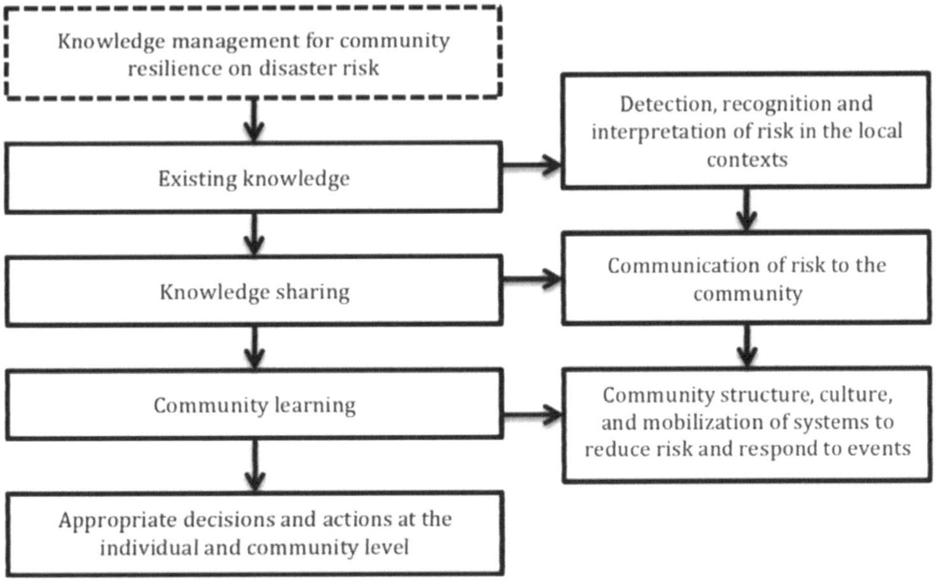

Figure 3: Knowledge utilization leading community in managing disaster risk

3.3 Overview of Indigenous Knowledge

In indigenous knowledge has emerged as a significant resource in dealing with many research and development issues. However, the increasing number of local and indigenous knowledge and practices documented in research on disasters have yet to lead to increased efforts to translate indigenous knowledge into initiatives that increase communities' resilience against disasters' impacts [13]. Indigenous knowledge has traditionally been regarded as inferior to science and technology [14], which has negatively affected communities' development processes [15].

The term "indigenous knowledge" arises from the experiences of many academics and practitioners. Some other terms are found in the literature referring to indigenous knowledge, such as folk, local, rural people's, indigenous technical, and traditional environmental knowledge.

Previous research justifies viewing indigenous knowledge not only as underlying technical solutions to daily problems or only information functioning as an early warning but also as containing non-technical insight, wisdom, ideas, perception, and innovative capabilities that deal with physical, environmental, ecological, biological, or geographical phenomena [15]. This is the reason that most academics and practitioners working in the field of indigenous knowledge have based their interpretations on geography, anthropology, and sociology. Indigenous knowledge could also be a more acceptable alternative to researchers of many backgrounds and areas of interest. Many definitions have been developed in a variety of fields in the literature to help understand indigenous knowledge and distinguish it from other concepts.

Indigenous knowledge has several essential characteristics that differentiate it from other types of knowledge, including that it is 1) embedded and originates in particular communities, 2) maintained through non-formal transmission, 3) collectively owned, and 4) developed over several generations [14] [16]. Indigenous knowledge is a repetition process from generation to generation and an unsystematic accumulation of new data over generations [17]. However, indigenous knowledge can potentially be disrupted through the breakdown of traditional, oral communication channels; the movement of communities to another place on a daily basis; and a lack of interest in learning indigenous knowledge [18]. In the present study, indigenous knowledge refers to communities' knowledge practices—formed over numerous generations—arising from their capacity to deal with and understand their environment in particular contexts and places.

3.4 Indigenous Knowledge in DRR

Indigenous knowledge has received increasing attention in the fields of DRR by both scientists and practitioners in recent years [7] [19]. Work on indigenous knowledge began to permeate the DRR discourse in the 1970s. However, this research expanded quite slowly, and the link between indigenous knowledge and DRR remained vague and indirect. By the 1980s, indigenous knowledge was still being ignored within the established disaster management discourse even as more attention was given to the role and value of indigenous knowledge in others field [16].

The recognition of indigenous knowledge's role in SFDRR now includes that indigenous peoples, through their experience and traditional knowledge, significantly contribute to the development and implementation of plans and mechanisms, including early warning systems [2]. Another major tsunami could happen at any time [20] in the future, so indigenous knowledge needs to be integrated into the entire early warning system through a local understanding of risks. A successful application of indigenous knowledge can allow locals to recognize the emerging threat in sufficient time to take informed action, thereby reducing risk and mobilizing appropriate responses and actions for the entire community, with the great benefit of reducing loss of life. Indigenous knowledge has a significant role in local community, including how to manage dynamic interactions in community systems that move from the level of individuals' "know-how" to individual and community-level action.

Most indigenous knowledge is intangible and, consequently, not easily codified [21]. It is hard to communicate real but tacit knowledge to people with a different level of knowledge [22]. Some approaches have been developed to manage knowledge effectively and successfully transfer tacit knowledge into explicit formats, but researchers need to have a clear understanding of the dynamic nature of knowledge itself [23]. Development initiatives that pay attention to local ways and perceptions are more likely to be relevant to people's needs and to generate sustainable interventions [17].

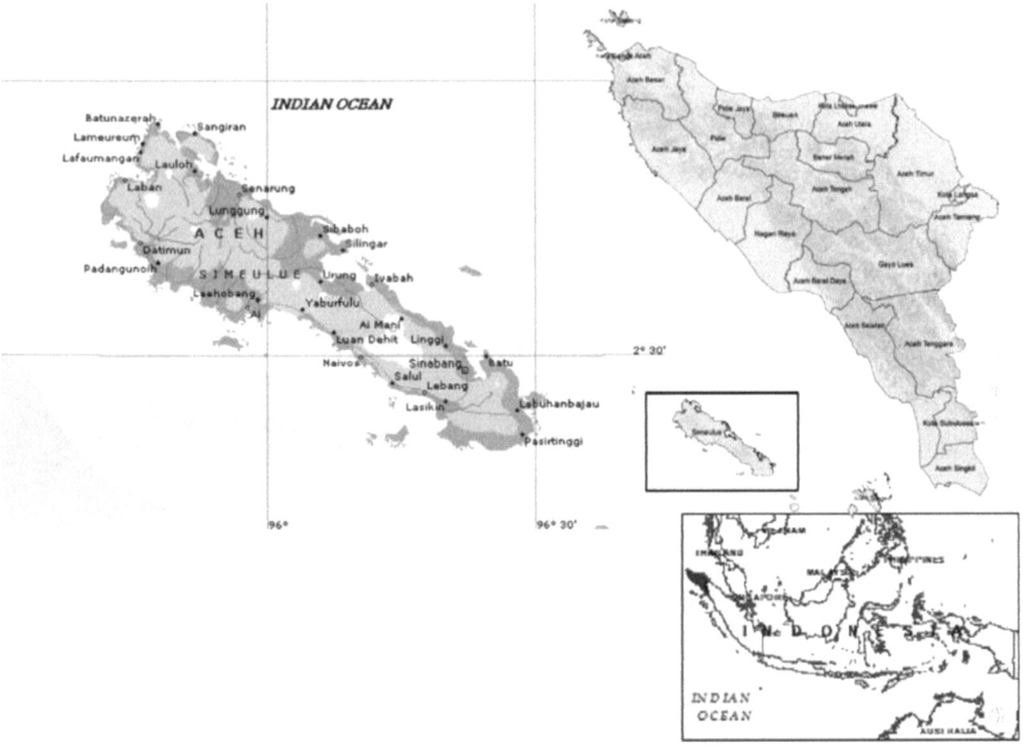

Figure 4: Map of the Island of Simeulue

Source: Adapted from Sumatransurfariis.com [24]

4. Underlying Lessons Learned from the Smong Tradition

The islands of Simeulue are located to the southwest in the province of Aceh, located 105 miles north of Meulaboh, West Aceh, or 85 miles from Tapak Tuan, South Aceh, at $2^0 15^{'}$ -$2^0 55^{'}$ north latitude and $95^0 40^{'}$ -$96^0 30^{'}$ [25] (see Figure 4.). Simeulue is a group of islands that comprises more than 147 large and small islands. The largest island is called Simeulue, and more than 90% of the archipelago's total population live there. Simeulue's total area is 1,838.09 km^2 or 183,809 ha.

The community on Simeulue is a heterogeneous society with several languages, namely, Devayan, Lekon, Sigulai, and Aneuk Jamee. Simeulue's Department of Population and Civil Registry recorded that its population increased from 84,704 people, in 2010, to 93,499 people, in 2014, with a gender composition of 47,969 males and 45,530 females. More than 99.74% of Simeulueans are Moslem, which means their culture is influenced by Islamic values [25].

4.1 Smong Success Story

On January 4, 1907, a 7.8 magnitude earthquake hit Aceh, and a tsunami developed in the Indian Ocean and caused the death of 70% of people on Simeulue [20]. The survivors shared their sadness, experiences, and knowledge with the next generation by creating the Smong tradition. The survivors recorded their experience in oral stories and passed these to the next generation from the oldest generations to the younger generations.

The word "Smong" derives from the Devayan language, which means the splash of water or tidal wave. Smong further refers to tsunami warnings when big earthquakes occur on Simeulue. Smong also explains the tsunami disaster phenomenon when a big earthquake is followed by a low tide of seawater on the beach and a giant wave that sweeps across the land (See Table 1) [16] [26].

Table 1: Smong lyrics in Devayan, Sigulai, and English

Devayan	Sigulai	English
Enggel mon sao curito	*Longola amba curito*	Please listen to this story.
Inang maso semonan	*Pado zaman nafe'e*	One day in the past,
Manoknop sao fano	*Tobanam amba desa*	A village was sinking.
Uwi lah da sesewan	*Nak daya feila la curitokan*	That's what has been told.
Unen ne alek linon	*Ya lunen afe dulu*	Starting with earthquakes
Fesang bakat ne mali	*Lentuk bakat yu ekhi eba*	Followed by a giant wave,
Manoknop sao hampong	*Tobanam amban gampung*	The whole country was sinking
Tibo-tibo mawi	*Tibo-tibo amak*	Immediately.
Anga linon ne mali	*Bo dulu ni abe le*	If the strong earthquake is
Uwek suruik sahuli	*Idane yu ata'a*	Followed by the lowering of seawater,
Maheya mihawali	*Rongkap akhuli*	Please find in a hurry
Fano me singa tenggi	*Banuami yu ala wa*	A higher place.
Ede smong kahanne	*Nak daya emong deini*	This is called Smong,
Turiang da nenekta	*Curito nenek moyang ta*	A story of our ancestors.
Miredem teher ere	*Enuge ekhi-ekhi*	Please always remember
Pesan dan navi da	*Amanah afe nasehatla*	This message and instruction.

Source: Adapted from Wikipedia.org [27]

The transmission of Smong as indigenous knowledge across generations of Simeulueans is a complex but basic process embedded within deep sociocultural patterns. Most Smong knowledge is tacit, and it is created and shared through community interactions. Smong has been mainly acquired from local sources such as family, older people, and community events. Since Smong is a matter of pride for Simeulueans and, predominantly, tacit knowledge shared within their cultures, rituals, and ceremonies related to disaster reduction, this knowledge already exists. However, it is not formally recognized in the official social system, for instance, a failure to integrate explicitly Smong information into disaster education curriculum.

A Simeulue community leader interviewed on August 15, 2016 mentioned the existence of older people who had experienced the 1907 tsunami, reporting that these community leaders have played an important role in coordinating different local cultures' awareness of this experience among community members. The elderly have contributed greatly to the government's access to the oral story practices through which Smong knowledge is conveyed, but most have died from old age (see Table 2).

However, another factor explains more clearly how Smong has successfully been translated into community action. The main ethnic groups living in the coastal areas of Aceh, such as Banda Aceh and Aceh Singkil, include in their cultures descriptions of tsunami. The tsunami story and messages based on what happened in the past are implied in oral literature referred to as Ie-Beuna in Banda Aceh and Gloro in Singkil. However, the Aceh Besar and Aceh Singkil people do not recognize these as a message from the past about how to reduce disaster risks. This tradition is perceived as merely beautiful poetry. Although the Smong, Gloro, and Ie-Beuna traditions originated in the same province, they have developed differently for several reasons.

Table 2: Smong knowledge processes

Knowledge Acquisition	Knowledge Development	Knowledge Sharing	Knowledge Preservation	Knowledge Application
1907–2004				
• Smong knowledge acquisition is predominantly local, from family and old people who had personal experiences with the 1907 tsunami.	• Smong developed as knowledge in oral stories.	• Oral transmission.	• Most Smong knowledge was preserved in oral stories. • Explicit sources were not used to maintain Smong knowledge.	• The right decisions and appropriate actions were made when the 2004 tsunami occurred. • Smong successfully alerted people when the tsunami occurred.
2004–present				
• Smong knowledge has been acquired from the experience of the 1907 tsunami and the successful Smong alert of people in 2004. • Access to Smong is through a variety of informal knowledge sources across communities.	• The term Smong is recognized within the community. • There are a few local initiatives based on Smong within efforts to raise community awareness of tsunamis.	• Access to Smong is influenced by oral story and cultural patterns such as nandong, buai-buai (lullabies) and nafi-nafi (advice). • Few local people are involved in transforming Smong into various cultural formats.	• No particular program includes Smong in formal disaster education. • Some documentation of Smong exists, but it is limited to the Smong success story rather than a plan for ways Smong could contribute to community knowledge.	• Most of the local people easily recognize the term Smong. • Few knowledge intermediaries have preserved Smong in printed formats.

The head of the Majelis Adat Aceh (Acehnese Adat Board) at the province level, who was interviewed on September 14, 2016, mentioned the armed conflict and massive military presence before the 2004 tsunami. The conflict between the Free Aceh Movement and the Indonesian Military continued for more than 30 years and created cultural changes on the Aceh mainland that have only slightly affected Simeulue. He suggested this could explain why the Smong tradition survived in the Simeulueans' memory and successful alerted them to danger when the 2004 tsunami occurred.

After the 2004 tsunami, many poets and writers created material that delivers messages to the next generation. Some authors tried to modify traditional song lyrics such as Nandong by adding some DRR messages to the community [26]. One author interviewed on September 3, 2016 observed that the initiative to modify Smong through new approaches has only come from individual and small group initiatives.

Key lessons learned from the Smong experience about community resilience in managing tsunami risk are as follows:

- The primary media storage for Smong knowledge is oral transmission through cultural patterns in which community members share their knowledge.
- Family, older people, and local leaders played a major role in disseminating and determining community members' actions when the tsunami occurred.
- The peaceful situation in this society has contributed to ensuring the sustainable development of culture patterns through which most Smong knowledge continues to exist.

4.2 Smong Knowledge Management Processes

The processes of a Smong knowledge management approach include the following: knowledge acquisition, development, sharing, preservation, and application of Smong information in community activities. The model proposes a transformation of information into an attitude based on the knowledge that the community can make the right decisions and take appropriate actions in every stage of disaster management. Figure 5 describes the elements of this indigenous knowledge management model to enhance community resilience to tsunami risk (see also Table 3).

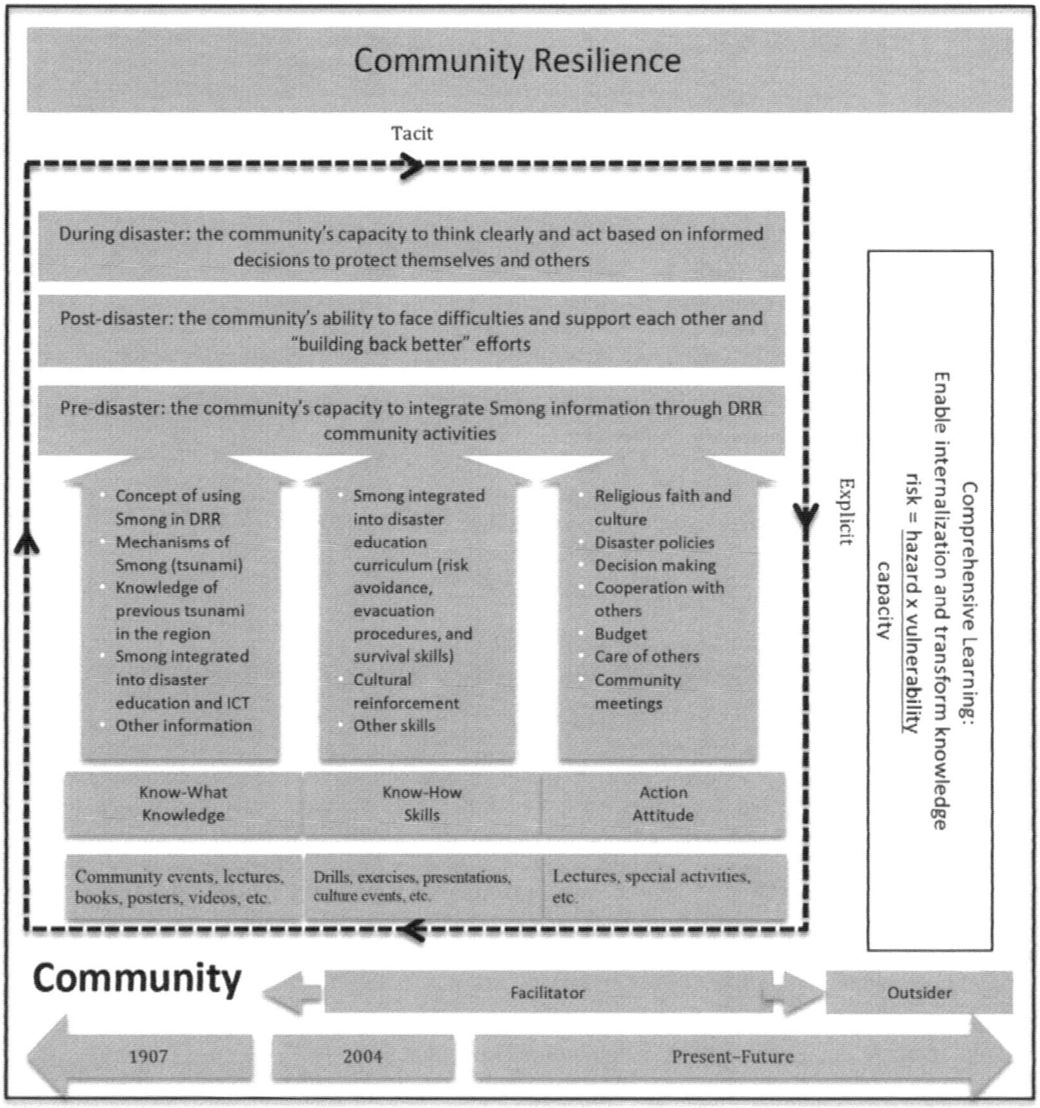

Figure 5: Smong knowledge management to enhance community resilience to tsunami risk

Table 3: Key terms in indigenous knowledge management to enhance community resilience to tsunami risk, on the island of Simeulue

Term	Description
Community Resilience	The community's capacity to adapt, reduce, manage, and recover from the worst impacts of hazards by utilizing local resources with the appropriate and actionable decisions in pre-disaster, disaster, and post-disaster phases
Local Community	The shared knowledge base for the relevant knowledge (i.e., Smong) because all members possess certain knowledge
Facilitator	A stakeholder who has the ability to manage the knowledge base, take responsibility for establishing and maintaining the knowledge base, and facilitate communication between the community and multiple stakeholders and outsiders.
Outsiders	The people who also possess relevant knowledge that could be used by the community, including technology, formal curriculum, and information and communication technologies
1907	The previous tsunami on January 1907, when more than 70% of Simeulueans were killed and lost: the basic source of knowledge through the local community's recognition of the Smong tradition
2004	The earthquake and tsunami on December 26, 2004, when less than five people were killed on Simeulue: a good lesson learned to construct further community knowledge
Present–Future	The conditions after the 2004 tsunami and ways the community prepares itself to face future tsunamis
Know-What	The knowledge resulting from interiorizing information
Know-How	The transformation of knowledge into skills or applications
Action	The abilities or attitudes needed to make decisions and take action using knowledge-based skills
Risk	The probability of tsunami events and their negative consequences affecting the community
Hazard	A tsunami event that may have negative consequences for the community.
Vulnerability	The community's assets, systems, circumstances or characteristics that make it susceptible to the damaging effects of tsunami hazards
Capacity	The combination of all the strengths, attributes, and resources available within the community that can be used to achieve goals for managing tsunami risk
Tacit Knowledge	The knowledge, ideas, and experiences that people have in their minds, which are difficult to access because this information is often not codified and may be hard to express
Explicit Knowledge	Knowledge that can be readily articulated, codified, accessed, and verbalized and then easily transmitted to the others.
Pre-Disaster	The appropriate decisions and actions before a disaster that reduce the potential for human, material, or environmental losses caused by tsunami hazards and ensure that these losses are minimized when disasters strike
During Disaster	The appropriate decisions and actions when disasters occur
Post-Disaster	The appropriate decisions and actions after tsunami events so that the community recovers effectively to face challenges and support each other and "building back better" efforts

The proposed approach begins with the community in which the platform for tsunami knowledge exists over numerous generations. The community provides the basic knowledge (see Figure 5 above).

Another important element is understanding ways that the community defines risks, recognizes and interprets risks, and communicates these risks to its members. Indigenous knowledge management can be used to manage Smong information. Its transmission can be encouraged with a combination of standard DRR approaches and the existing knowledge in the cultural material already respected by the community. The term Smong is widely recognized by locals, providing the root historical experiences of tsunamis that frequently devastate the area. Smong can be improved into new songs, stories, legends, proverbs, rituals, or ceremonies relating to DRR messages.

Although songs, stories, and cultural patterns relevant to the Smong tradition already exist, innovation is still needed to spread and ensure knowledge of Smong within the community. The model is concerned about not only transforming tsunami information into explicit knowledge but also, most importantly, ensuring that the knowledge can be transformed into community attitudes that help make right decisions and take appropriate actions when tsunamis strike. Smong knowledge should be encouraged and spread into formal education, to reinforce this cultural pattern. Some programs are described in this model including, among others, books, drills, lectures, posters, exercises, presentations, survival skills, and community meetings and cultural events.

The Smong tradition can be effectively managed and integrated to enhance community resilience when managing tsunami risk by appointing stakeholders who can manage the knowledge base, take responsibility for establishing and maintaining this knowledge, and facilitate the community's communication with other stakeholders and outsiders. Facilitators must work closely with community leaders to develop a good environment for knowledge sharing activities in local communities. This leadership has a significant role in creating cultural patterns that enhance knowledge sharing, utilization, and retention in local communities.

Finally, the proposed approach describes indigenous management processes that can achieve the main goal of enhancing community resilience to tsunami risk by making appropriate decisions and taking action. This goal can be reached in all phases of disasters. First, in the pre-disaster stage, the community could be willing to mainstream and integrate indigenous knowledge with DRR issues into their regular activities. Second, during disasters, Smong could make the community able to think clearly, act based on informed decisions, and protect themselves and others by using their Smong knowledge. Last, in the post-disaster phase, the community could be strong enough to face challenges and to support each other and "building back better" efforts, using local resources.

5. Conclusions

Smong provides valuable knowledge about enhancing community resilience when managing tsunami risk. The study's results indicate that the knowledge management approach can be used to manage the Smong tradition and introduce the involvement of knowledge facilitators, other stakeholders, and relevant outsiders. The findings support the conclusion that knowledge management is one alternative that can prevent the disappearance of Smong as indigenous knowledge on future tsunami risk, offering a way to strengthen the community's ability take the best decisions and take appropriate actions before, during, and after disasters. Finally, this study also found that, to achieve enhanced community resilience, indigenous knowledge needs to be linked to the entire range of disaster management activities in all disaster phases.

6. Future Avenues of Research

In the next stage of this research, this study will seek to formulate theoretical and empirical approaches to constructing a comprehensive and widely applicable indigenous knowledge management model for DRR. The next research should be conducted in participatory approach by involving the community to formulate the model.

7. References

[1] UN 2015 *GAR; Global Assessment Report on Disaster Risk Reduction; Making Development Sustainable: The Future of Disaster Risk Management* (United Nations)
[2] UNISDR 2015 *Sendai Framework for Disaster Risk Reduction 2015-2030* (Geneva: UNISDR)

[3] Spiekermann R, Kienberger S, Norton J, Briones F and Weichselgartner J 2015 The Disaster-Knowledge Matrix – Reframing and evaluating the knowledge challenges in disaster risk reduction. *International Journal of Disaster Risk Reduction, 13*, 96-108.
[4] Weichselgartner J and Pigeon P 2015 The Role of Knowledge in Disaster Risk Reduction *International Journal of Disaster Risk Science* **6**, 107-116
[5] North K and Kumta G 2014 *Knowledge Management: Value Creation Through Organizational Learning* (Springer International Publishing Switzerland)
[6] Probst G, Raub S and Romhardt K 2000 *Managing knowledge: building blocks for success* (New York: John Wiley)
[7] Hiwasaki L, Luna E, Syamsidik and Shaw R 2014 *Local & indigenous knowledge for community resilience: Hydro-meteorological disaster risk reduction and climate change adaptation in coastal and small island communities* (Jakarta: UNESCO)
[8] Norris F H, Stevens S P, Pfefferbaum B, Wyche K F and Pfefferbaum R L 2008 Community resilience as a metaphor, theory, set of capacities, and strategy for disaster readiness. *American journal of community psychology* **41** (1-2) pp 127-150
[9] Twigg J 2009 *Characteristics of a Disaster-Resilient Community A Guidance Note Version 2* (London: DFID)
[10] Aldrich D 2012 *Building Resilience Social Capital in Post-Disaster Recovery* (Chicago: The University of Chicago Press)
[11] Comfort L K, Boin A, and Demchak C C 2010 *Desaigning Resilience: Preparing for Extreme Events* (Pittsburgh: University of Pittsburgh Press)
[12] Berkes F 1993 Traditional Ecological Knowledge in Perspective in J Inglis *Traditional Ecological Knowledge Concepts and Cases* (Ottawa Canada: International Program on Traditional Ecological Knowledge, Canadian Museum of Nature) p1
[13] Shaw R, Uy N, and Baumwoll J 2008 *Indigenous Knowledge for Disaster Risk Reduction: Good Practices and Lessons Learned from Experiences in the Asia-Pacific Region* (ISDR)
[14] Agrawal A 1995 Dismantling the divide between indigenous and scientific knowledge *Workshop in political theory and policy analysis* (Bloomington: Indiana University)
[15] Gadgil M, Berkes F and Folke C 1993 *Indigenous Knowledge for Biodiversity Conservation* (Ambio, Springer) Vol. 22 pp 151-156
[16] Baumwooll J 2008 The value of indigenous knowledge for disaster risk reduction: A Unique Assessment Tool for Reducing Community Vulnerability to Natural Disasters *Master Thesis*
[17] Sillitoe P and Marzano M 2009 Future of indigenous knowledge research in development *Futures* **41** (1) pp 14-23
[18] Granier L 1998 *Working with Indigenous Knowledge, A Guide for Researcher* (Ottawa: International Development Research Centre)
[19] Inglis J T 1993 *Traditional Ecological Knowledge Concepts and Cases* (Ottawa, Canada: International Program on Traditional Ecological Knowledge Canadian Museum of Nature)
[20] McAdoo B G, Dengler L, M ERRI, Prasetya G, and Titov V 2006 Smong: How an Oral History Saved Thousands on Indonesia's Simeulue Island *Earthquake Spectra,* **22** (83) pp 1-9.
[21] World Bank 1998 *Indigenous knowledge for development: A framework for action* Retrieved June 4, 2016 from http://www.worldbank.org/afr/ik/ikrept.pdf
[22] Polanyi M 1996 *The Tacit Dimension* (New York: Anchor Day Books)
[23] McInerney C 2002 Knowledge management and the dynamic nature of knowledge *Journal of the American Society for Information Science and Technology* **53** (12), pp 1009-1018.
[24] Surfariis S n d Retrieved 09 6, 2016 from http://www.sumatransurfariis.com/images/simmaplarge.gif
[25] BPS Simeulue 2015 *Simeulue dalam Angka 2015* (Simeulue, Aceh, Indonesia: Badan Pusat Statistik Kabupaten Simeulue)
[26] Syafwina 2014 Recognizing Indigenous Knowledge for Disaster Management: Smong, Early Warning System from Simeulue Island, Aceh *Procedia Environmental Sciences* **20** pp 573-582
[27] Wikipedia.org n d Retrieved 09 6, 2016 from https://id.wikipedia.org/wiki/Kabupaten_Simeulue

IOP Conf. Series: Earth and Environmental Science **56** (2017) 012019 doi:10.1088/1755-1315/56/1/012019

Assessment of post-tsunami disaster recovery of Banda Aceh city of Indonesia as window of opportunities for sustainable development

E Meilianda[1,2,*], K Munadi[1,3], Azmeri[1,2], Safrida[1,4], Y Direzkia[5], Syamsidik[1,2] and R S Oktari[1,6]

[1]Tsunami and Disaster Mitigation Research Center (TDMRC), Syiah Kuala University, Jl. Prof. Dr. Ibrahim Hasan, Gampong Pie, Banda Aceh, 23233, Indonesia
[2]Civil Engineering Department, Engineering Faculty, Syiah Kuala University, Banda Aceh, 23111, Indonesia
[3]Electrical & Computer Engineering Department, Engineering Faculty, Syiah Kuala University, Banda Aceh, 23111, Indonesia
[4]Agriculture Faculty, Syiah Kuala University, Banda Aceh, 23111, Indonesia
[5]Aceh Mental Health Hospital, Banda Aceh, 23126, Indonesia
[6]Medical Faculty, Syiah Kuala University, Banda Aceh, 23111, Indonesia

*email: ella.meilianda@tdmrc.org

Abstract. Post-tsunami recovery process at Banda Aceh city of Indonesia were assessed in this study. Several actions and programs implemented during the recovery process were exercised and examined through several FGDs, to identify any windows of opportunities to change were captured in the aspects of infrastructure and housing, economic revitalization of the affected community, mental health and psychosocial condition and development, establishment and implementation of disaster risk reduction programs and community preparedness. Subsequently, whether or not those changes fit into the principle criteria of sustainability were examined. The results give insights on the dynamics of recovery process after more than a decade since the tsunami was affected the area. Some success and not-so-success stories of actions and program implementations during the recovery process were captured. On the aspect of livelihoods and public finance, the local government seems to have seen a window of opportunity and subsequently seize the opportunity to revitalize the administrative system of financing the micro-finance for communities. In contrast, on the aspect of socio-ecological systems integrity toward preserving the natural environment, the case of housing development at the coastal areas against the blueprint city masterplan exemplifies the failure in seizing the window of opportunity to "build back better".

1. Introduction

A natural disaster may be seen as windows of opportunity for change, with or without one realizing it. It may change people's perspective to see life with respect to the way of thinking and acting. To some extent, disaster can immediately be seen as a window of opportunity to create a better life. Although it may start from individual and communal actions, which is rather informal, it can be elevated into the level of institutional policies. Some of the actions are sustain, some are not, dependent to the internalization and commitment of those who involved in the process.

Postulated by [1] that separate streams of problems, policies, and politics come together at certain critical times, then solutions become joined to problems, and both of them are joined to favourable political forces. According to [2], windows of opportunity enabled institutions not previously engaged to become involved in the issue and often precipitated renegotiation of leadership on the issue within already engaged institutions. It is also an opportunity to an opening for new institutions to develop and become engaged, and shifting the network of actors involved. In the international community, the Action Learning Network of Accountability and Performance (ALNAP) in Humanitarian Action is predominant and undertakes annual reviews of response and relief work to encourage learning in the humanitarian sector. But again, focused on technical efficiency rather than the possibility that the political systems of impacted societies might themselves undergo formal and informal changes post-disaster [3].

The study aims at assessing important achievements of recovery in different aspects in life and to identify windows of opportunities and postulates a model of recovery from one of the largest disaster events in the modern human lifetime history. It seeks for lessons learnt that ultimately deliver recommendations for local and national level of government in implementing Disaster Risk Reduction programs in the future toward sustainable development. The assessment constitutes four aspects in life, which includes: 1) infrastructure and housing, 2) economic revitalization of the affected community, 3) mental health and psychosocial condition and development, and 4) establishment and implementation of disaster risk reduction programs and community preparedness towards disaster.

2. Data and Methods

2.1. Data Collection

The method of data collection in this study is a combination of qualitative analysis of data and information collected from several dedicated Focus Group Discussions (FGDs). A number of groups of beneficiaries and actors involved during the rehabilitation and reconstruction of Banda Aceh City, the capital city of Aceh Province of Indonesia, were invited during five sessions of FGDs, which further refined with in-depth interviews whenever necessary. Four working groups facilitating the focused discussions related to reconstruction of school infrastructure (WG1), community's economic revitalization (WG2), mental health & psychosocial (WG3) and disaster risk reduction program (WG4).

Five FGD sessions were conducted inviting no less than 80 participants representing different stakeholders, for instance, the 1st session were participated by school principals, teachers and school pupils from elementary, junior high school and highschools which schools are located at the tsunami-affected areas. The subsequent session were participated by representatives of communities located both inside and outside of the tsunami-affected areas. In the third session, governmental officers from several departments were invited, the fourth session was participated by academicians and the fifth session were participated by the media and NGOs representatives.

In each session of FGD, some general questions related to the topics of each working groups (WG) were delivered and discussed among group participants to capture different perspectives on seeing the recovery process. Conclusive remarks were made when information or data received most confirmations among groups were taken as the valid data, while those which remains inconclusive, would be further considered to be the focused questions set in the in depth interviews and/or questionnaire surveys following the FGDs. Figure 1 reveals the situation during one of the FGD sessions.

Figure 1. Situation during one of the FGD sessions to discuss the post-tsunami recovery in Banda Aceh from different perspectives.

2.2 Data Analysis Methods

Since the nature of the data is expected to be rather qualitative, data analysis were done through mind-mapping method formulated during the FGD in which statements or remarks delivered by the focused discussion participants were captured. The mind mapping were formulated by identification of several parameters including process leads to windows of opportunities for sustainability, output, outcomes and success factors.

The first parameter is output, which identifies what have been done in the effort of recovery since the tsunami occurred in 2004. This includes the aspects of infrastructure, product, services, networks, knowledge skills, rules be it formal or informal. Subsequently, we look into the process of how things were done and who was involved during the process. This may consist of government, International Organizations, Non-Government Organizations, Civil Society Organizations, private sector, individual actions, projects, programs, resources and skills and barriers and tactics. Part of the process, which is the main subject of interest in this article, is to whether such process, the way things done, would lead to windows of opportunity for sustainable development. How much attention have been put by all stakeholders involved during the recovery, how influential the political support gained during the process, resource support (money, knowledge, skills, networks, etc), creating new allies, openness, awareness and how much portion of the process ensure sustainability and seen as an opportunity by identifying those processes categories into the one that leads to sustainable development. This will be validated by justification of the outcomes from the beneficiaries which emerged and identified during the FGDs.

From those analysis, the success factor would be naturally identified as well. The success factors identify which of the factors involved in the process of recovery are critical leading to success. For instance, it can be from collaboration, vision/role model, perseverance, innovation in institutions, social learning & hands-on, and communication. From those analysis of parameters, we further categorize all the identifiable elements of each parameters which hold the principles of sustainable development.

Following the data collection from the FGDs, further analysis of data were done by individual working group facilitators to identify cases, implementation of programs or interventions that were seen as windows of opportunties for change toward sustainable development. This method was adopted from combination of work from [4] and [5] which established as a framework to map those cases, implementation of programs or interventions occurred during the recovery process into the principle criteria of sustainability. This includes all processes leads to recovering or even building back better environment, well-being, resources, citizenship which potential create equity and adaptability for not only the current generation but also for the next generation.

3. Results and Discussions

3.1. Recovery status of Banda Aceh City

Lessons learnt came out of the assessment are divided into two categories: 1) the success or potentially success stories; and 2) the not-so success stories of the Banda Aceh city's recovery progress which has been lasting for the last twelve years since the Indian Ocean tsunami of 2004. Table 1 describes the exerpts of success stories of the status of several aspects of life in Banda Aceh city of Indonesia after 12 years recovering from the mega tsunami disaster which struck the city on 26 December 2004.

Table 1. Case study of Banda Aceh City: Success stories

Aspects	Outputs	Process (WOOFS)	Outcomes	Success factors
School Infrastructure	Several disaster-prepared schools established	***Introduction of school-based disaster preparedness program***	Integrated effort for disaster preparedness in schools	School committee to involve parents, community and local government
Micro-economic Revitalization	Micro-finance institution for start-up and small businesses	***Start-up financial support donated by foreign donors, continued by government***	continued micro-economic revitalization	Continuous local government's attention for small-businesses micro-finance
Mental health and psychosocial problems	Improved mental health infrastructure and program	***Collection and updates of mental health data***	Increasing accessibility to mental health treatment, better understanding on psychosocial problems	Government's attention to improve infrastructure and programs
DRR Program implementation	National Disaster Management Agency down to district level established	***Capacity building for local government in DRR***	Increasing capacity and better coordination among stakeholders on disaster management	National government attention to support disaster risk reduction programs

On the aspect of school infrastructure recovery, several disaster-prepared schools have been established as the integrated effort for disaster preparedness in schools were the key factor of success. The effort came from the school committee who engaged the free-willing parents, the school-surrounding community and local government. All the evolving process for the past twelve years have seized the window of opportunity to establish potentially sustainable school-based disaster preparedness program in several schools in Banda Aceh. Nevertheless, such opportunity would soon be diminished if commitments from the local or central government was not strong enough to maintain or even better to formalize the program into, for instance, local-content-disaster-based school curriculum.

On the sector of community-economic revitalization, the initiation on incentives provided for start-up or small business donated by foreign donors, has surprisingly turned out into a beneficial micro-economic development in post-tsunami era. This was because the local government has increased its capacity in organizing typical micro-finance institution for start-up and small businesses, rather than investing to major business sectors in Banda Aceh. As a result, it provide the opportunity for the tsunami survivors to quickly start their new lives with their own start-up or small businesses. Such micro-economic revitalization model has been gradually increases as the government continuously supports the program up till now.

A rather special attention has be given for the recovery of mental health and psychosocial problems in this assessment. It was not so common such topic has ever been assessed thoroughly in cases of post-disaster event at the other places. Almost immediately after the tsunami event, the mental health institute started to realize that institutionally, the records and data of mental health as well as infrastructure are the upmost essential matters in creating programs for mental health and psychosocial recovery. However, those data and records had almost never been updated in the past. Embarking from this condition, an effort was done with support from the central government and foreign assistance to improve updating mental health data collection, infrastructure and programs. The outcome of the effort is that there has been increasing accessibility for the community to get mental health treatment, and at the same time to have better understanding on psychosocial problems as the excess of the mega disaster.

The local government has paid good attention on improvement of mental health institution and programs. Even more, currently the Aceh model for mental health program has become the reference at the national level. One of the most important success story from the post-tsunami recovery case of Indonesia is the realization and implementation of Disaster Risk Reduction Program. This is denoted by the establishment of the National Disaster Management Agency at the central government level down to district level in all provinces in Indonesia, including in Aceh Province. This is such a phenomenal governmental decision for the sake of increasing the capacity of all stakeholders to take part in disaster risk reduction program in good coordination. The remaining tasks are to take this opportunity to enhance the performance of the coordinating agency to develop sustainable programs which eventually increase the resilience of the community.

On the other hand, there were also some not so successful story of the recovery process at Banda Aceh city for the past 12 years. Table 2 summarized the implications of the recovery process on several aspects in life which have not yet so far indicating any changes leading to sustainability. For instance, on the aspect of school infrastructure, the building code regulation (e.g. earthquake resistance building standarization) were intensively introduced during the reconstruction process, either by national government or by foreign NGOs. However, not all new school buildings would eventually apply such code in their design and construction. The failing factor seems to be sourced from the loose control on building code regulation implementation and the inconsistency in construction supervision.

Another side of the micro-economic revitalization is that the development job by UN standard emerged during the rehab/recon process. This either attracted short-term or non-permanent profession changes, or creating sky-rocketed inflation, disproportionate work forces and eventually lack of job creation. Such instantaneous and temporary revitalization may sometime followed by no exit strategy after temporary development job, and more importantly, lack of self-dependency in business (entrepreneurship). On the case of implementing the Disaster Risk Reduction Program, one of the main agenda of the city development plan was to create disaster-based city masterplan. However, in practice, coordination among stakeholders seem to fail to put the city masterplan forward. Rather, whereas the coastal low-lying areas supposed to be restored as the 'green-belt' area up to a few kilometers inland, instead, the reality was that more houses were rebuilt towards the coastline.

Table 2. Case study of Banda Aceh City: Not So Success Stories

Aspects	Process (WOOFS)	Outputs	Outcomes	Failure factors
School Infrastructure	***Introducing building code regulation by foreign NGOs to government and construction company***	Not all new school buildings applying quake-resistant building code	Increasing awareness of building code application, but not strictly applied	inconsistency in supervision and loose regulation on building code
Micro-economic Revitalization	***Development job by UN standard for rehab recon emerged, attractive for profession changes***	Increasing unemployment vs new start-up and small businesses	Inflation, disproportionate work forces, lack of job creation	No exit strategy after temporary development job, lack of self-dependency in business
Mental health and psychosocial problems	***Introduction to different methods of trauma healing***	Undetectable early PTSD symptom	emerging different future mental illness manifestation	Sporadic and unsystematic trauma healing program
DRR Program implementation	***Introduction of disaster-based city masterplan***	Rehab and recon of the coastal areas not entirely done according to the city masterplan	Failure in implementing city development according to masterplan	Loose implementation of city masterplan

3.2. Windows of Opportunities towards Sustainable Development

There are ways of defining sustainability under different contexts. One would consider sustainability as the continuity of project over time, in which multi-faceted into economic viability to achieve social equity within the environmental integrity. While in a larger aspect of life, sustainability should be considered in a holistic way as an action that equally account for planet, people and profit. The latter would put forward the safe and just space for humanity which includes sustainable economic development. Overall, sustainable actions strive to enable economic and social welfare in an equitable manner (worldwide) without compromising the viability and integrity of the supporting social and ecological systems in the long term [6].

Accordingly, this article tries to map the recovery of Banda Aceh City based on key parameters which constitutes the principles of sustainable development. Four parameters proposed by [6] were considered herein as the criteria for sustainability: 1) environment, 2) resources, 3) livelihoods and public finance, and 4) citizenship and governance. Those criteria were subsequently cross-cutted by another two parameters which promote sustainability in public goods: 5) equity, and 6) adaptability to change.

The results of the mapping of the recovery actions which have been progressing during the recovery process of Banda Aceh city revealed in Table 3. Qualitative data gained from the discussions regarding how far has the recovery process in different life aspects fulfilled the four criteria for sustainability and whether those are seen as opportunities to be sustained in a long run for the current and the future generation, as well as adaptation to the current situation in order to seize the opportunities for change toward sustainability.

Table 3. Identification of Windows of opportunities for sustainable development in Banda Aceh City.

Enhancing the public goods		**Equity**	**Adaptation**
Natural Environment Ensuring socio-ecological systems integrity	Failure in implementing city development according to masterplan (-)	To continue socialization on the use of vertical evacuation facilities in the neighborhood	Revitalization of disaster-based coastal city development
Natural Resources Maintaining resources and using them efficiently	Increasing awareness of building code application, but not strictly applied (+)	To conduct consistent supervision and strict regulation on building code	Construction program for reinforcing existing buildings
	Emerging different future mental illness manifestation (-)	To provide access to mental health services	Systematic trauma healing program
Livelihoods & Public Finance Ensuring livelihood opportunities & local investments	continued government support for micro-economic revitalization (+)	To continue program of micro-finance for small-businesses	Increasing self-dependent economy
	Inflation, disproportionate work forces, lack of job creation (-)	Government to support creating new jobs	Returning to previous jobs
Citizenship & Governance Ensuring citizenship & governance systems integrity	Integrated effort for disaster preparedness in schools (+)	Proportionate participation of school committee, surrounding community and local government	To complete the disaster-prepared school program for the 'danger' zone
	Increasing accessibility to mental health treatment, better understanding on psychosocial problems (+)	To make a model for post-disaster mental health program	To change the old to new system for mental health
	Increasing capacity and better	To implement contingency plans for stakeholders	To improve Disaster Management Plan

	coordination among stakeholders on disaster management (+)	

In Table 3, the findings from the post-tsunami recovery assessment in Banda Aceh City were exercised and examined, by fitting the identifiable changing situations which may be seen as the window of opportunitites, into either of the four principle criteria of sustainability [6]. Taking into account for the recovery process that has been lasting for more than a decade, analysis were made to identify which actions or programs implemented during the recovery seen as the window of opportunities and how the recovery process within more than a decade has or has not seized those opportunities (e.g. "+" denotes for a seized opportunity, "-" denotes for a missed opportunity). Those orderly identifiable actions would subsequently encountered with further actions during the recovery process which would lead to sustainable development, underlying the importance of the equity to be benefited from the recovery intervention for the current and future generation, as well as to identify adaptation being done to maintain the sustainability in a long run. The followings are the samples of cases of three different situations can be identified during the post-tsunami recovery process in Banda Aceh city.

On the aspect of livelihoods and public finance, in the former time before the tsunami, micro-financing from the government was not administratively easy to access, particularly for the communities like farmers and fishermen. During the post-tsunami recovery, to ensure the livelihood opportunities and local investments after the tsunami hit Banda Aceh city, several donors both from national and international organizations delivered micro-finance schemes for the tsunami-affected communities along the coastline to revitalize the economy of the communities amid recovering from the mega disaster. For some practical reason, the financial distribution in the beginning was not coordinated with the local government, instead each organization would get in contact with the communities leaders for direct distribution. At a later stage during the recovery, as the local government resume to be operational, all finance assistance were well-coordinated. At this point, the local government seems to have seen a window of opportunity to revitalized the administrative system of financing the micro-finance for communities. This was shown by the continuation of the government support for micro-finance and micro-economic revitalization programs for small businesses until present.

In contrast with the aspect of livelihoods and public finance, on aspect of socio-ecological systems integrity toward preserving the natural environment, during the early stage of post-tsunami recovery, a new masterplan blueprint of the Banda Aceh city was made with the help from Japanese government. The masterplan set to have a green-belt area along the coastline with belt reached up to 200 m inland, at which location public infrastructure and housing were not supposed to be reconstructed. Low density housing area with elevated floors design should have been located at the landward of the green belt, and more dense housing areas and public facilities further inland. This is obviously a window of opportunity to build back the tsunami-affected area better than the former situation (e.g. before the tsunami). Nevertheless, the reality is that after a decade more houses were built than ever before just next to the coastline, ignoring the renewed masterplan. Apparently, the masterplan were not strictly followed during the rehabilitation and reconstruction process during which most of the houses were rebuilt. This exemplifies the failure in seizing the window of opportunity to build back better.

On the side not, not all the "negative" situations would necessarily lead to further negative outcome during the recovery process. For instance, in the case of enhancing public goods in the area of natural resources, there has been emerging mental health situation, where the untreated mental conditions of the surviving tsunami victims had not been detected and therefore not being properly treated much earlier. In addition, during that early stage and continued at a later stage during post-tsunami recovery, several trauma healing programs were implemented rather sporadically by different health-related humanitarian organizations. Along the line, new methods of mental health interventions were also found, introduced and tested. It was learnt that after more than a decade since the tsunami disaster, several mental illness had emerged among the survivors, which were actually the manifestation of the undetectable or untreated mental conditions (e.g. Post-Traumatic Stress Disorder / PTSD) at the early stage after the tsunami event. Inspite of that, by now access to mental health services has been relatively good, as the provincial government has invested a long-term program of mental health revitalization since the early stage of the post-tsunami recovery. Although it might have been late for the early-treatment of the trauma, nevertheless, with the current better access to mental health services, better non-sporadic formulated interventions can be readily available for those in need.

4. Conclusions

The article aims at assessing important achievements of recovery in different aspects of life and to identify windows of opportunities and postulates a model of recovery from one of the largest disaster events in the modern human lifetime history, the Indian Ocean's tsunami in 2004. It seeks for lessons learnt from Banda Aceh city of Indonesia as one of the most devastated area hit by the Indian Ocean tsunami.

The qualitative methods of the assessment fitting the identifiable actions and programs during the post-tsunami recovery process that were seen as the windows of opportunities to change toward sustainability were put into perspective of the theoretical critical principle criteria of sustainability. Those includes the aspects of infrastructure and housing, economic revitalization of the affected community, mental health and psychosocial condition and development, establishment and implementation of disaster risk reduction programs and community preparedness.

The results give insights on the dynamics of recovery process after more than a decade since the tsunami-affected area. Some success and not-so-success stories of actions and program implementations during the recovery process in the last decade post-tsunami were captured. There were changes which can be identified to have potential on sustainability since the beginning of the post-recovery process. There were also programs or actions which did not seem to work at first, but after intervention or continuation of programs throughout the recovery period they became sustainable. On the other hand, not all the "negative" situations would necessarily lead to further negative outcome during the recovery process.

The implication of this post-tsunami recovery assessment are that communities recovering from natural disasters have an important choice: rebuild damaged areas as they were, or to "Build Back Better". This decision should be able to improve communities's capacity to recover from future disasters, and also ensure that investments made today withstand the impacts that may come with natural hazards such as earthquakes and tsunamis.

Acknowledgements

Authors wishing to acknowledge all stakeholders invited during the FGDs, students volunteers, many colleagues in each working group of the assessment for their assistance and shared knowledge and expertise during the assessment and FGDs. A special thanks to Prof. Yasuo Tanaka for bringing up the idea of conducting several FGDs for data collection, and Katja Brundiers for introducing the concept of windows of opportunities for sustainable development at the final stage of the assessment which helps to put the assessment into perspective. The assessment of post-tsunami recovery was conducted by TDMRC of Syiah Kuala University in 2014 in commemorating the 10th year of the Indian Ocean tsunami disaster. We thank Badan Penanggulangan Bencana Aceh (BPBA) and Kobe University for financing the assessment of post-tsunami recovery of Banda Aceh City.

References

[1] Kingdon J W 1995 *Agendas, alternatives, and public policies* (New York: Harper Collins College Publishers)

[2] van Eijndhoven J, Clark W C, Jager J 2001 Risks *The long-term development of global environmental risk management* (The Social Learning Group, Learning to Manage Global Environmental vol 2) (Cambridge: The MIT Press)

[3] ALNAP (Action Learning Network for Accountability and Performance in Humanitarian Action) 2008 *List of publications in the internet* (Available via http://www.alnap.org/publications/rha.htm.)

[4] Birkmann J, Buckle P, Jaeger J, Pelling M, Setiadi N, Garschagen M, Fernando N, Kropp J 2008 Extreme events and disasters: a window of opportunity for change? Analysis of organizational, institutional and political changes, formal and informal responses after mega-disasters *Nat. Hazards* DOI 10.1007/s11069-008-9319-2

[5] Wiek A, Ries R, Thabrew L, Brundiers K, Wickramasinghe A 2010 Challenges of sustainable recovery processes in tsunami-affected communities *Dis. Prev. and Manag.* **19** 4 pp 423 - 437

[6] Gibson R B 2006 Sustainability assessment: basic components of a practical approach *Imp. Ass. & Proj. Appr.* **24** 3 170-182.

IOP Conf. Series: Earth and Environmental Science 56 (2017) 012020 doi:10.1088/1755-1315/56/1/012020

Disaster management curricula: strategy to create doctors with disaster resilience in Aceh, Indonesia

N Wahyuniati[1], R Maulana[1] and I Ichsan[1,2,*]

[1] Faculty of Medicine, Syiah Kuala University, Darussalam-Banda Aceh, 23111, Indonesia
[2] Tsunami and Disaster Mitigation Research Center (TDMRC), Syiah Kuala University, Darussalam-Banda Aceh, 23111, Indonesia
*Corresponding author: Faculty of Medicine, Kopelma Darussalam Banda Aceh 2311, Indonesia. Tel: +62 8116807781, Email: ichsan@tdmrc.org / ichsanmd_aceh@unsyiah.ac.id

Abstract. Faculty of Medicine of Syiah Kuala University has one special block called Disaster Management block (the last block on the curricula) on Competency-Based Curriculum of Medical Sciences with the Problem-Based Learning method. This block has four credits, allocated seven weeks learning period including one week for evaluation. The placement of disaster management block in the 7th semester (last semester) aims to allow students to implement more easily the complete basic and clinical medical knowledge and then have it integrated with the management capabilities during adisaster.This article evaluates two components: 1) Disaster management module, by comparing the content of modules used in three different academic years, the academic year 2013/2014, 2014/2015 and 2015/2016, 2) The final grade, by comparing the final grade of disaster management block in 4 years (comparing students class of 2008, 2009, 2010 and 2011). The results revealed that on every academic year there were additions and strengthening of the material to ensure that students achieve a complete learning experience, and there was a slight increase in student's grades where the number of students who receive grades A has increased while the number of students who receive grades E decreased.

1. Introduction

Indonesia is a disaster-prone country and experienced amajor loss of both victims (dead, missing, injured) and property loss. Aceh province is one of the areas in Indonesia who has suffered catastrophic 9.1 magnitude earthquake followed by atsunami on December 26, 2004. Most areas of Aceh was destroyed in a matter of minutes, more than 200.000 inhabitants were dead, missing or injured and countless properties lost or damaged. This tsunami has become a 'wake-up call' for the Indonesian nation and especially for the people of Aceh that has a very low disaster preparedness in both governments and communities level. Based on this reality, Indonesian Government then formed the Law No.27 of 2007 on disaster management, and later formed the National Disaster Management Agency (BNPB) and the Regional Disaster Management Agency (BPBD). The process of post-disaster continuous improvement in all aspects in Aceh after the 2004 tsunami included disaster education establishment. The disaster education in Aceh is a pioneer project in Indonesia and it is now implemented in school level to University level. Various strategies in the field of disaster education to increase the better understanding and preparedness in Aceh is still to be pursued, including the strategies implemented to medical students at the Syiah Kuala University. [1,2,3]

Faculty of Medicine of Syiah Kuala University has a several local curricula, one of them called 'disaster management' which are arranged based on the determination to produce graduates who are aware of and understand that disaster management is similar to day-to-day clinical emergency management but has more victims, and wider damage to facilities as well as health and psychosocial impacts in large communities, so they are expected to overcome problems arising from disasters and coordinate with relevant agencies in disaster management. Disaster management aims to reduce or avoid, potential loss from hazards, ensure fast and precise assistance for disaster victims, and achieve fast and effective recovery. Faculty of Medicine of Syiah Kuala University has one special block called Disaster Management block (the last block on the curricula) on Competency-Based Curriculum of Medical Sciences with the Problem-Based Learning method. This block allocated seven weeks learning period, including one week for evaluation and had four credits. The placement of disaster management block in the 7th semester (last semester) aims to allow students to implement more easily the complete basic and clinical medical knowledge and then have it integrated with the management capabilities during adisaster.This article aims to explain in detail about the disaster management block which is an improvement strategy for disaster preparedness that has been conducted in thefaculty of medicine of Syiah Kuala Kuala University since 2006 to produce future doctors with disaster resilience. [4]

2. Method

The method used in this article was conducted based on the evaluation of the disaster management curriculum implementation at the Faculty of Medicine, Syiah Kuala University. The author evaluates two components:

1. Disaster management module, by comparing the content of modules used in three different academic years, the academic year 2013/2014, 2014/2015 and 2015/2016. It is discovered that in every academic year there were additions and strengthening of the material to ensure that students achieve a complete learning experience. This article describe a comparison of the area of competence to be achieved by student, issues / topics competencies, topic tree, the number and qualification of college experts, lecture topics, the format of learning activities, the composition of the credits and the time allocated to each learning activities, the number and type of topic being taught, number and type of tutorial scenario, the theme/scenario of the table-top practices, the design and scenario of the disaster drill practice, and the supporting institutions involved in each academic year.
2. The final grade, by comparing the final grade of disaster management block in 4 years (comparing students class of 2008, 2009, 2010 and 2011) (as shown in figure 1). The final grade obtained from accumulate: 1) the 'process' score, consist of tutorial (discipline, attitudes, knowledge), table-top practice and disaster drill practice, 2) the cognitive score obtained from cognitive theory examination (computer based test).

The data in this article is presented in narrative form, tables and graphs. The main source of data used in this article came from disaster management module, primary data from assessment unit and academic guide book of thefaculty of medicine of Syiah Kuala University.

3. Results and discussion

Syiah Kuala University as the largest and oldest National University in Aceh is committed to developing excellence in science, technology and art in seven (7) themes: Food Security; Potential Development of Coastal, Marine and Fisheries; Disaster Mitigation and Management; Climate change; Renewable energy; Life Skill and Character-oriented Integrative learning; Law and human rights

IOP Conf. Series: Earth and Environmental Science **56** (2017) 012020 doi:10.1088/1755-1315/56/1/012020

Revitalization. Syiah Kuala University now has Tsunami and Disaster Mitigation Research Centre (TDMRC) featured as a study center that collects Syiah Kuala University's resources to addressing the issue of alocal disaster, national and global aspects of knowledge and application.In line with the development roadmap of the University, Faculty of Medicine of Syiah Kuala University also made a breakthrough in its curricula by adding a disaster-related local content. Faculty of Medicine aims to produce doctors who have the competence to implement primary health care which includes professional health care servicesof all ages, the detection of diagnosis of any disease as early as possible is implemented in perfection, holistic, sustainable and have coordination with other health professions. [5]

Doctors graduating from Faculty of Medicine of Syiah Kuala University would have major competencies built on a foundation consisting of noble professionalism, introspective and self-development, as well as effective communication, and is supported by pillars: information management, the scientific basis of medicine, clinical skills and management of health problems. Supporting competencies that will be owned by graduates are a local content curriculum developed in accordance with the vision, mission and local conditions of Syiah Kuala University in Aceh province and therefore translated into a vision, mission, goals and objectives of the Faculty of Medicine, Syiah Kuala University.Disaster Management block is the local curricula in Faculty of Medicine of Syiah Kuala University and it is still the one and only disaster-related medical education curricula implemented in Indonesia. This block will provide a thorough understanding and skills, practical and simple according to the academic level of students in the field of disaster management. This block also emphasizes the importance of good cooperation between the medical profession with the entire community in disaster management. Disaster management block will equip students with a basic understanding ofdisaster, provision and what he must possess, how they are able to communicate, coordinate, and synchronize with the victims, their families and with related parties that played a role in response to various types of disasters.[6]

3.1 Evaluation of disaster management module

Learning activities in disaster management block are designed in the form of Problem-Based Learning (PBL) as follows:

1. Tutorial: The three jump method
2. Lecture
3. Table-top and field practical session: manufacturing of hazard map and disaster simulation. The practical session will be held in the community and involve many supervisors from various professions.
4. Independent learning
5. Institutional visit

The content of disaster management module from 3 academic years were being compared based on 11 categories as shown in table 1. Over these three years there were a major changes in college expert involvement where there were addition of disaster management expert on academic year 2015/2016, addition of 'management of mass chaos' in lecture topic starting from academic year 2014/2015, addition of learning activities format (lectures provided through both interactive session and e-learning method), also increase number of supporting institution involved in this curricula. Those changes aim to ensure the enhancement of curricula quality and ensure the better understanding for students.

Table 1. Comparison of disaster management module in 3 academic years.[7,8,9]

Category	Academic Years		
	2013/2014	2014/2015	2015/2016
Area of competence to be	7 areas of competence	7 areas of competence	7 areas of competence

achieved			
Topics of competence	17 topics: - 6 topics with competence level 2 - 3 topics with competence level 3A - 4 topics with competence level 3B - 4 topics with competence level 4	17 topics: - 6 topics with competence level 2 - 3 topics with competence level 3A - 4 topics with competence level 3B - 4 topics with competence level 4	17 topics: - 6 topics with competence level 2 - 3 topics with competence level 3A - 4 topics with competence level 3B - 4 topics with competence level 4
Topic tree	Disaster management cycle	Disaster management cycle	Disaster management cycle
College expert	- Anaesthesiologist (2) - Forensic specialist (1) - Pulmonologist (1) - Public health practitioner (2) - Medical education expert (2) - ENT specialist (1) - Psychologist (1)	- Anaesthesiologist (2) - Forensic specialist (1) - Pulmonologist (1) - Public health practitioner (3) - Medical education expert (1) - ENT specialist (1) - Psychologist (1)	- Anaesthesiologist (3) - Forensic specialist (1) - Pulmonologist (1) - Public health practitioner (2) - Disaster management expert (1) - ENT specialist (1) - Psychiatrist (1)
Lecture topics	- Introduction and overview of disaster and hazard - Disaster management cycle - Disaster risk reducement - Disaster preparedness of health provider - Contingency plan for health facilities - Crisis management - Incident command system - Role of health provider in hydrometeorology disaster - Role of health provider in earthquake and tsunami disaster - Post-disaster infectious diseases' management - Disaster Victim Identification Management - Mental health in disaster	- Introduction and overview of disaster and hazard - Disaster management cycle - Disaster risk reducement - Disaster preparedness of health provider - Contingency plan for health facilities - Crisis management - Incident command system - Role of health provider in hydrometeorology disaster - Role of health provider in earthquake and tsunami disaster - Post-disaster infectious diseases' management - Disaster Victim Identification Management - Management of mass chaos. - Mental health in disaster	- Introduction and overview of disaster and hazard - Disaster management cycle - Disaster risk reducement - Disaster preparedness of health provider - Contingency plan for health facilities - Crisis management - Incident command system - Role of health provider in hydrometeorology disaster - Role of health provider in earthquake and tsunami disaster - Post-disaster infectious diseases' management - Disaster Victim Identification Management - Management of mass chaos. - Mental health in disaster
Learning activities format	- Tutorial: The 3 jump method - Lecture - Table-top and field practical session: manufacturing of hazard map and disaster simulation. The practical	- Tutorial: The 3 jump method - Lecture (E-learning and interactive) - Table-top and field practical session: manufacturing of hazard map and disaster	- Tutorial: The 3 jump method - Lecture (E-learning and interactive) - Table-top and field practical session: manufacturing of hazard map and disaster

	session will be held in the community and involve many supervisors from various professions. - Independent learning - Institutional visit	simulation. The practical session will be held in the community and involve many supervisors from various professions. - Independent learning - Institutional visit	simulation. The practical session will be held in the community and involve many supervisors from various professions. - Independent learning - Institutional visit
Allocation of credits and time	4 credits (SKS), 7 weeks	4 credits (SKS), 7 weeks	4 credits (SKS), 7 weeks
Tutorial scenario (number and type)	4 scenarios: new placement setting, flood, tornado, post-disaster management	4 scenarios: rainy season, volcano eruption, tornado, post-disaster management	4 scenarios: rainy season, volcano eruption, flood (freshet), post-disaster management
Table-top practices (theme/scenario)	- Creating disaster simulation - Creating hazard map and management plan - Utilization of instruments during disaster	- Creating disaster simulation - Creating hazard map and management plan - Utilization of instruments during disaster	- Creating disaster simulation - Creating hazard map and management plan - Utilization of instruments during disaster
Disaster drill practice (design and scenario)	Disaster victim identification	Disaster victim identification	Disaster victim identification
Supporting institutions	Institutional visit to: - Indonesian's red cross - Fire department - Search and Rescue department Supporting institutions: - Aceh's Provincial health office - Aceh's Disaster management's agency - Indonesia Red Cross - Search and Rescue depatment - World Health Organization - Fire Department - Non Government Organizations (2 NGOs) - Disaster resilience school	Institutional visit to: - Crisis management center - Search and rescue department Supporting institutions: - Aceh's Provincial health office - Aceh's Disaster management's agency - TDMRC - Search and Rescue depatment - World Health Organization - Crisis management center - Non Government Organizations (2 NGOs) - Disaster resilience school	Institutional visit to: - Health crisis management center - Search and rescue department Supporting institutions: - Aceh's Provincial health office - Aceh's Disaster management's agency - TDMRC - Search and Rescue department - World Health Organization - Crisis management center - Non-Government Organizations (2 NGOs) - Disaster resilience school

3.2 Evaluation of the final grade

At the end of the block, every student will be marked a final grade which are obtained from total value of the 'process' score and cognitive score. 'Process' score obtained from tutorial (discipline, attitudes, knowledge), table-top practice and disaster drill practice, whereas the cognitive score obtained from cognitive theory examination (computer based test).

The composition of the final grades as follows:

1. Process score (20% of total score), consist of:

- Tutorial (discipline, knowledge, attitude)
- Activities report
- Practices report

2. Final cognitive theory examination with computer based test method (80% of total score)

Figure shown below showed a comparison between students' final grades in disaster management block in student class of 2008-2011. It revealed that there were a slight increase in student grades where the number of students who receive grades A has increased while the number of students who receive grades E decreased.

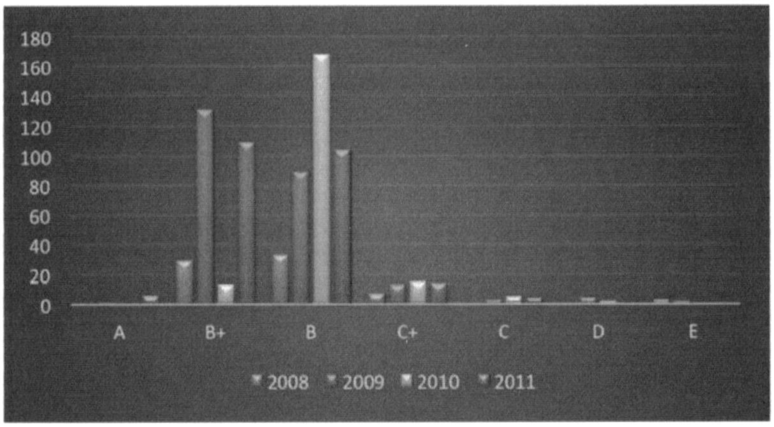

Figure 1. Comparison of students' final grades in disaster management block in student class of 2008-2011.

4.Conclusions

Disaster Management is a featured block which implement competencies: communication skills management, emergency and traumatology management, and across sectors coordination management. The concept of handling cases of emergency and trauma which is based on a priority basis to prevent death and disability would be the soul that always accompanies the medical profession when dealing with disasters.

With the provision of the above teaching concept, students are expected to have the same mindset in the future that in disaster management a medical profession may not work alone but should always be in a system and able to cooperate with everyone. Nonetheless, the medical professionalism should always be prioritized with the knowledge and skills update through varioustraining, so that doctors will be able to conduct their role as a major part in patient care and patient safety on every disaster.

The existence of the disaster management block at faculty of medicine Syiah Kuala University need to be strengthened in the form of block module optimization, the development of innovative learning activities, increase recruitment of professional/disaster practitioner as a lecturer/tutor, disaster drill optimization, the development of innovation in evaluation of learning outcomes, cooperation expansion with another institutions to optimize the implementation of the block.

Based on the results of internal SWOT analysis and faculty self-evaluation document, together with the academic monitoring and evaluation results, Faculty of Medicine of Syiah Kuala University confident of being able to answer the challenge to produce professional future doctors with disaster resilience through the implementation and optimization of the disaster management block.

5. References

[1] Markenson D, Dimaggio C and Redlener I 2005 Preparing health professions students for terrorism, disaster, and public health emergencies: core competencies*Academic Medicine*. 80(6):517-526

[2] Pfenninger EG, Domres BD, Stahl W, Bauer A, Houser CM and Himmelseher S 2010 Medical student disaster medicine education: the development of an educational resource*Int J Emerg Med*. 3:9-20.

[3] Ministry of Health RI 2007 Technical Guide for Health Crisis Response on Disaster : Guidelines for health workers that involve in health crisis responses on disaster in Indonesia

[4] Indah R, Mulyadi M and Maulana R 2013 Analysis on the quality of test items on disaster management block*Proceeding of the 3rd annual international conference Syiah Kuala University in conjunction with the 2nd international conference on multidisciplinary research (ICMR)* (Banda Aceh/Syiah Kuala University Press)

[5] Syiah Kuala University 2015 *Strategic Planning 2012-2017 and Long Term Development Plan, Revised edition May 2015* (Banda Aceh: Syiah Kuala University Press)

[6] Faculty of Medicine 2014 *Self Evaluation Document* (Banda Aceh: Faculty of Medicine, Syiah Kuala University)

[7] Medical education unit 2012*Disaster management module, fifth edition*(Banda Aceh: Faculty of Medicine, Syiah Kuala University)

[8] Medical education unit 2013 *Disaster management module, sixth edition* (Banda Aceh: Faculty of Medicine, Syiah Kuala University)

[9] Medical education unit 2014 *Disaster management module, seventh edition* (Banda Aceh: Faculty of Medicine, Syiah Kuala University)

Assessing school disaster preparedness by applying a comprehensive school safety framework: A case of elementary schools in Banda Aceh City

A Sakurai[1*], M B F Bisri[2], T Oda[3], R S Oktari[4] and Y Murayama[5]

[1] Associate Professor, International Research Institute of Disaster Science (IRIDeS), Tohoku University, 468-1-S304 Aramaki Aoba, Aoba-ku, Sendai-City, 980-0845 Japan,
[2] Ph.D. candidate, Graduate School of International Cooperation Studies, Kobe University, 2-1 Rokkodai-cho, Nada-ku, Kobe-City, 657-8501 Japan
[3] Associate Professor, Miyagi University of Education, 149 Aramaki Aoba, Aoba-ku, Sendai, 980-0845 Japan
[4] Lecturer and researcher, Tsunami and Disaster Mitigation Research Center (TDMRC) and Faculty of Medicine, Syiah Kuala University, Jl. Tgk Abdul Rahman, Gp. Pie, Meuraxa District, Banda Aceh, Indonesia
[5] Professor, Professional school of education, Yamagata University, 1-4-12 Kojirakawa, Yamagata-City, Yamagata 990-8560 Japan

*email:aikosak@gmail.com

Abstract. The study assessed the depth of school disaster safety at public elementary schools in Banda Aceh City, Indonesia in terms of comprehensive school safety, especially school location, disaster management and disaster education. The findings indicate that 56% of public elementary schools in Banda Aceh City are exposed to high tsunami risk, and most externally driven school disaster preparedness activities were not continued by the schools due to lack of ownership and funding. To realize comprehensive school safety, disaster preparedness programs should neither be brought in by external donors, nor be in a patchwork. Rather, it should be conducted jointly and sustainably by the local school and the community and supported by multi-sectoral support in the city. Comprehensive school safety of public elementary schools in Banda Aceh City could be realized by reviewing, updating and localizing school disaster preparedness programs by all the education partners in the city with strong political will and commitment.

1. Introduction

"Disaster education is the foundation of all the efforts for disaster risk reduction. The capacity of a society to accommodate itself to and recover from the effects of a natural disaster depends on how much each individual was able to learn from the past experiences and knowledge passed down from our ancestors, and how much they were able to prepare for a disaster in daily life through partnerships among households, schools, and the community" [1]. This is a preamble of the Sendai Declaration adopted at the International Forum for Promoting Education on Disaster Resilience, which was a public forum at the 3rd World Conference on Disaster Risk Reduction in Sendai, Japan. This sentences represents Japan's 20 years of experience in enhancing school disaster safety. It clearly states that disaster education involves efforts to increase the capacity of individuals and the society as a whole based on learning from previous disaster experiences, and that disaster education is a continuous community effort. In addition, the Sendai Framework for Disaster Risk Reduction 2015-2030 supports the idea of a comprehensive school safety [2] that includes an understanding of school disaster safety, particularly safe school locations and buildings, effective school disaster management and education to reduce disaster risk, disaster loss and disruption of educational services at the time of a disaster. In Banda Aceh City, several forms of external assistance have been promoting school disaster safety since the 2004 Indian Ocean Tsunami (IOT). A decade since the disaster occured, there remains questions of how much these forms of assistance could contribute to enhancing school safety against disaster, as well as how much these external supports have been institutionalized at the school level.

2. Purpose

The purpose of the study is to assess the depth of school disaster preparedness at public elementary schools in Banda Aceh City, and to examine ways for a sustainable approach for the schools to continue by themselves. The study aims at finding out the situation as to what extent the experience of the 2004 IOT is being integrated into school safety against tsunami disaster at all public schools.

3. Methods

The study was conducted over two years from March 2014 through December 2015 and included two field visits by the authors. There were three steps involved in the assessment of the depth of school disaster preparedness. The first step was to conduct a city-wide elementary school mapping with the 2004 tsunami inundation data in order to identify school locations and to determine which schools are still located in tsunami risk areas. In the second step, we conducted interviews and focus groups with school headmasters and teachers of 47 schools located at the inundated area. Structured interviews were arranged with headmasters of the selected schools. In addition, a focus group discussion was conducted with elementary school teachers. The third step involved collecting survey data from 113 students from four schools in grades 4-6.

During the second step, we retrieved the following data: 1) detailed personal information from the participants (i.e., headmasters and teachers), including their assignment history and teaching experience; 2) participants' experience with the 2004 IOT and participation in disaster trainings and other capacity building activities; and 3) the condition of the schools before and after the 2004 IOT, including status of the school's disaster preparedness program (SSB status), risk exposure of the school, disaster prevention, mitigation and preparedness activities conducted by the school, as well as nature of school-community collaboration. During each school visit, we also conducted an observation to assess the building type, primary structure, existence and condition of tsunami evacuation locations, as well as the existence and condition of tsunami evacuation maps, tsunami evacuation route and signage.

4. Results and Discussion

4.1 Safety of school location

The distribution of public elementary schools in Banda Aceh City in the 2004 IOT tsunami inundation area was found by school location mapping based on the inundation data estimated by the Japan International Cooperation Agency's (JICA) [3]. The results showed that 55 of the 98 public elementary schools in Banda Aceh City are located in the 2004 tsunami inundation areas.

4.2 Experiences of school disaster preparedness programs in the past
Among the 55 elementary schools located in the 2004 IOT tsunami inundation area, 47 public and Islamic elementary schools were surveyed. 29 of those schools had received at least one school preparedness program between 2009 and 2014 and were labeled as SSB schools, while 18 of the schools had not received any programs at all. Supporters use various program names to represent a "school disaster preparedness program". However, for analytical purposes, we labelled all of the schools that received school disaster preparedness programs as SSB (Sekolah Siaga Bencana). In total, 34 elementary schools in Banda Aceh City have received SSB programs in the past, and 61 have not received any programs [4]. This means that so far, the SSB program has been prioritizing the schools located in the previous inundation area. The sample in this study is representative of the schools in Banda Aceh City, as 29 of 34 schools that received SSB programs were included in the analysis. Among those schools, six of them have received the program more than once, with one school receiving the program four times from different agencies.

Each agency provides different types of disaster preparedness activities. For example, the school disaster preparedness program by the Tsunami and Disaster Mitigation Research Center (TDMRC) of Syiah Kuala University provides a school watching workshop, first aid training, an art workshop on disaster mitigation, provision of disaster related equipment, and disaster risk reduction (DRR) festivals. On the other hand, the SSB program of TDMRC and Bank of Indonesia assisted schools in preparing their tsunami evacuation maps, evacuation drills, school disaster preparedness task forces and school disaster management standard operating procedures. This was confirmed in survey to SDN 6 (SDN stands for *Sekolah Dasar Neger* or public elementary school) 6, SDN 17, SDN 41 and SDN 65. Schools that received the SSB program from the Indonesian Red Cross (PMI) were equipped with tsunami evacuation signs and "safe location" signs.

Table 1. Types of activities covered by SSB Programs in Banda Aceh City

Name of the SSB Program	**Types of activities included in the SSB Program**				
	Teaching content (in subject / curriculum)	**Extra-curricular activities**	**Evacuation map**	**Evacuation Drill**	**Provision of disaster equipment (e.g., signage)**
DRR-A Program	o	o	x	x	x
TDMRC, LIPI & UNESCO SSB Program	o	o	x	x	x
TDMRC & LIPI SSB Program	x	o	o	o	x
TDMRC, ACT, PMI SSB Program	x	o	o	o	o
TDMRC & Bank Indonesia SSB Program	x	o	o	o	o
TDMRC & ACT SSB Program	x	o	x	o	x
Education Agency Aceh province SSB Program	x	o	o	o	o

Source: Compiled by authors, from various sources
Remarks: DRR-A (Making Aceh Safer through Disaster Risk reduction in Development, UNDP & Government of Indonesia, 2009-2012), TDMRC (Tsunami and Disaster Management Research Center, Syiah Kuala University), LIPI (Indonesian Science Institute), ACT (*Aksi Cepat Tanggap*, a local NGO), PMI (Indonesian Red Cross).

4.3 Disaster Management
Some of the schools with SSB recognition are more aware of their designated tsunami evacuation site, are equipped with an evacuation route map and sign and have conducted at least one disaster drill during the

IOP Conf. Series: Earth and Environmental Science **56** (2017) 012021 doi:10.1088/1755-1315/56/1/012021

program. However, based on the statistics, the number of schools demonstrating more awareness was not significant. For example, only 12 of 29 SSB schools are equipped with evacuation route maps. In addition, there were only two non-SSB schools that were equipped with evacuation route maps. Only 9 of 29 SSB schools were aware of their designated evacuation sites, and only five teachers actually had the experience of visiting and assessing the site. Some schools were equipped with an illustrative image indicating the location of the schools, access roads, and names and locations of the designated evacuation sites. Other schools used an aerial photo to indicate the access road for the evacuation place. However, all of the disaster evacuation maps, regardless of the type, lacked scaling information to indicate the distance or time required to access the designated evacuation place. This is a fundamental feature of a map and the evacuation maps need to be updated so that they are in compliance with the 2014 National Disaster Management Agency (BNPB) Planning Guidelines on Tsunami Evacuation Routes and Signage [5].

It is also critical to notice that the "safe location" sign installed at several SSB schools and assembly point for evacuation were not hazard specific (i.e., fire, earthquake or tsunami). Thus, they could give a confusing message to the school community regarding the safety of the assembly point or safe location. Moreover, in the case of a tsunami, the Indonesian tsunami evacuation guidelines does not recognize the term "safe location" (or "*tempat aman*" in Indonesian). Instead, it introduces the concept of a "temporary evacuation place" (*TES: Tempat Evakuasi Sementara*) and "final evacuation place" (*TEA: Tempat Evakuasi Akhir*) [ibid]. Accordingly, the school community must be trained to recognize their designated evacuation place, either TES or TEA, and evacuate to those places rather than waiting or stopping at the "safe location". BNPB's 2014 standards on the signage for TES and TEA also includes specific dimensions and colors, as well as guidelines on its placement.

There is also little correlation between preparation of an evacuation route map and having a designated evacuation place actually used during a disaster simulation or drill. Only 6 of 29 SSB schools actually used the evacuation route map for the simulation. Another issue is that schools generally do not update their evacuation route maps or make efforts to assess the situation along the route. Only one SSB school reported having updated its evacuation route map. Nevertheless, in terms of relevance of the disaster type, almost all surveyed schools with evacuation maps stated that the information is relevant for an evacuation in the event of an earthquake and tsunami, and some added that it is also intended for an evacuation in the event of a fire and landslide. We also found that generally, the schools do not designate additional resources for maintenance of the tsunami-related signage. Thus, in some schools, their tsunami evacuation maps become completely unreadable.

Figure 1. Various Types and Conditions of Tsunami "safe location" (left and center columns, survey, October 2015), in comparison with BNPB Guideline (right column)

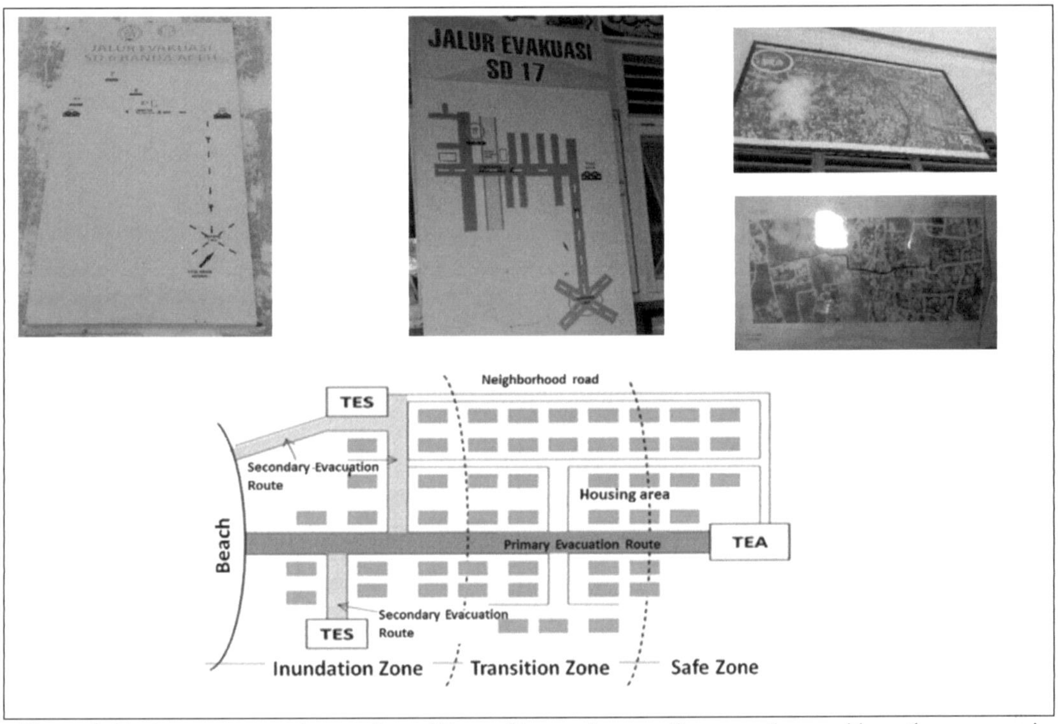

Figure 2. Various types, versions and conditions of tsunami evacuation maps (upper side, authors survey in October 2015), in comparison to BNPB's guideline (lower side, Rahayu et al, 2014)

We also observed problems with the availability and placement of evacuation route signs and "safe location" signs. Only 11 SSB schools had evacuation route signage available at the time of the survey in 2015. In those schools, the tsunami evacuation route direction and "safe location" were placed without consideration for the appropriateness of the location, whether it was visible or being blocked by other objects, or whether it gave the same message when being looked upon from both sides. Most of the tsunami evacuation route signs also needed to be updated so that they would be in compliance with the 2014 BNPB's Planning Guidelines on Tsunami Evacuation Routes and Signage, particularly the signage standards addressing the name and distance to the designated evacuation place. The most recent standards also regulate color (green background with the text and information written in white), dimensions and layout so that the signage is readable from any direction. Figure 3 showcases various types and conditions of tsunami evacuation routes available in several elementary schools in Banda Aceh City, in comparison with the national standard.

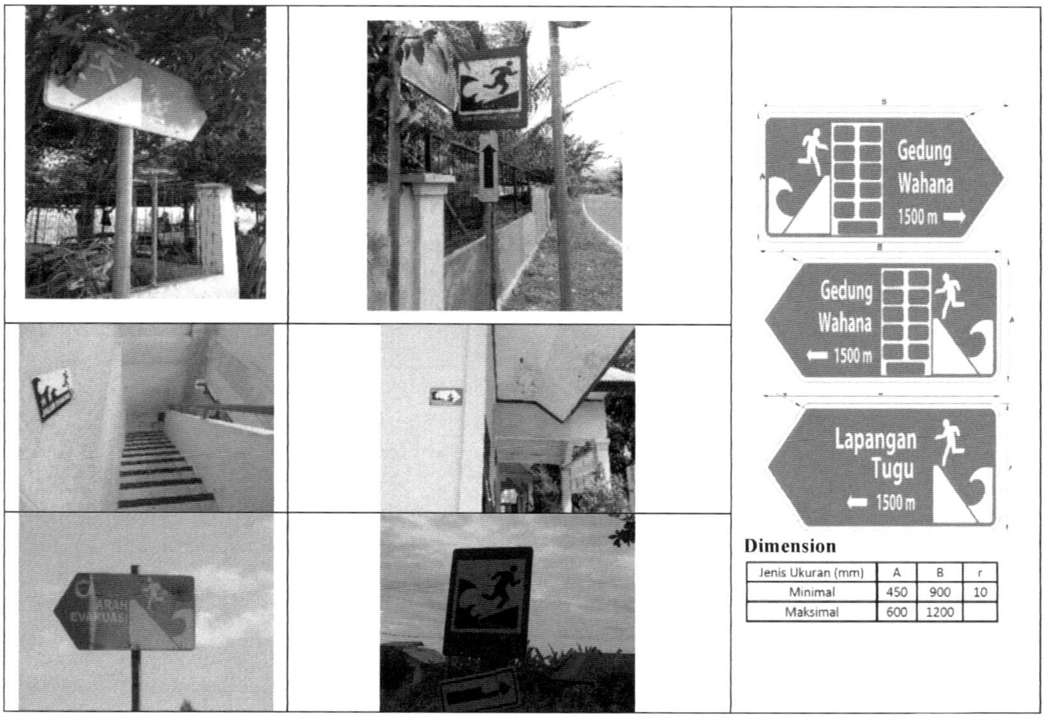

Jenis Ukuran (mm)	A	B	r
Minimal	450	900	10
Maksimal	600	1200	

Figure 3. Various types, versions and conditions of tsunami evacuation route (left and center columns, survey, October 2015), in comparison to BNPB Guideline (right column)

In general, SSB schools have more experience in conducting evacuation drills or simulations. However, this does not mean that non-SSB schools do not receive or conduct disaster drills. Among the 25 SSB schools and 7 non-SSB schools that reported conducting evacuation drills or simulations, there was only one school that consistently organized disaster drills from 2005-2014 period, with at least one and sometimes three drills per year. After the SSB program ended, the other SSB schools did not assume the responsibilities of conducting school evacuation drills or regularly visiting and assessing the designated evacuation locations. This implies an issue of sustainability. Since school principals and teachers are regularly transferred to other schools, routine implementation of evacuation drills at schools are critical for transferring knowledge and skills to newly assigned teachers and ensure the safety of the students in the case of tsunami, especially since they are located in the tsunami inundated areas. The headmaster or teachers who received the SSB training may have a basic idea of disaster preparedness and evacuation, since non-SSB schools have very limited understanding of this. For example, one of the non-SSB school teachers interviewed stated, "*(We will evacuate to) the school ground (if there is) an earthquake and (to) school hall on the third floor in the case of tsunami...*" (a teacher, female, interviewed in September 2015). This indicated that they were unaware that there is no guarantee of safety in the school hall. The results also show that four SSB schools reported that they have never conducted evacuation drills. This implies that even though the school experienced the drill, none of the current teachers had ever participated in a drill.

In addition, there is little correlation between having a SSB-status and having the means for the school to access the disaster early warning information and evacuation order. A total of 34 schools, 18 SSB and 16 non-SSB, admitted that they had no means to receive official evacuation decisions from either the Education Agency or BPBD. Only nine schools reporting having heard a tsunami siren during an earthquake in April 2012. One of the SSB programs did provide an earthquake alarm for the schools. However, all of the beneficiaries said that the alarm was no longer working. The other schools said that they created their own early warning instrument using bamboo or listened to announcements from local mosques. During the interviews, one of the headmasters described the school's situation as follows: *"(There is) no official information received. I had to decide by myself for evacuation basically. But, when in 2014 there was a risk of big flood, I participated in a briefing organized by Education Agency for (elementary schools) headmasters, on how to manage the flood. Also, we had received earthquake alarms from Indonesian Red Cross, but (they are) broken now. Also, when in 2014 we have an earthquake during school day, the students just escaped and left behind all their bags and shoes"* (a headmaster, female, interviewed in September 2015).

4.4 Disaster Education

Disaster Risk Reduction Project for Aceh (DRR-A) developed guidelines, entitled '*Sahabat Siaga*', on how schools can prepare for disasters, and integrate disaster content into the elementary school curricula for grades 1-3 and grades 4-6. Outputs of the initiative were also endorsed by the Governor of Aceh Instruction 2/2012 on integrating disaster content into the curriculum [4]. However, from the trained schools (SDN 1, SDN 2, SDN 31, SDN, 36 and SDN 39) in DRR-A among the 47 surveyed schools, only one school clearly still referred to the curriculum guideline. In addition, since a new national curriculum was introduced in Indonesia, these guidelines should be also updated accordingly. There is little correlation between having received the SSB program and including disaster education content during the teaching process. Both SSB and non-SSB schools reported that they provided disaster content during teaching hours through multiple methods and time allocations. Most schools from the SSB and non-SSB groups stated that they teach students disaster education. The pattern in both groups is similar in that the disaster education teaching material was owned by the schools, the same grades received disaster education, and the same types of hazards were being taught.

Both SSB and non-SSB schools stated that they have additional reading material for teaching disaster content, supplementing the use of an ordinary teaching book. However, some SSB-schools had more material or utilized various means for teaching disaster content. For example, 2 schools used the DRR-A materials, student projects or assignments, printed materials (e.g., maps, posters, disaster photos) and a mock, model or video of disaster. Other similarities include the student grade level selected for receiving disaster content and the type of hazards being taught, which included earthquake and tsunami. Common types of hazards taught include flood, landslide, and volcanic eruption. Both SSB and non-SSB schools preferred to target grades 4, 5 and 6, as can be seen from the Figure 4.

There were variations in terms of the teaching subject or activity used to deliver disaster content. At the non-SSB schools, the content was being taught in either natural science, social science, citizenship or physical education. However, SSB schools taught the content a wider range of subjects. including natural science and social science, mathematics, Indonesian language, and extracurricular activities (e.g., Boy Scouts/Girl Scouts). In terms of teaching hours, the pattern is also similar for both SSB and non-SSB schools with the same portion of schools that taught disaster content irregularly: 1-5 teaching hours/semester or 6-10 teaching hours/semester. However, some SSB schools allocated more than 15 hours/semester to teach disaster content. To this point, SSB schools still have the upper hand for teaching disaster content and have tried various means. As the vice headmaster at a SSB school stated, "... *(we teach) in the subject of natural science and social science. Each at around 2 class hours per in a week, especially with the new thematic teaching style. In total, it might have accounted for 12 hours per month. We indeed tried to connect with local content as well"* (a vice headmaster, female, interviewed in September 2015).

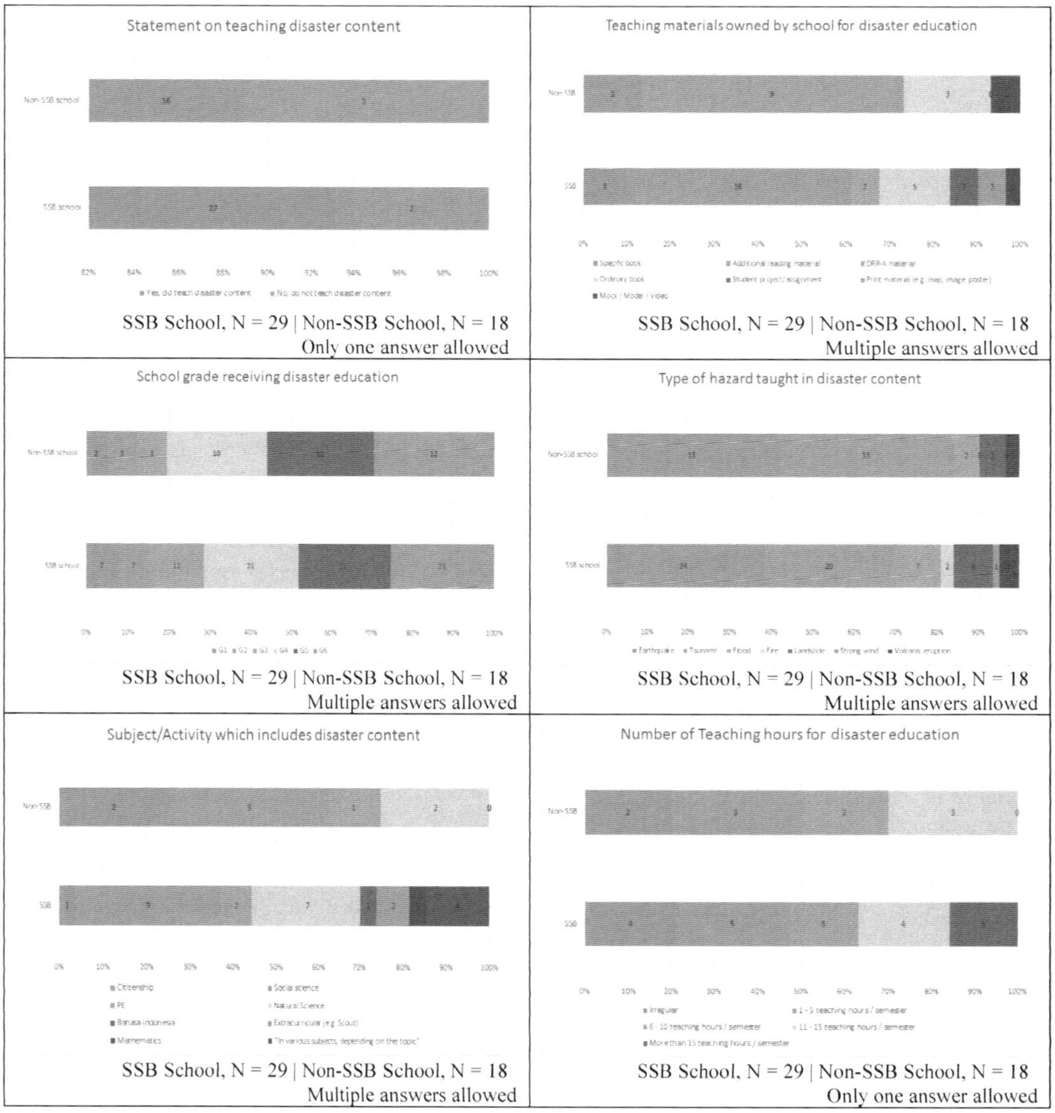

Figure 4. Variation on SSB and Non-SSB schools teaching characteristic on disaster content (Source: created by authors)

Generally, schools also did not keep records or documentation of the physical conditions before and after the 2004 tsunami. Thus most of the schools used resources from after the IOT and general information, as opposed to resources specific to the school's locale, as a means to teach students about the tsunami itself. There were only 6 schools (4 SSB schools and 2 non-SSB schools) that kept memories of the 2004 IOT, which were mostly photos and videos, and usually belonged to the personal collections of the teachers themselves. Similarly, only four schools (1 SSB and 3 non-SSB) kept records and documentation of the school situation before the 2004 IOT. However, almost all of the schools' headmasters and teachers knew the general oral history of physical damage, people's loss of life in the school due to the 2004 IOT and year of schools' reconstruction as well as the name of organization/donors that built their school. Only 2 non-SSB schools and 1 SSB school were found to have a complete absence of records of the history of the school due to the 2004 IOT. Nevertheless, there was no evidence that the documentation of the schools' records and documentation of the 2004 IOT were being used for teaching about the disaster or DRR education. However, having tsunami poles at the school site helped them to have a point of discussion for disaster education or to commemorate the 2004 tsunami, as 11 of 19 schools with tsunami poles used them for this purpose. In contrast, 24 schools reported organizing a visit to Aceh Tsunami Museum as part of education about the 2004 IOT. Other means for education include newspaper articles, photos and videos about the 2004 IOT, although these materials were not necessarily about the school's experience. In cases in which the school was located in proximity to a preserved memorial facility of the 2004 IOT or tsunami-related infrastructure (e.g., spot of the "floating boat" or tsunami evacuation building), we found that there was a tendency for the school to organize a trip to visit those locations.

Figure 5. Example of a record of the school's status before and shortly after the IOT 2004

Figure 6. Example of tsunami poles built at schools (left, SDN 31; right, SDN 27)

5. Conclusion

The study found that although most of the schools were totally damaged by the 2004 Indian Ocean Tsunami, 56% of public elementary and Islamic schools are still located in the 2004 tsunami inundated area. According to the Banda Aceh City Education Office, as of 2014, the lives of 21,301 students are exposed to a great potential tsunami disaster risk. Through post-disaster recovery assistance, massive externally-driven disaster risk reduction activities have been implemented to enhance the school disaster safety in the Banda Aceh City, including reconstructing concrete and two-story school buildings, making evacuation route maps and signs, conducting evacuation drills, preparing for teaching and developing learning materials. However, the study results indicated that these activities were mainly donor-driven and were implemented without full involvement from the local education agency. After the external supports ended, these activities were not assumed by local education agencies and schools. Tsunami evacuation information and disaster education curricula have not been updated. Since most of these disaster preparedness projects at the schools were supported by external donors during the past decade, a "sense of dependency" on external actors was observed among teachers at many schools. Schools and headmasters were not aware that they were responsible for ensuring the continuation of disaster education.

To realize comprehensive school safety, disaster preparedness programs should neither be brought in by external donors, or be a patchwork. Rather, it should be conducted jointly and sustainably by the local school and the community and supported by multi-sectoral support in the city. Comprehensive school safety of public elementary schools in Banda Aceh City could be realized by reviewing, updating and localizing existing school disaster preparedness programs among all the education stakeholders in the City with strong political will and commitment.

In the Sendai Framework on Disaster Risk Reduction 2015-2030, "increasing public education and awareness of disaster risk" in the post-disaster recovery, rehabilitation and reconstruction phase is newly emphasized. Since school disaster preparedness in Banda Aceh has been a precedent case in a post-disaster recovery process, continuous follow-up efforts should be made to make schools a center for creating a culture of safety against the next potential disaster. It is recommended to promote collaboration with other tsunami affected areas, such as that of the 2011 Great East Japan Earthquake and Tsunami to exchange experience, which could help avoiding the erosion of memories and experiences of the tsunami disaster.

References

[1] International Research Institute of Disaster Science (2015) "The Sendai Declaration, adopted at international forum for promoting education on disaster resilience: Development of a resilient community and improving disaster education and regional disaster preparedness", held on March 14, 2015 at Tohoku University, Sendai City, Japan. (http://drredu-collabo.sakura.ne.jp/cms/wp-content/uploads/20151101-3_en.pdf)
[2] UNISDR and Global Alliance for Disaster Risk Reduction and Resilience in the Education Sector (2014) Comprehensive School Safety. (http://gadrrres.net/resources/comprehensive-school-safety-framework)
[3] Japan International Cooperation Agency (2005) The study on the urgent rehabilitation and reconstruction support program for Aceh province and affected areas in north Sumatra, Final Report (1).
[4] Bisri, M. B. and Sakurai, A. (2016). Disaster education and school safety governance after the 2004 Indian Ocean Tsunami in Indonesia: from national policy to local implementation, in Djalante, R., Thomala, F., Garschagen, M., and Shaw, R., *Disaster Risk Reduction in Indonesia: Progress and Challenges in Managing Risks, Reducing Vulnerability and Building Resilience*.
[5] Rahayu, H., Latief, H., Anita, J., Riawan, E., Bisri, M. B., & Rizka, S. (2014). Pedoman Perencanaan Jalur dan Rambu Evakuasi Tsunami (Planning Guideline on Tsunami Evacuation Route and Signage). Jakarta: BNPB (National Disaster Management Agency) and ITB (Bandung Institute of Technology).

Acknowledgement
This study is funded by the 2014 Project Grant from International Research Institute of Disaster Science, Tohoku University and MEXT KAKENHI Grand (No. 26510008).

IOP Conf. Series: Earth and Environmental Science **56** (2017) 012022 doi:10.1088/1755-1315/56/1/012022

Disaster risk reduction policies and regulations in Aceh after the 2004 Indian Ocean Tsunami

Syamsidik[1,2,*], I Rusydy[1,3], S Arief[1], K Munadi[1] and E Melianda[1,2]

[1]Tsunami and Disaster Mitigation Research Center (TDMRC), Syiah Kuala University, Jl. Prof. Dr. Ibrahim Hasan, Gampong Pie, Banda Aceh, 23233, Indonesia.
[2]Civil Engineering Department, Syiah Kuala University, Jl. Syech Abdurrauf, Banda Aceh, 23111, Indonesia.
[3]Geological Engineering Department, Syiah Kuala University, Jl. Syech Abdurrauf No. 7, Banda Aceh, 23111, Indonesia.

*Email: syamsidik@tdmrc.org

Abstract. The 2004 Indian Ocean Tsunami that struck most of coastal cities in Aceh has motivated a numerous changes in the world of disaster risk reduction including to the policies and regulations at local level in Aceh. This paper is aimed at elaborating the changes of policies and regulations in Aceh captured and monitored during 12-year of the tsunami recovery process. A set of questionnaires were distributed to about 245 respondents in Aceh to represent government officials at 6 districts in Aceh. The districts were severely damaged due to the 2004 tsunami. Four aspects were investigated during this research, namely tsunami evacuation mechanism and infrastructures, disaster risk map, disaster data accessibility, perceptions on tsunami risks, and development of tsunami early warning at local level in Aceh. This research found that the spatial planning in several districts in Aceh have adopted tsunami mitigation although they were only significant in terms of land-use planning within several hundreds meter from the coastline. Perceptions of the government officials toward all investigated aspects were relatively good. One concern was found at coordination among disaster stakeholders in Aceh.

1. Introduction

People often regard their government and the disaster management agencies as bodies to provide safety and welfare [1]. The expectations are hard to meet in the case of unprepared institutions. Failure to accommodate such expectations may lead to another restive situations, such as what were found in the 2010 Haitian earthquake [3] and the 2008 Cyclone Nargis [2]. The 2004 Indian Ocean Tsunami should strongly motivate the increasing quality of the government agencies to cater the needs of the disaster-affected communities. The questions of the lessons learned on the institutional changes and disaster risk reduction contribution in the affected region are yet to be answered. This includes Aceh as the most affected area due to the 2004 Indian Ocean Tsunami.

Most of coastal cities in Aceh were severely damaged due to the 2004 Indian Ocean Tsunami. The tsunami caused almost 140,000 people lives and numerous of damaged infrastructures were resulted. The damages caused large disruption on local government services and operations in this area. In term of environment, the tsunami changed large area of coastal morphology [11]. Port and coastal structures were severely damaged and not functioning for several years after the tsunami. The vast impacts of the tsunami also gave opportunities to the government to re-plan its coastal areas and its regulation related to disaster management. To rehabilitate the tsunami affected area, one agency was established where its mandates include re-arranging the coastal area land-use and to help to re-activate the function of local government. The agency was named as Aceh Nias Rehabilitation and Reconstruction Agency (BRR Aceh Nias). The operation of the agency was set to run the recovery process in four years, started from April 2005 until March 2009. The changes at the level of institution after a severe impact of disaster were seen in several countries, such as Indonesia and Sri Lanka. Both of the countries were among the most tsunami affected countries. The changes from the perspective of institution were also followed by a series of regulations and policies afterward. At national level, the 2004 Indian Ocean tsunami did trigger massive changes in disaster management and multi-level arrangements and disaster governance in Indonesia [4]. The previous decentralization process after the resign of new order led by Soeharto and many disasters before the 2004 tsunami did not motivate any significant changes in disaster management in Indonesia. Although the changes are clear at national level, the local level preparedness in term of institutional capacity and disaster related policies have not been mapped for Aceh context.

This research is aimed at capturing institutional changes in Aceh and regulations made after the 2004 tsunami, which are related to disaster risk reduction. The research is expected to partly respond to the questions of the lessons learned of the 2004 tsunami adopted at policies and regulations sectors. This would help disaster practitioners to a better understanding towards creating a more resilience region.

2. Study area

Aceh is located at the northern part of the Sumatra Island. It is situated between the Andaman Sea at the north, the Malacca Strait at the east, and the Indian Ocean on the west. **Figure 1** shows the study area. The study area comprises of six districts, i.e. Banda Aceh, Sabang, Aceh Besar, Aceh Jaya, West Aceh, and South Aceh Districts. All of the districts were affected by the 2004 tsunami with various levels of damages. The worst affected district was Banda Aceh. Prior to the 2004 tsunami, the population of the city was about 230,000. About 70,000 people were died in Banda Aceh or about one third of the city population. In comparison, the total reported death human casualties from the 2004 Indian Ocean were about 230,000 lives.

As the largest city in the Aceh Province, many of provincial government facilities and their organization were located in Banda Aceh. The massive damages due to the 2004 tsunami also affected the function of the provincial government and the city office too.

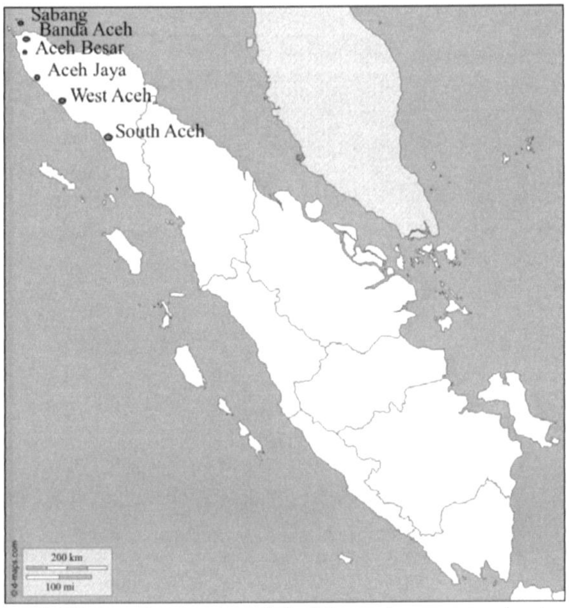

Figure 1. The study area (modified from d-maps.com/carte.php)

3. Methods

To assess the change in regulation and policies related to disaster risk reduction in this area, we employed a set of key questions, to capture the changes. There about 245 respondents were involved in this survey. The respondents come from disaster risk reduction related stakeholders, such as Aceh Disaster Management Agency (BPBA), Districts Disaster Management Agency (BPBD), Public Works Department (PU), Health Department, Hospitals, Education Department, District Development and Planning Agency, Transport and Communication Department (DISHUBKOMINTEL), Water Resources Department, Marine Affairs and Fishery Department. The targeted organizations were part of provincial or districts government units. The questions addressed to the respondents cover aspects as follows:

- Quality and Accessibility to disaster data.
- Openness of disaster data
- The uses of disaster risk map
- Tsunami risk understanding.

Assessment was also conducted to evaluate a number of disaster related documents produced between 2005 until early 2015. This research was conducted as part of the 10-Year Tsunami Recovery Process Assessment in Aceh performed between October 2014 until March 2015. The policies produced by the provincial government that adopt disaster mitigation were also studied using inventory procedure and mapping them in the disaster management cycle to identify the fulfilment and the gaps.

Focus of the analysis in term of the institutional changes was on the roles of Aceh Disaster Management Agency (BPBA) as a leading organization for disaster risk reduction in Aceh. This organization was part of the significant reform at national level on disaster management. It was established in 2010 based on a *Qanun* (by law product). The institutional set-up, regulation or Standard-Operating-Procedures (SOPs) and activities performed by the institution after the establishment are part of the logical framework analysed in this research. Figure 2 shows the logical framework employed in this research.

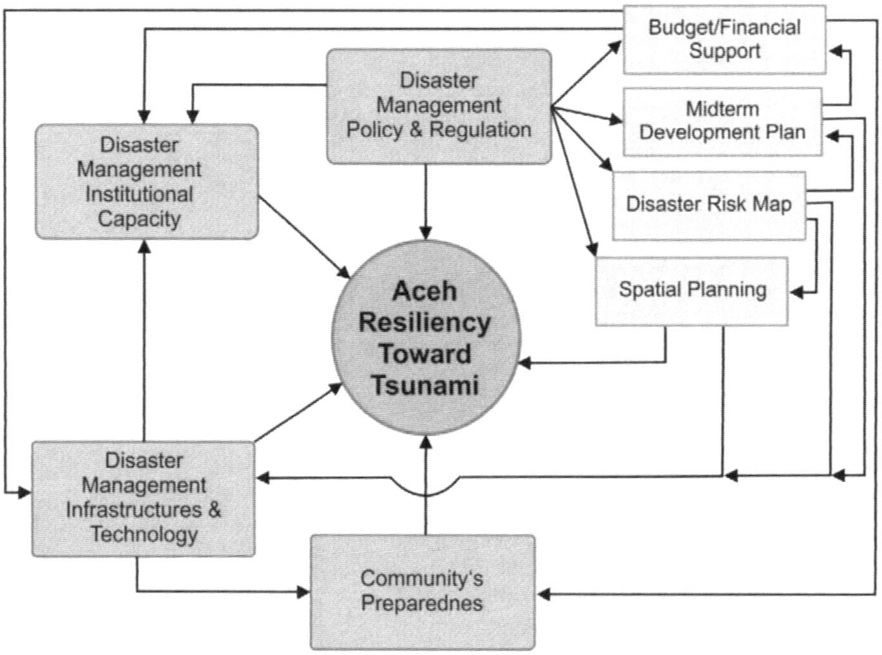

Figure 2. A logical framework employed in this research to assess the policies and regulations changes in Aceh related to DRR.

4. Results

Aceh has declared 10 development missions in its Mid-Term Development Planning for 2012-2017. One of the missions is on environmental quality and disaster management. To achieve this mission, the Aceh Provincial Government indicated three priorities, i.e.:

- Improvement of infrastructures that consider environmental quality and disaster mitigation.
- Increasing protection, critical area recovery, conservation and utilisation of natural resources and environment as part of development capitals in order to increase the quality of life.
- Disaster management including pre-disaster, emergency responses, rehabilitation and reconstruction process.

Provincial government units responsible to each of the activities in the priorities can be seen in Figure 3. The figure shows that BPBA has a central role in disaster management at Provincial level. Therefore, observing the roles and the operation of BPBA throughout the 10-year of the tsunami recovery process is pivotal to understand the policies and regulations changes related to DRR in Aceh.

4.1. Institutional Set-Up and Operation

BPBA was set-up on June 22, 2010. The legal basis of the BPBA establishment was referred to National Law No. 24 Year 2007 on Disaster Management, Aceh Qanun (by law product) No. 5 Year 2010 on Disaster Management, and Qanun No. 6 Year 2010 on The Establishment of BPBA Working Procedure and Organizational Set-up. Beside the Qanuns and the National Law, a more detail explanation on the operation and the functions of BPBA was stipulated in Aceh Governor Decree No. 7 Year 2011 on Details of Duties, and Function of Structural Officers in BPBA. In pre-disaster phase, BPBA is expected to conduct disaster management planning, DRR, to implement spatial planning, to coordinate efforts in conserving environmental function, to coordinate development planning, to conduct disaster education activities, to compose disaster risk analysis, and to compose disaster management technical standards. A complete list of regulations produced at provincial level to govern disaster management in Aceh can be seen in **Table 1**.

Table 1. Provincial Disaster Regulations in Aceh released between 2005-2015

	Regulation Number	Title of Regulation
Year 2008	Governor Decree No. 74/2008	Implementing of Loans to Empower Acehnese Business Persons Due to Earthquake and Tsunami in Aceh
	Governor Decree No. 93/2008	Amendments to Governor Decree No. 74/2008
Year 2010	Qanun No. 5/2010	Disaster Management
	Qanun No. 6/2010	The Establishment of Aceh Disaster Management Agency (BPBA)
	Governor Decree No. 43/2010	Tsunami Early Warning System and Emergency Responses in Aceh
	Governor Decree No. 48/2010	Aceh Local Actions Plan for Disaster Risk Reduction (2010-2012)
Year 2011	Governor Decree No. 7/2011	Details of Duties and Functions of Structural Officers at BPBA
Year 2012	Governor Decree No.49/2012	A Guideline for Floods Contingency Plan at Disaster Management Prioritized Zones in Aceh
	Governor Decree No. 87/2012	Results of Analysis on Structural and Functional Officers at BPBA
Year 2013	Governor Decree No. 64/2013	Actions Plan on Rehabilitation and Reconstruction After Earthquake in Central Aceh and Bener Meriah Districts (2013-2015)
Year 2014	Governor Decree No. 61/2014	Amendments to Governor Decree No. 64/2013
	Governor Decree No. 87/2014	Details of Main Duties and Functions of Steering Committee at BPBA

The expected activities are inter-related to other agency/government unit's working area. For an example, spatial planning is a mandated to Provincial Development and Planning Agency (BAPPEDA Aceh). Although BPBA was included in consultation process, the role of BPBA is not dominant. In practice, the role in conserving environmental function is run by Environmental Impacts Control Agency (BAPPEDAL) and by Forestry Department (DISHUT). Although there are duties of BPBA mentioned in Strategic Planning of BPBA for 2013-2017, it is not clear how all the duties have been performed and being communicated to other agencies/organizations. To perform its operation, BPBA has organizational structure as shown in **Figure 3**.

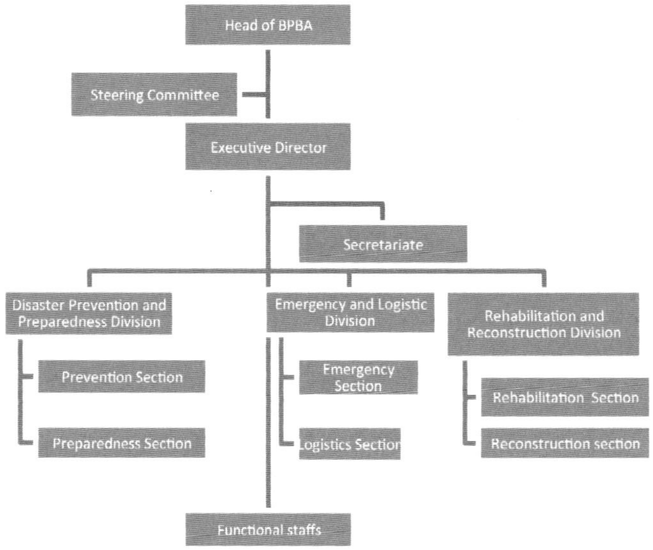

Figure 3. Organisational chart of BPBA according to Qanun No. 6 year 2010.

During 10-year of the tsunami recovery process, BPBA has encountered a number of disasters. Some of significant disasters were:

- Twin earthquakes, 8.5 Mw and 8.0 Mw, occurred on April 11, 2012.
- A 6,2 Mw earthquake on July 2, 2013.
- Massive floods in December, 2015 along the western coasts of Aceh
- Floods in December, 2014 around eastern coast of Aceh.

The disasters were real tests to the functions of BPBA after its establishment. In 2012 twin eartquakes, BPBA was not involved in activating tsunami alarm punctually [8]. The earthquakes created massive panic in the study area. Although the giant earthquake did not generate any tsunami, the central system of tsunami early warning in Jakarta operated by BMKG sent message of potential tsunami. During the first five minutes after the earthquake, BMKG successfully performed its function to send the message to the potential affected area. However, the massage was not responded properly. Part of the reasons was untrained staffs of BPBA at that period to align with the tsunami early warning system. This will be discussed in later section.

A 6.2 Mw earthquake in central region of Aceh that occurred on July 2, 2013 caused about 34 people died, 92 people were severely injured, and displaced about 48,000 people from two districts, i.e. Central Aceh and Bener Meriah districts. The impacts of the earthquakes were identified in the two districts [9]. It is interesting to learn that despite three years after the establishment of BPBA, the rehabilitation and the reconstruction process in the affected areas was led by BAPPEDA instead of BPBA. This was stipulated in Aceh Governor Decree No. 64 Year 2013 about Action Plan on Rehabilitation and Reconstruction Process After Earthquake in Central Aceh and Bener Meriah District. The Provincial BAPPEDA was expected to run the process within 2013 until 2015. Further confusing matters were seen when National Disaster Management Agency (BNPB) put a similar action plan for the affected area with the duration of the process between 2013 and 2014 [BNPB 2013]. Notwithstanding with the action plans, the role of BPBA in this process was not central yet.

At the end of 2010, all districts in Aceh have established their District Disaster Management Agency (BPBD). All BPBDs were established based on District Qanun. However, by the early 2010, 7 out of 23 BPBDs did not have a strategic planning to perform their operation and only two of them have detail regulation on their functions [6]. After decentralization of the Indonesia government, the role of district levels government units are important and more powerful. Therefore, a sufficient capacity to perform disaster management by BPBD at district level is crucial.

4.2. Tsunami Evacuation Infrastructure and Mechanism

Indonesia tsunami early warning system is basically divided into two parts, i.e (1) structural dimension consisting of early detection, analysis, broadcasting, and warning messages system; (2) cultural dimension comprising public announcement at city level, siren towers, and evacuation procedure [10]. Here, in this study, we studied the cultural dimension, which is under local government management. In this case, the local government were the Governor of Aceh and BPBA. All of the six districts studied in this research all of them have developed their tsunami evacuation procedures. However, infrastructures to facilitate tsunami evacuation differ from one district to another. Banda Aceh has the most complete facilities compared to others. It has 6 designated evacuation buildings, three major evacuation roads perpendicular to the coastline of Banda Aceh, and has about 7 siren towers that can cover most of coastal areas of the cities. However, other districts, such as West Aceh District only has two siren towers and one evacuation building. It is difficult for most of the districts to finance their facilities using the district's annual budget.

Some initiatives were made as attempts to suffice the needs of the tsunami evacuation buildings. For example, West Aceh District Government ask double storeys buildings along tsunami affected area to allow their roofs to be functioned as evacuation points, as temporary alternatives before the district has enough financial resources to build a new one. **Figure 4** show examples of tsunami shelters developed at double storeys building in Meulaboh city, the capital of West Aceh District. The West Aceh District has assigned its evacuation mechanism since 2012. As same as the other districts who have established their SOPs for Tsunami Evacuation, the clarity of the procedures can be seen in evacuation routes, evacuation location/buildings, and job description of staffs involved in the evacuation process. However, all the evacuation mechanisms have yet to have clarity on what would be the tsunami estimated time of arrival expected for each districts. Syamsidik et al. [5] published a data base for tsunami estimated times of arrival (ETA). They were developed based on a number of generated tsunami events around Aceh Province and the ETA recommended to be considered in the evacuation was based on the shortest ETA. Gaps are identified at all districts included in this study. For example, Banda Aceh still has about 25 minutes of a city-wide tsunami evacuation duration compared to the shortest ETA for the city, which is about 35 minutes. With the existing condition, the overall situation in the study area is still facing serious challenges to accommodate and to facilitate a smooth and fast tsunami evacuation process.

Figure 4. Double Stories Buildings in Meulaboh City of West Aceh District were made as alternatives to Tsunami Evacuation Building.

4.3. Government staffs perceptions on DRR

Based on the 245 returned questionnaires distributed in the six districts, they revealed that the perceptions of the government officials at provincial and at district levels are generally at good levels. **Figure 5** shows the perceptions of the government staffs on four aspects described in section 3.

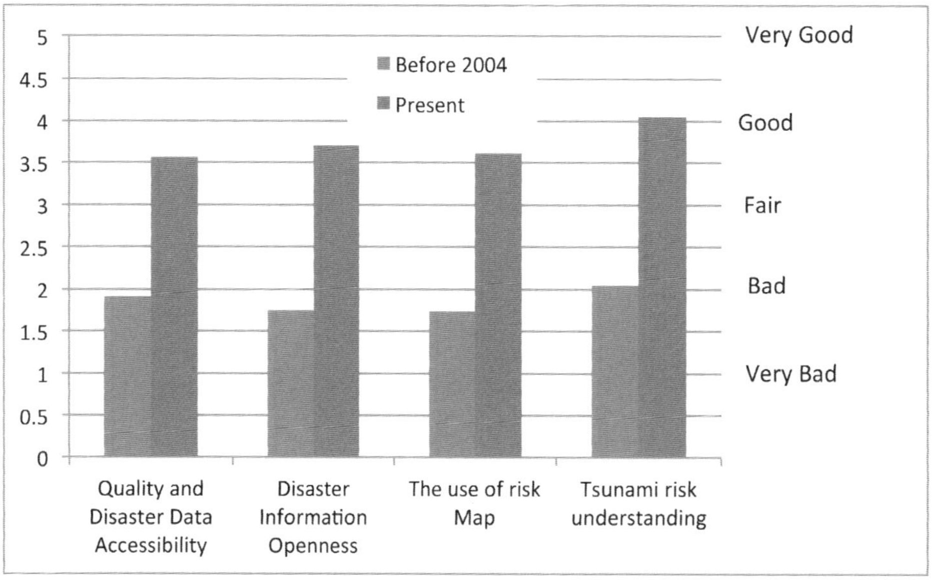

Figure 5. Perception of government staffs in Aceh on four elements in disaster.

Comparisons of the condition before 2004 and at the present were made individually by the respondents. Selected respondents were interviewed to understand reasons behind the perceptions. Quality and disaster data accessibility has been improved significantly after the introduction of Aceh Disaster Information and Data (DIBA) in 2011. The DIBA system was later implemented nationally by BNPB through its Indonesian Disaster Information and Data (DIBI). It has been made accessible to public through an online system. Efforts made by media and government agencies to report fast and accurate information on disaster events have also made the disaster information openness become better compared to the condition before the 2004 tsunami. Due to fast the development of internet infrastructures in Aceh, online media outlets have made the report of disaster event reach wider communities. Local printed newspapers in Aceh are not many. However, the circulation of the printed media every day has made possible to include disaster related news to communities who have no access to internet.

In 2011, Tsunami and Disaster Mitigation Research Center (TDMRC) of Syiah Kuala University, funded by Multi Donor Trust Fund through its program named Disaster Risk Reduction for Aceh (DRR-A), has collaboratively composed the first disaster risk map for Aceh. The map was later named as Aceh Disaster Risk Map which is now accessible through online (tdmrc.unsyiah.ac.id/drmis). The risk map has been considered in a number of development planning documents, such as Aceh Spatial Planning and Aceh Mid-Term Development Planning (RPJMA).

The perceptions on tsunami risk at the government staffs are considered at good condition. The 2004 Indian Ocean tsunami has taught many lessons learned to the staffs. Some significant changes in the government organizations and working flows have made them aware of tsunami risk posed by this region.

5. Conclusions

This study reveals that the 10-year tsunami recovery process has brought Aceh and its districts new practices in disaster management. The institutional changes and a number of policies/regulations made have enabled this region to demonstrate disaster risk reduction efforts. Notwithstanding the achievements, there are gaps identified at evacuation mechanism and facilities in several districts. The capacity of BPBA as the focal point agency for disaster management in Aceh has been upgraded significantly. Yet, serious challenges need to be highlighted at BPBA capacity to handle reconstruction and rehabilitation process should any disaster struck again. The rehabilitation and reconstruction process led by other agencies than BPBA is put under questions since the core function of BPBA also includes rehabilitation and reconstruction process. Overall present perceptions of Aceh Government staffs at provincial and district levels are in good level compared to the condition before the 2004 tsunami. Process of integrating disaster risk map into some of development planning documents has been performed. The accessible disaster risk map has also helped a number of government units to include the risk information into their working area.

6. References

[1] Coppola DP 2015 Participants-Governmental Disaster Management Agencies. *Introduction to International Disaster Management*. Elsevier, http://dx.doi.org/10.1016/B978-0-12-801477-6.01001-7

[2] Acht M, Mahmoud TO, and Thiele R 2015 Corrupt government do not receive more state-to-state aid: Governance and the delivery of foreign aid through non-state actors. *Journal of Development Economics* **114** p 20-33

[3] Hooper M 2014 Priority setting amid the rubble: Organisational approaches to post-disaster reconstruction in Haiti. *Intl. Journal of Disaster Risk Reduction* **10**(A) p 270

[4] Seng DSC 2013 Tsunami Resilience: Multi-level institutional arrangements, architectures and system of governance for disaster risk preparedness in Indonesia. Environmental & Policy **29** p 57-70

[5] Syamsidik, Rasyif TM, and Kato S 2015 Development of accurate tsunami estimated times of arrival for tsunami prone cities in Aceh, Indonesia. *Intl. Journal of Disaster Risk Reduction* **14**(4) p 403-410.

[6] Aceh Disaster Management Agency (BPBA) 2013 *BPBA Strategic Planning 2013-2017.* Banda Aceh, Indonesia

[7] Indonesia Disaster Management Agency (BNPB) 2013 *Actions Plan for Rehabilitation and Reconstruction Process after Earthquake in Central Aceh and Bener Meriah Districts*. BNPB, Jakarta, Indonesia
[8] Rapid Assessment Team (TIM) 2012 *Evaluation of Tsunami Early Warning System during Earthquakes and Tsunamis in Aceh on April 11, 2012*, in: E. Yulianto I. Rafliana(Eds.), published by GITEWS, Jakarta, Indonesia
[9] Rusydy I, Syamsidik, and Pagihariadi P 2015 Macroseismic mapping to estimate an earthquake source, A study case of Central Aceh Earthquake on July 2, 2013. *Proceedings of National Conference of Higher Education Forum for Disaster Risk Reduction*, Solo, 2015
[10] BMKG (Indonesia Geophysical, Climate and Meteorological Agency) 2012 *Tsunami Early Warning Service Guideline* (in Indonesian), Jakarta, Indonesia
[11] Syamsidik, Fahmi M, Al'ala M, and Tursina 2016 Tsunami wave impacts on coastal morphological changes around the Ulee Lheue Bay of Aceh, *Indonesia Proceedings of the International Offshore and Polar Engineering Conference*, p 796-800

Acknowledgments
Authors are grateful with the financial supports received from Aceh Disaster Management Agency (BPBA) to conduct 10-Year Aceh Tsunami recovery process in 2014. At the same time, authors acknowledge Syiah Kuala University supports to this research through its *Penelitian Unggulan Universitas (PUU)* Research Grant in 2015.

Inter-organizational network in Indonesia during disasters: Examples and research agenda on disaster management

M B F Bisri[1]

[1] PhD Candidate, Graduate School of International Cooperation Studies, Kobe University, 2-1 Rokkodai-cho, Nada-ku, Kobe-City, 657-8501 Japan. Contact: mizanbfbisri@stu.kobe-u.ac.jp

Abstract. Indonesia is facing various type of disaster risks, each with its own nature (sudden or slow onset, purely natural or man-made) and coverage of affected areas. Whereas science, technology and engineering intervention requires different modalities for each hazard, little has been known on whether the institutional setup and organizations involvement requires a different or similar types of intervention. Under a decentralized disaster management system, potential involvement of international organizations in response and growing diversified organizations involved in responding to disaster, it is important to understand the nature of inter-organizational network during various type of disasters in Indonesia. This paper is mixture of in-depth literature review and multiple case studies on utilization of social network analysis (SNA) in modelling inter-organizational network during various disasters in Indonesia.

1. Introduction

Indonesia is facing various type of disaster risks, each with its own nature (sudden or slow onset) and coverage of affected areas, and it is under an increasing trend under the influence of climate change. Record from 1900 until 2016 shows that there are 429 occurrences of disasters in Indonesia, resulted in total affected people of 29,011,349 and economic losses in 44,360,308 USD [1]. For Indonesia, geophysical disaster occurrences are less compare to hydro-meteorological disaster (40% against 60%), however geophysical disasters are more deadly although in terms of total affected people and total damage hydro-meteorological disaster is more dangerous [2]. Towards the future, from 496 cities/regencies, which being assessed on their disaster risks against nine type of hazards, 322 have been classified of having high risk and 174 have moderate risk to disasters [3]. Accordingly, Indonesian society requires advancement in disaster management to mitigate and lessen the risk, as well as to better responding in emergency situation induced by disaster.

Whereas the engineering mitigation approach requires different modalities for each hazard, little has been known on whether the institutional setup and organizations involvement requires different or similar types of intervention. Under a decentralized disaster management system, potential involvement of international response and growing diversified organizations involved in responding to disaster, it is important to understand the nature of inter-organizational network during various types of disaster in Indonesia. Djalante [2] has explored the trends of research on natural hazards, disaster risk reduction (DRR) and climate change adaptation (CCA) in Indonesia, by looking at 744 publications in the period of 1978 until 2015. It was found out that major research stream is on hazards, risks and disaster assessment, in comparison to DRR and CCA, with the most-frequent studied hazards are volcanic eruption, tsunami and earthquake [2]. Most of the research are also published in geography, earth sciences or inter-disciplinary journals on disaster, and less on social science, public administration, emergency management, and political science journals. Clearly, there is a need to investigate more on research of Indonesian disasters from the perspective of social sciences. Thus, Author would like to further promote the need to intensify socio-political research and analysis on disaster management in Indonesia.

Similar situation was observed in the United States over three decades ago. However, started from 1984 situation has change whereas through a joint effort between the Federal Emergency Management Agency (FEMA) and Network of Schools of Public Policy, Affairs and Administration (NASPAA), research in the field of disaster and emergency management become one of the mainstream in public administration, and at the same time there is growth in social science research in disaster studies [3]. There are five issues in disaster management which requires contribution from public administration field; i.e. inter-organizational coordination and collaboration, interoperability in communications, integrated approach, response and recovery, and recognition of the continuing vulnerability. It is in those five issues the scholars in public administration in the US mainly published their scholarly works. Thus, in the period from 1985 to 2009, the general categories of their publications fall into the following: flood hazard, hurricane, and earthquake mitigation; disaster management information technology; policy, communication, coordination, organizational learning; decision making under uncertainty, collaborative leadership, and presidential disaster declarations; intergovernmental relations; and national policy directions. Interesting for us in the developing countries to look at their future research agenda for public administration in relation to disaster studies. In addition to the aforementioned five research issues, future research agenda in the US includes the following; integration of geographic information system (GIS) with public administration architecture for disaster response and recovery; legal and organizational analysis for comprehensive disaster management; network analysis to detect functional and dysfunctional links in inter-organizational response and recovery systems following disaster; intensification of comparative case studies for understanding recurring hazards and reducing risk; sociotechnical systems and decision support which also include the area of incorporating social networking platforms; and complex adaptive systems [4].

2. Objective

The purpose of this paper is to assess whether or not type of hazard that trigger disasters has influence to the structure and characteristics of inter-organizational network pertaining to the disaster management. Alternatively, it intends to showcase that availability of policy and response framework, capacity of local government to respond specific disaster, status of disaster declaration that affect the presence of national government, and involvement of response from international organizations are more detrimental to the inter-organizational network during disaster. Pragmatically, this paper also would like to provide example of social network analysis (SNA) application in several disasters in Indonesia.

3. Inter-organizational network analysis on disaster situations using SNA

In nature, the paper is using a multiple case studies approach, although without an objective to directly compare one case to another. The selected cases are representing various disaster cases triggered by multiple hazards (earthquake, tsunami, flood, and volcanic eruption) and nature of its suddenness, i.e. sudden onset or slow onset. Another criterion on the selection of a case is on the availability of comparable inter-organizational network model from past research, since it is outside of the scope of this paper to generate new model. From the criteria, the following cases are assessed: the 2004 Aceh / Indian Ocean Tsunami, the 2009 West Java Earthquake, the 2009 West Sumatra Earthquake, the 2010 Mount Merapi Eruption and the 2014 Jakarta Flood.

This paper is limited on inter-organizational network modelling that employ social network analysis (SNA) as the main research tool. The SNA has the ability to represent the structure of relations between actors and to analyse a large number of relations within a network [5], which one of the fundamental properties is the ability to determine, through mathematical algorithms, whether network members are connected one to another, and to what degree, in various relationships [6].

This approach has been known for its versatility to answer many research queries in political and public administration aspects of disaster management; e.g., on the 2005 Hurricane Katrina [7], the 2010 Haiti Earthquake [8], the 2011 Great East Japan Earthquake and Tsunami [9] and the 2016 Nepal earthquake [10]. Table 1 exhibit the landscape of inter-organizational network analysis using SNA on disasters in Indonesia.

Table 1. Landscape of Inter-Organizational Network Analysis using SNA on disasters in Indonesia

Disaster case (Phase)	Key research question/objective	SNA measurements and software used
West Java Earthquake 2009 (Emergency response) [11]	In what ways, inter-organization cooperation creates network? How was the character of the network? To what extent the network can improve institutional arrangement of disaster management?	Degree centrality, betweeness centrality, closeness, density and cliques' identification SNA Software: Ucinet & Netdraw
Mount Merapi Eruption 2010 (Emergency response) [12]	How is the coordination structure between organizations during the emergency response of the 2010 Mount Merapi Eruption? Is there any organizational grouping during the emergency response and how was the relationship between organizations? Which are the key organizations during the emergency response period?	Average degree, network diameter, density, average path-length, network modularity, degree centrality and betweeness centrality SNA Software: Gephi
West Java Earthquake 2009, focus on housing reconstruction (Recovery) [13]	To understand the stakeholder's interaction process in housing reconstruction in Pangalengan district	Degree centrality, betweeness centrality, closeness centrality SNA Software: Ucinet & Netdraw
Great Jakarta Flood 2014 (Emergency response) [14]	To map groups and actors of government agencies during emergency response and grasp the coordination processes.	Average degree, network diameter, density, average path length, degree centrality, betweeness and modularity SNA Software: GEPHI
Aceh Tsunami 2004 (Recovery) [15]	What does the complexity landscape for a typical network of humanitarian aid for large-scale disasters look like? What does it mean for managing complexity in post-disaster governance?	Degree centrality, Betweeness centrality, k-core, diameter, average path length SNA Software: Pajek
Comparison between West Java & West Sumatra earthquakes 2009 (Emergency response) [16]	In what ways, did inter-organizational cooperation function and create networks during the emergency response of both earthquakes? What were the characteristics of cooperation in humanitarian operations? Which type of network performed better in the emergency response of both earthquakes?	Degree centrality and betweeness centrality, which weighted and modelled based on Principal-component, density and cliques' identification SNA Software: Ucinet & Netdraw

Aceh Tsunami 2004 (Recovery) [17]	To identify who populates the community networks and how they might be integrated with the formal, government networks responsible for a region's disaster management	Betweeness centrality (focus: current flow) and Conductance / Modularity; as proxy to measure social capital in community SNA Software: Visone
West Sumatra Earthquake 2009 (Emergency response) [8]	What is the role of transitivity, homophily and brokerage on interorganizational network formation during crises and how does a country's disaster management capacity influence the patterns of response network formation?	Transitivity, homophily and brokerage, with utilization of quadratic assignment procedure (QAP) in its network regression model and the use of conditional uniform graph (CUG) SNA software: Statnet

From the table, it can be seen that there are some inter-organizational network models that available for all five of the disaster cases. However, some of the models were done for community level and only focus one sub-sector, instead of organizational level. Accordingly, the inter-organizational network models analysed in this paper are the 2004 Indian Ocean Tsunami [15], the 2009 West Java Earthquake and the 2009 West Sumatra Earthquake, which the Author has conducted own network modelling [11] [16], the 2010 Mount Merapi Eruption [12] and the 2014 Great Jakarta Flood [14].

As such, for each case study, Author is observing comparable SNA measurements that produced by the original research: degree centrality, betweeness centrality, closeness centrality, and density. In simple term, centrality analysis refers to positions of a node (in the context of this paper an organization) within a network whereas the usual measurements are degree centrality and betweeness centrality. Degree centrality is calculated by counting the number of ties (or links) connected to a given node [5]. On the other hand, the betweeness centrality is calculated by counting the number times where a particular node locates in between another two nodes, which this indicate the tendency of functionality as coordinator or leader in a network [5]. For both measurements, the value may fall between 0 and 1, which the greater the value means higher centrality. On the visualization of the inter-organizational model of each case, author is using the original network for soft-comparison purpose. In addition, relevant SNA measurements such as density, average-path-degree and network diameter from the original research are also being mentioned for the interest of the readers.

4. Inter-organizational networks from selected disasters in Indonesia

4.1. Inter-organizational network on the 2004 Aceh Tsunami

The 2004 Indian Ocean Tsunami was one of the deadliest mega-disaster in modern history, which resulted in death of 250,000 people in Indonesia alone, mostly in coastal cities in Aceh and North Sumatra Province. For Indonesia, and Aceh province in particular, the tsunami stricken in the area where domestic conflict took place between the Free Aceh Liberation Movement (GAM) against the Indonesian Government. Accordingly, at the time of the tsunami there was minimal governmental apparatus operated in normalcy, in particular there was no disaster management agencies at local government levels nor (local) disaster response framework put in place at that time. In simple term, at that time the challenge was the absence of authority after such a large-scale disaster [15].

On this, the 2004 Indian Ocean Tsunami was also becoming one of the oldest disasters in Indonesia ever assessed [15] [17]. Jonathan Lassa's work [15] focus to understand the complexity through the use of network theory on post-disaster reconstruction at the organizational level, while Guarnacci [17] rather focuses at network at community level. Thus, this paper will mostly refer to the former one.

Lassa (2015) visualized the networks of organizations involved in the post-disaster recovery stage from 2005 to 2007.

Lassa (2015) used Gephi and Pajek's network analysis in modelling the network following the 2004 Aceh Tsunami. The network comprises of 797 nodes (organizations), 977 ties (links), which was derived from the database of Aceh-Nias Rehabilitation and Reconstruction Agency and particularly Lassa was analyzing the relationship between donor-partner data. As recorded, there were 472 international NGOs, 147 local NGOs, 25 multilateral organizations and 36 bilateral organizations involved in post-disaster reconstruction of Aceh between 2005-2007.

The inter-organizational network model for the 2004 Indian Ocean Tsunami was carried out using Pajek in mode 1 and treated as a directed network. It was found out the following network characteristics: average path length of 1.715, network diameter of 5, 28 number of loops. According to the analysis United Nations Development Program (UNDP) has the highest degree centrality and betweeness centrality, mainly due to their intermediary roles between donors, governments and civil society. Lassa also provided the sums of degree, outdegree and indegree centrality whereas the value of degree is significantly high for international NGOs (0.695), local/national NGOs (0.180) and multilateral organizations (0.133) in comparison to that of local governments (0.024) and national government organizations (0.012).

4.2. Inter-organizational network on the 2009 West Java Earthquake

The 2009 West Java Earthquake occurred on 2 September 2009 at 14:55 local time with magnitude 7.3 on the Richter scale. It affected 15 municipalities in West Java Province and resulted in death of 81 people, serious injuries to 1,287 people and displacement of 194,805 people. Moreover, according to post-disaster assessment, the the damage and loss was estimated of 7,905 billion IDR, with 89.79% being non-government assets and 10.21% being government assets [18].

On the 2009 West Java Earthquake, the author conducted an inter-organizational network modelling based on 343 emergency response activities that took place during the first-month after the earthquake, which compiled from situation reports published by National Disaster Management Agency (BNPB), United Nations Office for Coordinating Humanitarian Affairs (UNOCHA) and Indonesian Ministry of Health. It should be noted that officially the Governor of West Java, and supported by BNPB, set a two-weeks period of emergency response, and yet it was extended for another two weeks given the poor performance of the initial response, which also include acceptance of international aid and UN-lead activation of humanitarian clusters. In the end, there were 192 organizations involved in the emergency response, comprises of 50 local/national NGOs, 45 local government agencies, 22 international NGOs, 21 universities, 18 provincial government agencies, 10 central government agencies and 16 private companies.

Based on the data, the Author generated inter-organizational network model using UCINET version 6.4 that comprises of 192 nodes and 729 ties [16]. On the network, degree centrality and betweeness centrality measurements were performed to analyze the characteristic of actors within the network. In addition, SNA also calculated the network density, i.e. 5.005, and identified 37 inter-organizational cliques in the emergency response of the West Java Earthquake 2009.

On both degree centrality and betweeness centrality, the Satkorlak West Java ranked the highest (normalized degree centrality of 6.702 and normalized betweeness of 69.793). From central government side, the BNPB ranked second on degree centrality and fourth on betweeness centrality, which indicates that it has relatively high cooperation and yet lower level of coordination function, and Ministry of Health was ranked fifth on degree centrality and thirteenth on betweeness. Health Agency of West Java, on the other hand, ranked third from both degree centrality and betweeness centrality values. From the UN agencies side, UNOCHA was ranked fourth on degree centrality and eleventh on betweeness centrality while UNICEF was ranked eight on both measurements. Furthermore, in the remaining ranks of the top-fifteen organization with high value of degree centrality and betweeness centrality, it has more less representatives from international and national NGOs, local government agencies and military. Based on principal-component analysis, it was clear that the inter-organizational network in West Java was a lead-agency type of response [16].

4.3. Inter-organizational network on the 2009 West Sumatra Earthquake

The 2009 West Sumatra Earthquake occurred 30 September 2009 and measured on 7.6 Richter scale, it was followed with two major aftershocks and a subsequent landslide which destroyed three villages in Padang Pariaman Regency. The first aftershock measured 6.2 on the Richter scale 22 minutes later; and the second one measured 6.8 on the Richter scale with an epicenter located 225 km southeast of Padang city in the early morning of the following day. In total, the earthquake affected 13 cities and districts in the province. The government reported that the earthquake caused 1,117 deaths and 2,902 injuries [19]. In addition, there were 250,000 families (around 1,250,000 people) being internally displaced and affected by the earthquake. In total, there were 249,653 houses damaged; i.e. 114,797 homes severely damaged, 67,198 moderately damaged, and another 67,658 lightly damaged [19]. The total damage due to the earthquake is estimated at 19,369 billion IDR and total economic losses stand at 2,198.7 billion IDR.

On the 2009 West Sumatra Earthquake, the author conducted an inter-organizational network modelling based on 431 emergency response activities that took place during the first-month after the earthquake, which compiled from situation reports published by BNPB, UNOCHA and Indonesian Ministry of Health. It should be noted that officially the BNPB announced that the emergency response was set for two-months period with acceptance of international assistance and involvement of international organizations to co-lead the humanitarian clusters with provincial government agencies. Much of the decision was created based on performance in the emergency response of the 2009 West Sumatra Earthquake and the fact that the latter earthquake gave additional burden to the state resources. De facto, however, the emergency response was effective for one month, which is faster than the original plan. In the end, there were 223 organizations involved in the emergency response, comprises of 29 local/national NGOs, 40 local government agencies, 63 international NGOs, 12 agencies of the United Nations, 3 universities, 18 provincial government agencies, 17 central government agencies and 6 private companies.

Based on the data, the Author generated inter-organizational network model using UCINET version 6.4 that comprises of 223 nodes and 865 ties [16]. On the network, degree centrality and betweeness centrality measurements were performed to analyze the characteristic of actors within the network. In addition, SNA also calculated the network density, i.e. 5.259, and identified 40 inter-organizational cliques in the emergency response of the West Sumatra Earthquake 2009.

The result of degree centrality and betweness centrality in the 2009 West Sumatra Earthquake emergency response seems to support that humanitarian clusters approach was functional. In addition to UNOCHA (normalized degree of centrality at 4.762 and normalized betweeness centrality at 11.601), BNPB (2.902 and 8.054), and Satkorlak West Sumatra (2.530 and 7.652), among the top-fifteen of organizations with highest degree centrality, almost all of them are the lead agency in each cluster, i.e. UNICEF, International Organization of Migration (IOM), International Federation of Red Cross (IFRC), Save the Children (STC), United Nations Development Program (UNDP), Bappeda West Sumatra (provincial development planning agency), World Food Program (WFP), and Food and Agriculture Organization (FAO). Furthermore, by observing the normalized degree centrality and normalized betweeness centrality value of those organizations, it can be seen that the difference was not significant. Analyzed in its principal-component mode, the inter-organizational network in the 2009 West Sumatra Earthquake was in lead-partnership type of network whereas the network has several lead agencies represented by lead and co-lead agencies of humanitarian clusters, instead of single and powerful lead agency [16].

4.4. Inter-organizational network on the 2010 Mount Merapi Eruption

The 2010 eruption was the largest and most powerful in the past 100 years of Mount Merapi, with the affected areas of within 15 km radius. It resulted in the death of 277 people and the numbers of internally displaced people were 125,563 people [12]. The status of Mount Merapi was alleviated from the beginning of October 2010, and in particular large-scale evacuation was organized from 26 October 2010 and reached its peak on 5 November 2010. Officially the emergency response was defined for one month from 26 October until 25 November 2010, however due to the fluctuated and prolonged eruptions, thus the emergency response was extended until 9 December 2010.

Esterly (2012) created an inter-organizational network model of the 2010 Mount Merapi Eruption, which mainly derived from attendance list, coordination minutes of meeting and who-where-what matrix of emergency response activities documented by FPRB Yogyakarta (DRR forum of Yogyakarta province) that covers the emergency period from 26 October until 9 December 2010. It was found out that the network comprises of 177 organizations, with 39.98% of them are local/national NGOs followed by 19.21% local government agencies, 14.12% international NGOs, 6.21% education institutions, 4.52% central government agencies, 3.95% of UN agencies and private companies each, as well as 9.04% other organizations.

According to her model, from the interaction between 173 nodes, there were 340 ties identified. At the network level, it was found out that the network diameter is 9, average path length at 3.12, network density at 0.011 and average degree around 1.782. This implies that the network was rather large, only around 11% of total possible ties were formed and in order to coordinate an organization must through around three intermediaries. Esterly also identifies that 62.71% of the responders are extending organizations, meaning those which present at the community on a daily basis even though does not assume main responsibilities to handle disasters, and 29.38% are established organizations, i.e. those which established before the disaster and does have main responsibility in responding to a crisis.

Furthermore, to understand the structure of the network calculation on degree centrality and betweeness centrality was also provided. Based on degree centrality, the top-ten of organizations with high value were established organizations such as BNPB, Yogyakarta health agency, UNOCHA, Save the Children. The BNPB, as it was expected, has the highest degree centrality. However, interestingly, based on betweeness centrality, Oxfam GB has the highest value and followed by extending (local) organizations that was rather new to disaster emergency response, such as PSB, SAPDA, Plan, Hijau GPL and others. As a whole, the network shows that the disaster created a complexity where formal organizations that have been established before gave to adapt by structural change or alternation of tasks in interacting with new organizations in carrying emergency response [12].

4.5. Inter-organizational network on the 2014 Jakarta Flood

Jakarta Capital Region has a long-history of seasonal flood, whereas in the last ten years alone there were flood occurrences in 2007, 2013, 2014 and 2015. The 2014 Jakarta Flood reached the peak on 21 – 22 January 2014 where several rivers that flows through the capital could not handle the volume of water and created inundation area in 33 districts in Jakarta. In total, according to Jakarta BPBD, the flood affected 119,397 people with 89,334 of them have to take refuge in 307 evacuation points [14]. Various news reports also suggested that the economic losses reach IDR 5 trillion.

On its emergency response period, one network model during the 2014 Jakarta Flood was analyzed by Esterly and colleagues (2014) who focuses on inter-organizational and inter-personal coordination instead of cooperation during emergency response, of which the model was generated using survey result to individual responders [14]. This in itself is different and important network data compilation process in comparison to previous cases. However, albeit the production network models in several modes, the research of Esterly and colleagues (2014) does not clearly identified the number of nodes, both at organizational and individual levels.

Based on their measurements, the average degree of the network is 1.29, which implies actors only coordinate with one or two other actors during emergency response period. The network was also rather large, diverse and has a rather vertical relationship, indicated with the network diameter size, 8, and network density of 0.002. They also measured the average path length, 3.47, which implies for an information to travel to all nodes in the network, thus it generally travels through 3 – 4 intermediaries. Furthermore, in the inter-organizational network during the 2014 Jakarta Flood, there were 10 sub-groupings identified with modularity value at 0.485, which indicates that grouping during coordination was not random phenomena, yet representative to the field situation.

Based on the degree centrality that being calculated, the top-ten organizations with highest value are comprise of four local government agencies, Jakarta Disaster Management Agency (BPBD), Social Agency, Health Agency, Firefighter Agency, in addition to three district (kecamatan) offices and three sub-district (kelurahan) offices. While betweeness centrality value for organizations was not provided, Esterly and colleagues (2014) calculated degree centrality and betweeness centrality of individual actors within the network. They found that from both calculations, officers from BPBD Jakarta were dominating the individuals with high degree centrality and betweness centrality value. However, they noted that the results were rather based on "friendship and personal" relationship, rather than formal coordination channel.

4.6. Comparative perspective on inter-organizational network modelling from disasters in Indonesia
To this point, we have learned that there are several disasters in Indonesia of which inter-organizational network has been assessed. The network images can be found in Figure 1.

Prell (2012) noted that direct comparison of network is difficult to objectively conducted, if not impossible, due to rare identical two or more networks with same size. Therefore, to identify learning points from networks described in this paper, it is not on the real or normalized value of degree centrality and betweeness centrality, but rather on which organizations are recorded of having relatively high value on both measurements relative to the other nodes. Also, to what extent those organizations were established or recognized within the disaster management regulations and frameworks existed at the time of those disasters.

Among the five cases, only the 2004 Tsunami that occurred before the Law 24/2007 on disaster management enacted, which at that time disaster management in Indonesia was basically coordinated through the National Coordinating Agency on Disaster Management (Bakornas PB) and at the time of actual disaster there was no single agency that assume ultimate responsibility. However, as Lassa (2015) pointed out, it was not the Bakornas PB nor the BRR that recorded having the highest centrality and rather the UNDP. This clearly shows that at the time absence of national framework made no domestic organization capable of managing a mega-disaster at such magnitude.

On Aceh case, as Lassa himself noted, the inter-organizational network does not reflect the whole picture of how a coordinating agency played roles in the field as its much derived from the database of "who does what supported by whom". However, it does confirm the significant distance and indication that at least five intermediary agencies are needed in order to nurture a cooperative tie between organizations, i.e., as indicated that the diameter of humanitarian organizations network is 5. This post serious concern on how to reduce the structural gaps as well as to foster closer and foster ties between humanitarian organizations immediately after mega-disaster, even with the existence of humanitarian clusters. Nevertheless, Lassa's work (2015) was mostly on post-disaster reconstruction period took place in 2005-2007, and not the emergency response period itself. Thus, it is incomparable with other disaster cases due to a difference on disaster management stage.

IOP Conf. Series: Earth and Environmental Science 56 (2017) 012023 doi:10.1088/1755-1315/56/1/0120

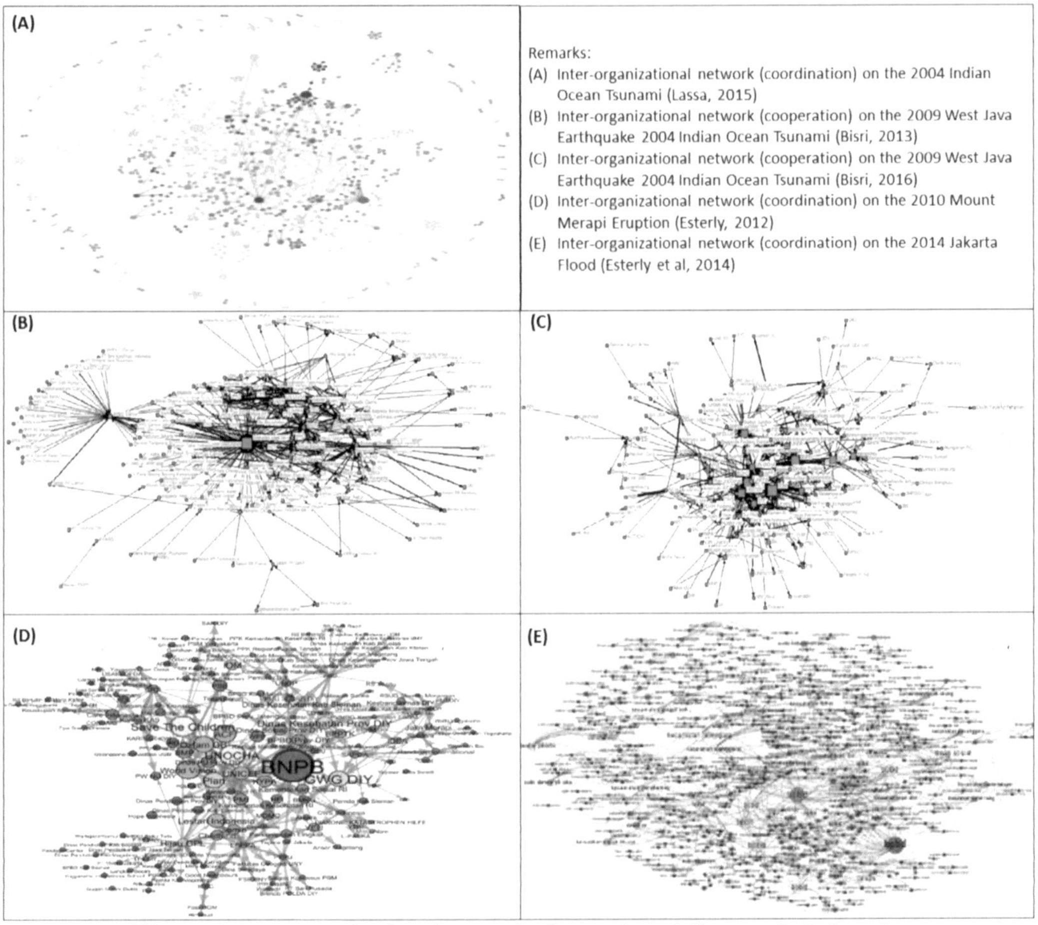

Figure 1. Inter-organizational networks from selected disasters in Indonesia

On the other hand, the other four disasters occurred after the Law 24/2007 was established and its subsequent detailed regulations, which mostly enacted through Head of BNPB regulation, including on acceptance of international aid (Head of BNPB Regulation 21/2008). Thus, the detrimental factor is whether or not the local government has established their BPBD and the political decision taken by both central and local governments on whether accepting international assistance and how it is managed. The West Java and West Sumatra earthquakes in 2009 and Mount Merapi Eruption in 2010 best explained this, particularly when the centrality values of BNPB are scrutinized. In all three disasters, the provincial governments have not established BPBD or its BPBD was still infant, yet BNPB decision was different in all three. In West Java, BNPB was ranked second due to its decision to support provincial government, which already determined response plan through West Java Governor Decree SK 360/kep.1260-Hukham/2009. Yet, both Satkorlak West Java and BNPB can be seen as the centre of the network, due to its significant difference with other organizations. On the other hand, when the 2009 West Sumatra Earthquake occurred the BNPB and provincial government agreed to welcome international assistance immediately after the earthquake and consequently the UN-led humanitarian clusters were activated. Accordingly, as shown the network was rather formed a lead-partnership type where humanitarian clusters lead agency and co-lead agency led the network, mostly one government agency and one non-government (UN agency or international NGOs). In the case of the 2010 Mount Merapi Eruption, the BNPB took leadership, and consequently ranked with

IOP Conf. Series: Earth and Environmental Science **56** (2017) 012023 doi:10.1088/1755-1315/56/1/012023

high centrality values. However, still many local organizations and international humanitarian organizations have high centrality values above the local government agencies, due to their high level of engagement in the emergency response.

The 2014 Jakarta Flood was at a different scale in comparison with the other cases whereas the repetitive disaster classified as a local (provincial) disaster. On this, with more mature organizations and better local government capacities, as expected, Jakarta BPBD and three other agencies (health, social and fire fighter) recorded high values on both centrality measurements. This indicated a working emergency response framework and social trust to those local government agencies in responding to disaster. For a more locally managed disaster, it is important to further strengthen the capacity of BPBD and other agencies in the local government for managing emergency situation and enhance ties with other non-government organizations that potentially contribute in emergency response.

In the end, none of the example above map the inter-organizational network within the total disaster management cycle, which in reality may provide hints on the societal and institutional capacity of the area throughout the overall disaster management cycle. Author in this paper is showcasing preliminary attempt to showcase such approach on inter-organizational network in West Java Province. Here, author built a database of all DRR programs and projects in the province two years before the 2009 Earthquake, use the initial model of inter-organizational network during emergency response and mapped post-disaster reconstruction activities two years after the earthquake, in similar fashion of that of Lassa (2015). The result can be seen in Figure 2. It clearly can be seen that before the earthquake, due to absence of provincial BPBD, the network was not fully connected and there are two big hubs on DRR programs and projects, i.e. the West Java Province Development Planning Agency (Bappeda) and Bandung Institute of Technology (ITB). On the other hand, the network during recovery stage, interestingly does not indicate a pure continuation of the network during emergency response. Rather, it was centred around UNDP (agency of the UN), BPBD West Java, which eventually formed in 2010, and BNPB (central government agency), thus similar to that of in the 2004 Tsunami (Lassa, 2015). While, the emergence of BPBD West Java shows a change in institutional setup, it is important to further assess on next big disaster in West Java whether the BPBD continue assume the leadership.

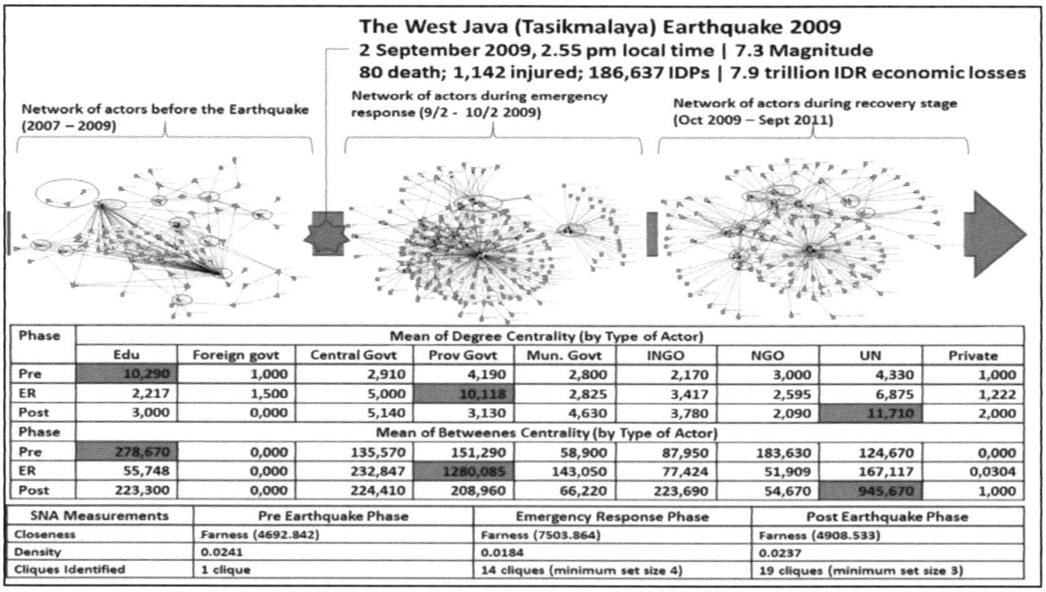

Phase	Mean of Degree Centrality (by Type of Actor)								
	Edu	Foreign govt	Central Govt	Prov Govt	Mun. Govt	INGO	NGO	UN	Private
Pre	10,290	1,000	2,910	4,190	2,800	2,170	3,000	4,330	1,000
ER	2,217	1,500	5,000	10,118	2,825	3,417	2,595	6,875	1,222
Post	3,000	0,000	5,140	3,130	4,630	3,780	2,090	11,710	2,000
Phase	Mean of Betweenes Centrality (by Type of Actor)								
Pre	278,670	0,000	135,570	151,290	58,900	87,950	183,630	124,670	0,000
ER	55,748	0,000	232,847	1280,085	143,050	77,424	51,909	167,117	0,0304
Post	223,300	0,000	224,410	208,960	66,220	223,690	54,670	945,670	1,000

SNA Measurements	Pre Earthquake Phase	Emergency Response Phase	Post Earthquake Phase
Closeness	Farness (4692.842)	Farness (7503.864)	Farness (4908.533)
Density	0.0241	0.0184	0.0237
Cliques Identified	1 clique	14 cliques (minimum set size 4)	19 cliques (minimum set size 3)

Figure 2 Inter-Organizational Network in West Java Province before, during emergency response, and after the 2009 Earthquake (Created by author)

5. Conclusions and Research Agenda for Disaster Management in Indonesia

This paper confirms that existence of policy and response framework, capacity of local government to respond specific disaster, status of disaster declaration that affect the presence of national government, and involvement of response from international organizations are detrimental factors to the inter-organizational network during disaster; and not solely the type of hazard. However, the time dimension that bring about by each type of disaster may provide different implications to the network formation. Therefore, in addition to the need to continuously develop local (organization) capacity to multiple hazards, specific resources, policy and organizational rosters need to be prepared for the most urgent and frequent hazard of each localities in Indonesia.

Throughout the paper, Author has briefly discussed the development and utilization of SNA as analytical tool for analysing inter-organizational network, which further grow as a mixture of political science, public administration, and sociology. SNA as a tool can now be utilized by political scientist and public administration scholar to manage and visualize big data of inter-organizational relationship in many development sectors, including disaster studies and disaster risk management sector. To repeat Berry and colleagues (2004), the future importance is to contextualize SNA result on various development sector, including disaster studies [20]. On the other hand, modification and improvement on the way public administration practices (such as in reporting and data management) are needed to feed quality data for SNA.

To further strengthen an evidence-based policy making, SNA can also contribute significantly with its characteristics, as proven in the three examples provided here; i.e. 1) it can be done to analyse nationwide network for shading lights on institutional trend organizational involvement at the time of disaster, 2) it can be used to track dynamic of inter-organizational relation in a particular development sector over time, as in the case of West Java, 3) it can be used for a comparative study within a country, and possible extended to a cross-country comparison, given the objective view of the network, as in the case of West Java and West Sumatra in Indonesia. All five research examples in this paper gave evidence that SNA can be used for policy making on disaster risk management, particularly on governance design, both at local and national level. For a more evidence-based policy making and institutional design, Author would like to call upon the importance of intensifying political science and public administration research using SNA, and network modelling in general for disaster studies in Indonesia.

6. References

[1] EMDAT, "The International Disaster Database," 2016. [Online]. Available: http://www.emdat.be/country_profile/index.html. [Accessed 21 September 2016].

[2] R. Djalante, "Recent trends on hazards, disasters, risk reduction and climate change in Indonesia: a systematic literature review," *Natural Hazards and Earth System Sciences,* pp. 1-48, 2016.

[3] BNPB, "Indeks Risiko Bencana Indonesia (Indonesian Disaster Risk Index)," BNPB (Badan Nasional Penanggulangan Bencana - National Disaster Management Agency), Jakarta, 2013.

[4] L. K. Comfort, W. L. Waugh and B. A. Cigler, "Emergency Management Research and Practice in Public Administration: Emergence, Evolution, Expansion, and Future Directions," *Public Administration Review,* vol. 72, no. 4, pp. 539-548, 2012.

[5] C. Prell, Social Network Analysis - History, Theory, and Methodology, London: Sage Publications, 2012.

[6] D. Varda, R. Forgette, D. Banks and N. Contractor, "Social Network Methodology in the Study of Disasters: Issues and Insights Prompted by Post-Katrina Research," *Population Resolution Policy Review,* vol. 28, pp. 11-29, 2009.

[7] N. Kapucu and T. Arslan, "Examining Intergovernmental and Inter-organizational Response to Catastrophic Disaster: towards a Network-Centered Approach," *Journal of Administration and*

Society, vol. 42, pp. 222-247, 2010.

[8] M. D. Siciliano and C. Wukich, "Network Formation During Disasters: Exploring Micro-Level Interorganizational Processes and the Role of National Capacity," *International Journal of Public Administration,* pp. 1-14, 2016.

[9] M. B. F. Bisri, "Observing partnership innovation through Inter-organizational Network Analysis on Emergency Response of the Great East Japan Earthquake and Tsunami 2011," *Japan Social Innovation Journal,* vol. 6, no. 1, pp. 27-41, 2016.

[10] M. B. F. Bisri and S. Beniya, "Analyzing the National Disaster Response Framework and Inter-Organizational Network of the 2015 Nepal/Gorkha Earthquake," in *Procedia of Engineering* , Boston, 2016.

[11] M. B. F. Bisri, "Examining Inter-Organizational Network during Emergency Response of West Java Earthquake 2009, Indonesia," *Procedia Environmental Sciences,* vol. 17, pp. 889-898, 2013.

[12] M. Estrely, "Analisis koordinasi antar organisasi pada fase tanggap darurat erupsi merapi 2010 dari perspektif jejaring sosial," Universitas Pertahanan Indonesia, Jakarta, 2012.

[13] F. Situngkir, "Karakteristik jejaring interaksi antar aktor dalam rekonstruksi rumah pasca bencana di Kecamatan Pangalengan (in Bahasa Indonesia)," *Jurnal Perencanaan Wilayah dan Kota A,* pp. 117-125, 2014.

[14] M. Estrely and W. Cahyono, "Mapping groups and key agents in emergency response system," 4 10 2014. [Online]. Available: http://www.peringatanbulanprb.net/wp-content/uploads/2014/10/04-Koordinasi-Tanggap-Darurat-dari-Perspektif-SNA.pdf. [Accessed 1 October 2016].

[15] J. Lassa, "Post Disaster Governance, Complexity and Network Theory: Evidence from Aceh, Indonesia After the Indian Ocean Tsunami 2004," *PLoS Current Disasters,* 2015.

[16] M. B. F. Bisri, "Comparative study on inter-organizational cooperation in disaster situations and impact on humanitarian aid operations," *Journal of International Humanitarian Action,* pp. 1-14, 2016.

[17] U. Guarnacci, "Joining the dots: Social networks and community resilience in post-conflict, post-disaster Indonesia," *International Journal of Disaster Risk Reduction* , pp. 180-191, 2016.

[18] Bappenas, "Rencana Aksi Rehabilitasi dan Rekonstruksi Wilayah Pasca Gempa bumi di Provinsi Jawa Barat dan Kabupaten Cilacap, Provinsi Jawa Tengah 2 September 2009 (Action Plan Rehabilitation and Reconstruction of Post West Java Earthquake 2 September 2009)," Bappenas (Ministry of Development Planning), Jakarta, 2009.

[19] BNPB, "West Sumatra and Jambi Natural Disasters: Damage, Loss, and Preliminary Needs Assessment," BNPB, Jakarta, 2009.

[20] F. S. Berry, R. S. Brower, S. O. Choi, W. X. Goa, H. Jang, M. Kwon and J. Word, "Three Traditions of Network Research: What the Public Maangement Research Agenda Can Learn from Other Research Communities," *Public Administration Review,* vol. 64, no. 5, pp. 539-552, 2004.

Acknowledgments
The author would like to acknowledge all organizations that supported financially to the overarching research of this paper, the Japan Ministry of Education Sport and Technology (MEXT) scholarship and Ryoso Award of Kobe University Alumni Association of Social Sciences (Ryosokai). For the content of the paper, the author would like to thanks Ms. Martina Esterly, who has share the SNA results of the 2010 Mount Merapi and the 2014 Jakarta Flood, and reviewer from the Secretariat of the 2016 AIWEST-DR.

Disaster budgeting of Banda Aceh's local government: Trends and analysis of post-tsunami Aceh 2004

R S Oktari[1,*], H Fahlevi[2] and W Irawati[3]

[1]Tsunami and Disaster Mitigation Research Center (TDMRC) and Faculty of Medicine, Syiah Kuala University, Jl. Tgk. Syech Abdul Rauf, Darussalam, Banda Aceh, 23111, Indonesia
[2] Accounting Department, Faculty of Economics and Business, Syiah Kuala University, Jl. Tgk. Syech Abdul Rauf, Darussalam, Banda Aceh, 23111, Indonesia
[3] Management Department, Faculty of Economics and Business, Syiah Kuala University, Jl. Tgk. Syech Abdul Rauf, Darussalam, Banda Aceh, 23111, Indonesia

*email: okta@tdmrc.org

Abstract. This study aimed to analyze Disaster Risk Management (DRM) related funds that budgeted by the government of Banda Aceh. The specific objectives of this study were: i) to assess and analyze the budget allocation related the DRM investment in the government of Banda Aceh , ii) to provide an update on achievements and key trends in DRM investments in government of Banda Aceh, iii) to evaluate the implementation of DRM budget and challenges on mainstreaming with DRM perspectives into policy of local budgeting, and iii) to propose appropriate recommendations for improvement. The study utilized both primary and secondary sources of data to achieve the objective. The method in this study included the following steps: searching/ collecting, checking, compiling, classifying, measuring, accounting and analyzing the existing budget document. The results of the data analysis showed that a large proportion of DRM investments was allocated after 2011. This study recommended some actions to be taken by the government and related stakeholder to increase the quantity and quality of DRM investments.

1. Introduction

Indonesia has witnessed numerous events of natural disasters. Indonesia Disaster Data and Information (*Data dan Informasi Bencana Indonesia*/ DIBI) has reported that between 1990 – 2016, around 9.7 million peoples were displaced as a result of natural disasters, including droughts, tsunami, earthquakes, floods, volcanoes, epidemics and social conflict [1, 2]. DIBI also reported that Aceh has the vulnerability to disasters in almost all regions across it. Not only prone to earthquakes and tsunamis, but Aceh also suffers from floods and landslides. Floods continue to be the most frequent catastrophes occur annually [2].

Banda Aceh, the capital city of Aceh, is also prone to several natural disasters, including flood, earthquake, cyclone and tsunami. During the last major tsunami in the Indian Ocean, Banda Aceh was one the most affected areas hit by the giant wave of 26 December 2004. The massive surge destroyed around 21.751 unit houses, 169 units of education facilities, 25 units of health facilities, 302 km roads, 63 units of governments offices and 46 social facilities [3].

Considering the fact that Banda Aceh is prone to natural disasters and also was one of the most affected areas during tsunami 2004, greater investments in Disaster Risk Management (DRM) is highly required. The importance of DRM investments has been highlighted in the Sendai Framework for Disaster Risk Reduction (SFDRR) 2015-2030, under the Priority 3. The SFDRR endorses the government and other stakeholders to pay more attention to the DRM investments for resilience [4]. Governments need to invest in DRM to avoid losses and reduce the impacts of a disaster. All development planning should integrate proper disaster resilience measures [5].

2. Purpose

This study aimed to analyze DRM related funds that budgeted by the government of Banda Aceh between 2007 and 2015. The specific objectives of this study were: i) to assess and analyze the budget allocation related the DRM investment in the government of Banda Aceh, ii) to provide an update on achievements and key trends in DRM investments in government of Banda Aceh, iii) to evaluate the implementation of DRM budget and challenges on mainstreaming with DRM perspectives into policy of local budgeting, and iii) to propose an appropriate recommendations for improvement.

3. Methods

The study utilized both primary and secondary sources of data to achieve the objective. The primary sources of data were collected from interviews with key government officials and stakeholders to obtain an insight of disaster budgeting practice, mechanism and its problem and challenges. The secondary sources of data consisted of the annual local government budget (APBD) of Banda Aceh government from 2007 to 2015 and other related documents. The method in this study included the following steps: searching/ collecting, checking, compiling, classifying, measuring, accounting and analyzing the existing budget document.

The classification of the activity in the budget data was reviewed and accounted as DRM investments based on the Government Regulation No. 21 of 2008 on the Implementation of Disaster Management. The study explores two different types of investments: stand-alone DRM investments (such as mitigation, risk assessment, preparedness, early warning, etc.) and mainstreamed or embedded DRM investments in the sector (such as health, education, agriculture, public works, etc.) [7, 8].

4. Findings and Discussion

4.1. Legal and Policy Framework for DRM

Laws and regulations are a necessary foundation for building disaster resilient communities and nations. They are essential to creating an enabling environment for reducing risks posed by natural hazards, preventing new risks from arising and making communities safer [8]. The importance of legal and policy framework to support DRM program has been highlighted in the Hyogo Framework for Action (HFA) 2005-2015 [9]. The SFDRR 2015-2030 has also called for a renewed focus on strengthening legal frameworks for DRM [4].

Indonesia has achieved significant progress on legal frameworks for disaster management, by enacting Disaster Management Law No. 24/ 2007 and several regulations both at national and local level. Indonesian Disaster Management Law No. 24/ 2007 stipulates the importance of integrating DRR into the policy framework of government for all phases of disaster management, including mitigation, preparedness, emergency response and recovery [10].

As shown in Figure 1, Disaster Management system in Indonesia has a comprehensive regulating system which includes laws/ regulation, planning, institution, budgeting, capacity development, including science and technology, to ensure a well-integrated and coordinated implementation of DRM program [11].

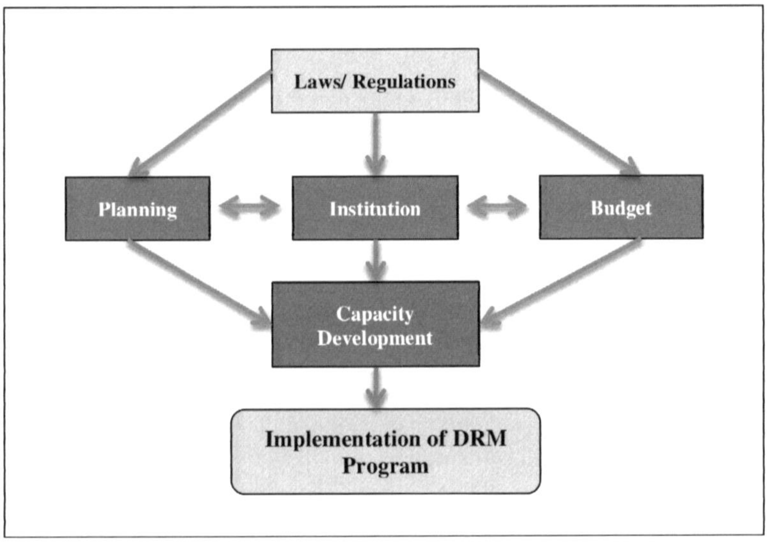

Figure 1. Disaster Management System in Indonesia
(Sources: Adopted from BNPB, 2011)

In order to provide more specific operational detail on the application of the general principles, BNPB currently has more than 80 additional specific regulations. There are also several Ministerial Regulations have been enacted to strengthen DRM measures in Indonesia (please see Table 1).

Table 1. National Regulations to strengthen DRM Measures

No	Regulation	Content
1.	Disaster Management Law No. 24/ 2007	Sets out important legal regulations on disaster management
2.	Government Regulation No. 21/ 2008	Stipulates the implementation of disaster management and the enforcement of Local Action Plan for DRM
3.	Government Regulation No. 22/ 2008	Funding and Management of Disaster Assistance
4.	Government Regulation No. 23/ 2008	Participation of International Institutions and Foreign Non-Government Institutions in Disaster Management
5.	Presidential Regulation No. 8/ 2008	Establishment of BNPB
6.	Ministry of Home Affairs Regulation No. 131/ 2004	Stipulates the disaster management in the Local Level
7.	Ministry of Home Affairs Regulation No. 46/ 2008	States that every region should develop a BPBD and provides manual on tasks, organizational structure, and operations of BPBD
8.	Head of BNPB Regulation No. 3/ 2008	Provides Guidelines for the Establishment of BPBD

(Sources: Extracted from BNPB, 2011)

The local regulations on DRM in Aceh have been established and endorsed by Governor Regulation and Governor Decree as shown in Table 2 below. These regulations provide legal frameworks for the

IOP Conf. Series: Earth and Environmental Science 56(2017) 012024 doi:10.1088/1755-1315/56/1/0120

Provincial Government to implement DRM and make DRM measures as part of development planning instrument. These regulations also provide institutional arrangements, supports the government to manage pre, during and post disaster, and helps to develop better mitigation and preparedness efforts to face future disaster.

Table 2. Local Regulations to strengthen DRM Measures

No	Regulation	Content
1.	Qanun No.5/ 2010 on Disaster Management	Adds important information access and advisory role of the communities in the DRM decision making process
2.	Qanun No.6/ 2010 on the Establishment, Organizational Structure and Management of the BPBA	Provides detail information on the Establishment, Organizational Structure and Management of the Aceh Disaster Management Agency (BPBA)
3.	Governor Regulation No.43/ 2010 on Standard Operating Procedures for Tsunami Early Warning System (TEWS) in Aceh	Equips the government with standard mechanism for increasing tsunami preparedness
4.	Governor Regulation No. 51/ 2011 on Provincial Disaster Management Plan for 2012-2017	Provides more strategic planning for Disaster Management in Aceh
5.	Governor Regulation No.48/2010 on Local Action Plan on DRM for 2010-2012	Provides comprehensive approach to DRM measures in dealing with various hazards in Aceh
6.	Governor Decree No. 360/6a/2011 on the establishment of the DRM Coordination Forum	Provides legal basis for the establishment of DRM forum in the form of multi-stakeholder forum mandated to assist the Government of Aceh in advancing DRM measure
7.	Governor Instruction No. 2/INSTRK/2012 on 28 May 2012	Provides a legal basis to implement DRM curriculum in schools
8.	Qanun of Banda Aceh No. 3/ 2011 on the Establishment, Organizational Structure, and Management of the BPBD Banda Aceh	Provides a legal basis to establish Local Disaster Management Agency (BPBD) in Banda Aceh

(Sources: Extracted from UNDP, 2012)

4.2. National Budget Allocation for DRM

At the national level, total government investment for DRM activities has increased significantly from IDR 2.6 trillion in 2006 to almost IDR 9 trillion in 2011. In fact, the actual DRM investments at the national level are greater since some activities are embedded in other sectoral programs. In the period 2006 to 2012, 75% of DRM investments were allocated for disaster mitigation and prevention activities, followed by disaster preparedness activities (13%) and research, education, and training (6%). Also, a budget classification for stand alone DRM has been developed to facilitate the monitoring of the DRM investments expenditure and outcomes [8,11].

Although most of the central government institution have invested some funding for DRM purposes, the average of DRM investment is less than 1 %, which is still lower than the internationally accepted ratio (1,2% of national budget) [8].

DRM has been integrated into policy framework at the national government level. Policies related to DRM have also been instituted in the Medium Terms and Annual Development Plans (2004-2009 and 2010-2014). However, some Ministries have not included DRM principles in their Strategic Plan or their Annual Working Plan [8].

4.3. Local Government Budget Allocation for DRM

This paper analyzed the budget allocation related to DRM investment in the local government of Banda Aceh between 2007 and 2015. The data was collected from revised budgets of the local government of two departments. Previously, DRM-related program was only found in Office of Development Welfare and Civil Protection *(Badan Kesejateraan Pembangunan dan Perlindungan Masyarakat/* Kesbanglinmas. After 2011, the government created Local Disaster Management Agency (*Badan Penanggulangan Bencana Daerah*/ BPBD). This department was originally a fire department.

BPBD Banda Aceh was established in 2011 through Local regulation or Qanun No. 3/ 2011 on the Establishment, Organizational Structure and Management of the BPBD [12,13]. The BPBD organizational structure as shown in Figure 2, support the statement in the Article 24 of Disaster Management Law No.24/ 2007 which stipulates that the local disaster management agency shall carry out the functions during a pre-disaster period, emergency response and post-disaster period [10].

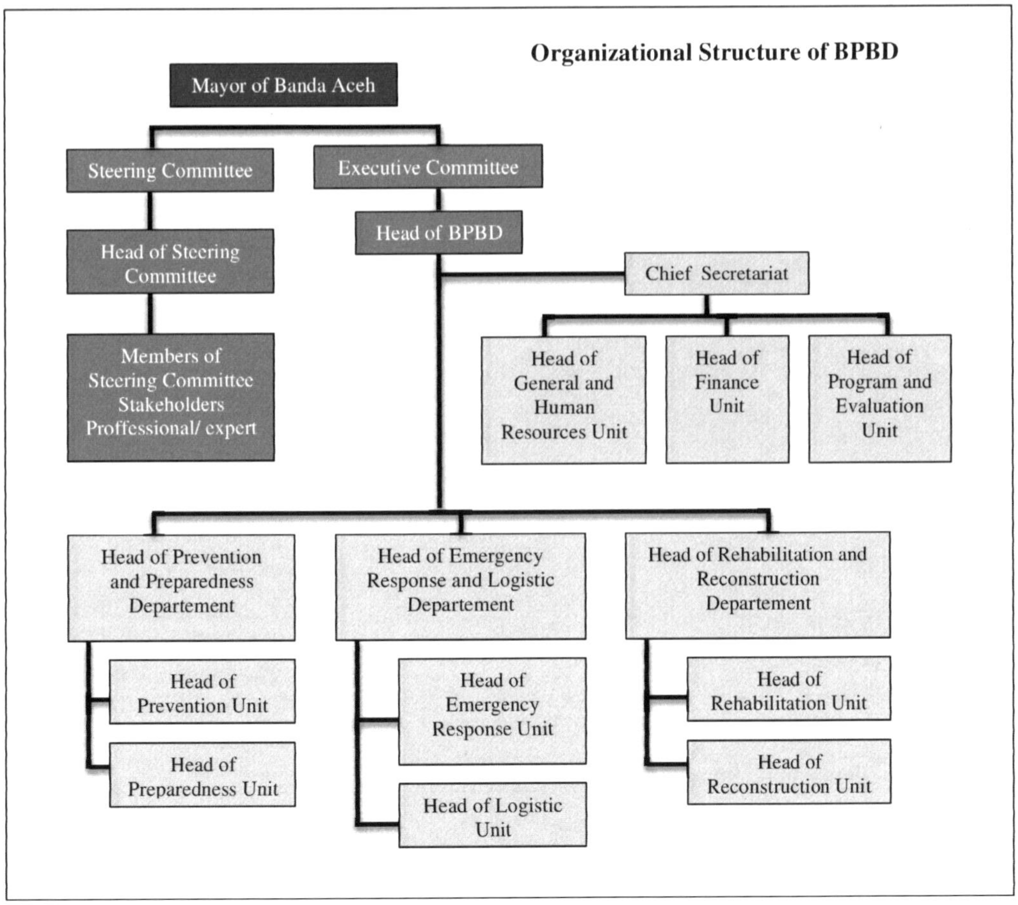

Figure 2. Organizational Structure of BPBD
(Sources: Adopted from Qanun No.3/ 2011)

4.4. Budget Allocation related to DRM in Banda Aceh

According to the data analysis shown in Table 3, the budget portion for DRM investment has been relatively very small compared to the total local budget or other department budgets. Before 2011, the DRM related programs were accounted for less than 0.06% of the total government budget. After 2011, the budget for disaster-related programs has not been changing substantially. It accounted not more than 0.60% of the total budget of Banda Aceh.

Table 3. DRM related program allocation as a percentage of the total local government budget

Year	Total Local Government Budget (in IDR)	Total DRM-related Budget (in IDR)	Percentage
2007	532.046.733.984	152.898.000	0.03 %
2008	500.040.754.837	245.544.250	0.05 %
2009	555.362.366.232	188.813.800	0.03 %
2010	589.258.259.907	78.246.200	0.01 %
2011	761.313.118.850	76.667.500	0.01 %
2012	825.234.713.764	1.034.180.976	0.13 %
2013	985.975.685.419	1.518.104.776	0.15 %
2014	1.194.481.738.232	5.226.884.820	0.44 %
2015	1.329.366.899.560	7.871.688.250	0.59 %

(Source: Extracted from APBD of Banda Aceh)

As a percentage of the total budget, the local government funds allocated for DRM amounted only 0.03% in 2007, and increased steadily to the level of 0.59% in 2015 (see Figure 3). Overall, the figure show an increased local government investment in DRM programs and activities.

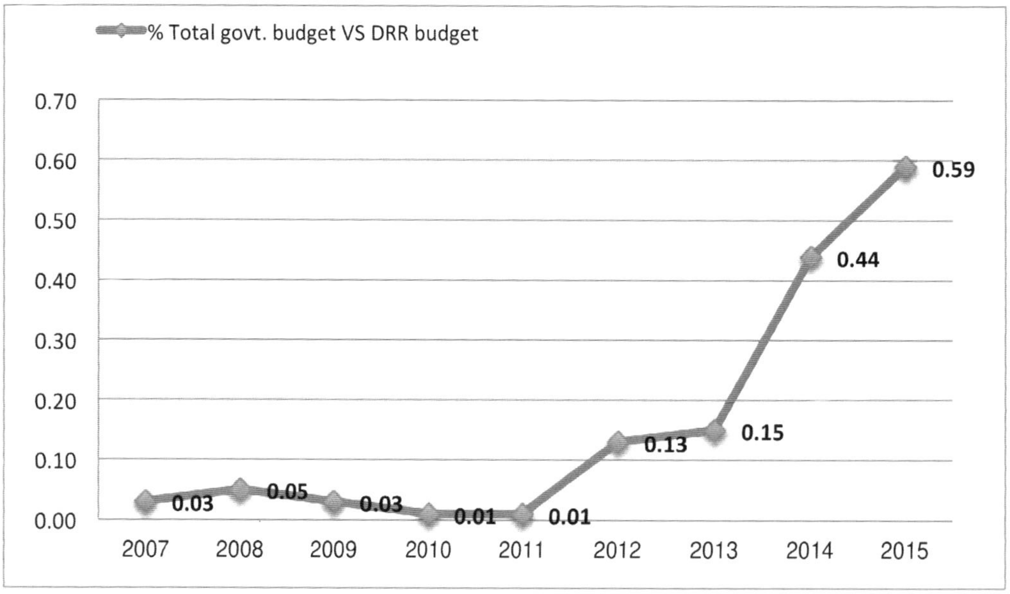

Figure 3. Trend in DRM investments in Banda Aceh
(Sources: Extracted from APBD of Banda Aceh)

IOP Conf. Series: Earth and Environmental Science 56(2017) 012024 doi:10.1088/1755-1315/56/1/01202

However, the DRM-related programs can only be found in *Kesbanglinmas* or BPBD budget as it is one of their core functions/ responsibility. This condition indicated that the disaster budget was allocated only in one department, rather than cross-department. For example, DRM related program was not found in health or education department. Although theoretically, these departments could play a significant role in DRM risk mitigation.

In addition, the DRM-programs that have been proposed were mainly after disaster/ recovery-oriented than preventive. Most of the budgeted program are from the procurement of temporary shelters, evacuation process during a disaster and early warning on disaster for the community. After 2011 (BPBD in charge for DRM program), the government focused more on the fire disaster as BPBD was originally a fire department. In this period, the term 'disaster' in the government budget was related only to fire disaster preparedness and prevention. The term 'disaster' was not found in other departments' budget except in BPBD budget. However, the term 'emergency' and 'extraordinary situation/ catastrophes' can be found in social department. In the social department, there was a program related to disaster e.g. response for emergency and extraordinary situation program. In this program, the social department budgeted IDR 15.000.000 to IDR 17.000.000/ year for procuring clothing for disaster victims.

Through some interviews with key officers in BPBD, insight of disaster budgeting in Banda Aceh government has been obtained. The budgeting mechanism was started from budget request/proposal prepared by each department. These budget request will be collected by local budget preparation team which consist of representatives from Development Planning Agency (Badan Perencanaan dan Pembangunan Daerah/ BAPPEDA) and department of finance. The budgeting team reviewed the budget request of each department before combining and preparing a local budget draft. The local budget draft will be discussed further with the head of government prior to the approval process by the legislative.

Mostly, BPBD received a smaller budget that it is proposed due to the financial constraint of local budget. Indeed, the BPBD approved budget can only be used for equipment maintenance and other routine programs. As a result, BPBD has not conducted any DRM program particularly DRM mitigation program for the last few years. According to the interviewees, most of DRM program that has been implemented were financed by Disaster Management Agency at National and Province level (BNPB and BPBA) or international NGOs. These bodies have more financial sources than the local BPBD. In fact, only few DRM related program that has been implemented after 2004 Tsunami. During the Tsunami recovery period, in fact, it was the NGOs played a vital role on DRM program, not the local government. Thus, the DRM issues and concerns seem failed to penetrate to local government budgeting and other departments so far.

Another issues uncovered is the lack of capacity of the programmer in BPBD. Most of the key officers have no educational background or prior training on disaster science. Consequently, they do not have the required-capacity to propose or create DRM related program particularly in preventing and mitigating DRM risk.

In a nutshell, disaster budgeting in Banda Aceh government is still not a budget-priority. Not only a small number of disaster budget, but also the proposed program is not adequate and appropriate for a region prone to disasters.

5. Conclusions

Disaster is difficult to predict and allocating a speculative amount of money in the budget system may lead to a lower budget's credibility and usefulness [14]. This argument could help us to understand why the government reduces its budget allocation for the disaster-related program. However, the analysis showed that a large proportion of DRM investments was allocated after 2011. The establishment of BPBD of Banda Aceh contributed to the increasing government investment in disaster management activities. As stipulated by the law, all local governments are required to establish a disaster management agency. After 2011 the figures show an increased budget allocation for DRM, particularly as a stand-alone DRM investment. Indeed, the BPBD approved budget was only be used for equipment maintenance and other routine program. BPBD has not conducted any DRM program particularly DRM mitigation program for the last few years using their own budget.

Investing in DRR yields multiple benefits as it builds a resilient future. It helps avoid losses when disaster strikes. Therefore, decision makers need to prioritize investments to build resilience, although these do not produce immediate benefits. This study provided useful data and highlighted the trends related to resource allocation for DRM investments in Banda Aceh. This study recommended some actions to be taken by the government and related stakeholder to increase the quantity and quality of DRM investments, including: i) capacity development in planning and budgeting for BPBD personnel, ii) formulate a sectoral policy on DRM to be adopted by other sector in the local governments (such as health, education, agriculture, public works, etc.), and iii) strengthen coordination with related to the sectoral institutions.

References

[1] BNPB 2015 *Indonesia's Disaster Risk Mangement Baseline Status Report 2015*. Jakarta: National Disaster Management Agency (BNPB). Available at: http://www.preventionweb.net/files/50832_5083220161031indobaselinereportfina.pdf
[2] Indonesia Disaster Data and Information (*Data dan Informasi Bencana Indonesia/* DIBI). Available at: http://dibi.bnpb.go.id/
[3] BAPPEDA Banda Aceh 2013 *Referral of Regional Spatial Plans (RTRW) Banda Aceh 2009-2029.* Banda Aceh: Regional Development lanning Agency (BAPPEDA). In Indonesian
[4] UNISDR 2015a Sendai Framework for Disaster Risk Reduction 2015–2030.
[5] UNISDR 2015b Economic Aspects of DRR. UN World Conference on Disaster Risk Reduction. 14-18 March 2015, Sendai, Japan. Available at: http://www.wcdrr.org/wcdrr-data/uploads/871/B.%20Brief%20issue%20paper.pdf
[6] Government Regulation No. 21 of 2008 on the Implementation of Disaster Management
[7] Gordon M 2013 Exploring Existing Methodologies For Allocating And Tracking Disaster Risk Reduction In National Public Investment. *Geneva: UNISDR.* Available at: http://www.preventionweb.net/files/globalplatform/51967c57c3c9fMarc_Gordon,_2013.pdf accessed on May 30, 2016.
[8] Darwanto H 2012 Preliminary examination of existing methodologies for allocating and tracking national government budget for disaster risk reduction (DRR) in Indonesia. *Study commissioned by UNISDR in collaboration with ADPC under the IAP project—Regional Stocktaking and Mapping of Disaster Risk Reduction Interventions for Asia and the Pacific.* Available at: http://www.preventionweb.net/english/hyogo/gar/2013/en/bgdocs/Darwanto, 2012.pdf accessed on May 30, 2016.
[9] UNISDR 2005 Hyogo Framework for Action 2005 – 2015
[10] Indonesian Disaster Management Law No. 24/ 2007 on Disaster Management
[11] BNPB 2011 *Disaster Management in Indonesia.* Available at: http://www.soi.asia/data/event/20111027-disastermng/pdf/4. session1 Disaster Management in Indonesia.pdf
[12] Qanun No. 3/ 2011 on the Establishment, Organizational Structure and Management of the BPBD Banda Aceh
[13] UNDP 2012 *Making Aceh Safer Through Disaster Risk Reduction in Development (DRR-A) Project 2009-2012.* Project Evaluation Report. Jakarta: UNDP Indonesia.
[14] Anderson, B (2009). Budgeting for Disaster. Available at: http://www.cepal.org/ilpes/noticias/paginas/7/34687/barry_anderson_Budgeting_for_Disasters.pdf accessed on May 30, 2016.

Acknowledgments
The author would like to thank Hibah Lab Unsyiah 2016 and all those who made its compilation and publication possible.

Rethinking "Build Back Better" in housing reconstruction: A proposal for "People Centered Housing Recovery"

Elizabeth Maly[1]

International Research Institute of Disaster Science (IRIDeS), Tohoku University, Sendai, Japan
maly@irides.tohoku.ac.jp

Abstract Since 2004, when "Build Back Better" (BBB) was used to call for post-tsunami recovery in Aceh that reduces risk and improves people's lives, BBB has become widely-used in disaster risk reduction and recovery, and has been featured in post-disaster recovery plans of several countries and the 2015 Sendai Framework for Disaster Risk Reduction. Improving on pre-disaster conditions is a shared goal for recovery, but in relation to housing reconstruction the term BBB has become both too broad to offer meaningful direction. It is also used to narrowly describe safer construction without holistic consideration of what is "better" for people's lives. This paper argues that "People Centered Housing Recovery" (PCHR) could provide a more meaningful and comprehensive set of principles to guide post-disaster housing reconstruction: housing design and form that meet people's needs; genuine participation of empowered residents in decision making and construction; and related policies that are accountable to all residents. This paper applies the framework of PCHR to three cases of post-disaster housing recovery--Mt. Merapi in Indonesia, Typhoon Yolanda in the Philippines, and the Great East Japan Earthquake and tsunami in Japan.

1. Introduction

The term "Build Back Better" was used to describe goal for disaster recovery after 2004 Indian Ocean Tsunami, after UN Special Envoy for Tsunami Recovery William Clinton used it to call for recovery that: reduces future risk; supports a "better and faster" transition from relief to recovery; and helps communities towards "a path to development" with improved safety and economic vitality [1]. In the decade since, the term Build Back Better (BBB) attained a prominent role in the Sendai Framework for Disaster Risk Reduction [2] adopted at the World Conference on Disaster Risk Reduction held in Sendai, Japan in March 2015, and has become more broadly used as part of recovery plans [4,5].

Although it is easy to agree on the goal of "building back better"—for recovery to improve upon the previous/pre-disaster situation—BBB is used to describe various aspects of recovery, making the concept into a vague umbrella term for "good practices" in recovery [3]. For example, Clinton's 10 propositions included reducing future risk, as well as improving inter-agency coordination and emphasizing roles of the private sector along with people's own roles [1]. BBB has also been used to emphasize linking of relief and development [3]. In the Sendai Framework, intended to guide policy for disaster risk reduction (DRR), Build Back Better is used to describe recovery measures that contribute to reduction of future disaster risk [2]. Beyond potential vagueness, BBB can also be interpreted narrowly, such as in the context of disaster risk reduction, when BBB can be used to justify requirements for certain levels of structural design, or the choice to relocate households away from a designated risk area without considering other life or livelihood impacts.

2. Objective and Methodology

This paper considers the use and limitations of term Build Back Better, through an examination of three recent international examples of post-disaster housing reconstruction and relocation. The three cases studies include residential relocation as part of post-disaster recovery in: Mt. Merapi, Indonesia, after the 2010 volcanic eruption; Tacloban City after the 2013 Typhoon Yolanda; and the Tohoku region of Japan after the 2011 Great East Japan Earthquake and tsunami.

Drawing on a review of relevant literature and the author's past research the three areas, this paper analyzes the three case studies of post-disaster recovery and housing reconstruction, and considers the relationship of the concept of BBB at the multiple scales of individual houses, community and neighborhood, and disaster-wide policy. Based on this analysis, the concept of people-centered housing recovery (PCHR) is proposed as an underlying principle to guide housing reconstruction that is indeed 'building back better' for affected people.

3. Build Back Better and People Centered Housing Recovery

As discussed, the term Build Back Better has been used to describe a variety of goals for recovery, ranging from broad integration of development ideas to specific improvements of structural safety. The three cases discussed in the following section all include the provision of strong housing in safer locations; BBB is often used to describe this kind of approach, and it has been argued by Kennedy et al. that building back *safer* might be a more appropriate term [6]. However, while providing stronger structures in a safer area is building back better buildings, it is not necessarily building back better lives for the affected people. In addition, in the provision and construction of housing, BBB does not offer any direction for thinking about housing design or people's participation in the reconstruction process.

In recent years, the ideas of "putting people at the center," or People-Centered Housing Recovery, have been increasingly supported in the fields of housing recovery. Several key documents [7][8][15] called for putting people at the center of their own housing decisions since the 1970s, and people centered approaches are also supported in international development [16,17]. Several key guidelines for housing reconstruction and and have called for people-centered approaches, including publications from the World Bank, UNHabitat, UNDP, and the Red Cross [9][10][11][12][13][14][18]. The idea of 'putting people at the center' of recovery has been increasingly supported in recent years [9][10][11][12][13], as have principles emphasizing residents' participation in recovery decisions and/or reconstruction [9][11][12][13][14][19][20]. Based on an analysis of these and related documents, this paper uses the concept "People-Centered Housing Recovery" to include: 1) housing recovery policy that supports livelihood recovery; 2) a recovery process that includes residents' involvement in decision-making; and 3) houses whose design and construction meet residents' needs. The following sections describe the three international housing reconstruction case studies and consider them in terms of BBB and PCHR ideas.

4. Post-Merapi REKOMPAK program in Indonesia

In October 2010, the volcanic eruption of Mt. Merapi, north of the city of Yogyakarta, caused heavy damage to communities living near Mt. Merapi. Almost 3000 houses were destroyed by the pyroclastic flow, buried under the lahar, or flooded as material from the volcano caused river floods. Drawing on past disaster recovery experiences, the Indonesian government implemented the Community-Based Rehabilitation and Reconstruction Settlement Project (REKOMPAK). Using a process with significant participation of residents from the early planning stages, site and community design, and community-based construction, this project also provided technical experts and facilitators to support each phase of the project, which led to the construction of close to 3000 new houses [21].

Based on a 6x6 meter expandable core house design, residents could choose from several different configuration of rooms, and continue to modify/expand their houses. House design and configuration was also developed in response the various sites, along with site planning in consultation with technical experts/facilitators. With intense participation throughout the process, the REKOMPAK program exemplifies the focus on decision making and resident involvement at the community scale. During the implementation, the largest number of houses provided through the REKOMPAK program became those in collective resettlement sites. However, the program provided support for a number of options: rebuilding individual houses on former or newly acquired sites outside of the hazard area; support for small groups of residents who found their own resettlement sites; and combining support with other non-government donors.

5. Post-Yolanda housing reconstruction in Tacloban

In November 2013, Typhoon Yolanda caused devastation across the Philippines and Tacloban City was especially hard hit. With almost 30,000 houses destroyed, the majority from informal coastal communities, Tacloban City's housing reconstruction focused on the provision of new housing in resettlement sites in the northern part of the city [22]. As of 2016, the City-coordinated plan included more than 13,000 housing units to be built by the National Housing Authority (NHA) and another 2,500 houses to be built NGOs on land coordinated by the City [23]. Independent from the City, two other resettlement sites are planned by NGO/donors [23].

The city-coordinated housing recovery process includes various combinations of different support to provide safer housing to the city's most vulnerable residents; the vast majority of houses are built by the National Housing Authority (NHA). With 22.5 square meter are, NHA houses are row houses built of reinforced CMU block, and are constructed by developers, according the NHAs established methods to provide socialized housing. At the scale of the house, the design is standard, without much options for modifications by residents. Whereas the residents' voices were not included in the planning process to designing houses or settlement planning, coordination from Tacloban City Dept. of Housing and Community Development attempted to facilitate community-building within the process of relocation. The City Dept. of Housing and Community Development office has attempted to move residents together as groups from temporary to permanent housing, and support this transitional process through staff support and promotion of community leadership.

In contrast to the standardized housing units provided through the NHA, there are also examples of resettlement sites organized by non-profit organizations. Including members Development and Peace and Urban Poor Associates, the FRANCISCO consortium has started construction of 500 houses was part of a model community that includes livelihood and farming. Along with local partners, the international NGO Catholic Relief Services (CRS) is also planning a resettlement area for 900 families. UPA and CRS are each supporting the construction of a housing resettlements site on land they have acquired. Using community facilitation, these projects include the future residents in the planning for the site; residents are also involved in the building construction [23]. Limited by scale, future residents of these projects represent only a small fraction of all post-Yolanda housing beneficiaries in Tacloban, but these projects provide a meaningful demonstration of how people-centered housing recovery can be done in this context.

6. Great East Japan Earthquake

On March 11, 2011, the Great East Japan Earthquake and tsunami devastated communities along Japan's northeastern Tohoku coast; almost 400,000 houses were damaged or destroyed [24]. Based on past precedents for disaster recovery, government support for housing reconstruction includes two main approaches: the construction of new Disaster Recovery Public Housing units; and provision of new residential lots as part of Collective Relocation for Disaster Mitigation projects, which move residents to high land areas. Including single family detached public housing and collective multi-family apartment style public housing, about 30,000 units of public housing are planned for the disaster-affected municipalities; as of July 2016, construction had been completed for about 20,000 units, or 2/3 of the total [25]. Site preparation is complete for almost half of the 19,500 lots planned for residents to use for private house reconstruction in relocation projects.

Residents have only several options in regards to housing reconstruction and the use of government provided housing support; the main decision is between private rebuilding or public housing. For those who decide to rebuild their own houses, options include renting or buying a residential lot provided through the collective relocation project, or choosing not to wait for the government sponsored project, and finding their own land [26]. Of course, residents rebuilding their own houses can select their own contractor and house design.

Residents who choose to move into public housing have fewer options, but some may have some housing choices for which area they want to move to. Similar to the case of Tacloban mentioned above, the various housing units available as part of the recovery project do not necessarily mean that individual residents can make individual choices. In many rural fishing village areas, single family public housing units are provided; in many municipalities, larger multi-family units are provided in areas with more urban density. The size and configuration of public housing is decided by family size; some efforts have been made to accommodate the needs of local residents—for example large sinks in houses for residents of fishing communities who need to be able to clean fish. However, from a PCHR point of view, residents are not involved in the design and planning of housing units or sites; these projects are carried out without residents' participation in decision making, and remain a commodity provided as part of social welfare support.

7. Comparison and Analysis

Although the focus of all three cases is government-driven housing provision as part of relocation and resettlement, as shown in Table 1, the aspects of PCHR vary applied at the three scales of policy, community, and houses. Of the three examples, the support for housing recovery in Japan is the most monolithic—relying exclusively on the national government budget and recovery project support, without any other complementary support from sources. In contract, the examples from Merapi and Tacloban include some flexibility and variation of options through combinations of government and non-government support. Since residents are not always able to chose which of these options they would like for their family on an individual basis, there may some unevenness in access to different kinds of housing support. In Japan, on the other hand, although there is some variety in how municipalities chose to implement housing reconstruction, there is a greater degree of standardization across the disaster area.

Although housing advocates and international guidelines have for decades promoted the idea that housing recovery should be implemented in a way that it can become a tool for residents to improve their lives, the three cases show that there is still a prevalence of housing support that takes the form of objects being provided to passive residents. Reasons contributing to this situation vary between the different local context in these cases. In the case of Japan, a wealthy nation, the majority of residents are not used to being directly involved with the construction of their houses—rather they pay for the finished product--although households receiving support (especially public housing) in Tohoku are not wealthy. Whether it condominium or detached housing, housing is a commodity in the non-disaster housing market. Yet the government is acting on their obligation to provide support (in the form of housing) to disaster survivors.

In the case of Tacloban, the economic situation in the Philippines is the opposite—housing built for the disaster recovery targets extremely poor and vulnerable residents from informal settlement, who are used to building their own shelters from whatever materials and on whatever land they can access. Yet, similar to Japan, when the national government is the primary housing provider, for the majority of permanent housing provided by the National Housing Authority in Tacloban, the voices of the residents are not included in the decision-making process, and the hands of the residents are not included in the construction process.

Table 1. Comparison of Housing Recovery cases and relationship to PCHR

Disaster and Housing Reconstruction	Housing recovery program	PCHR at house scale	PCHR at community scale	PCHR at policy scale
Oct. 2010 volcanic eruption of Mt. Merapi, near Yogyakarta, Indonesia. Around 3000 houses damaged; designated areas at risk from future volcanos.	Community-based housing recovery program (REKOMPAK) supported new housing on site (not risk) or resettlement areas. Close to 3,000 houses built.	Residents involved in choosing the design (room configuration) of their houses; directly involved in the construction.	Intensive resident participation in planning process from the early stage, to site planning and construction, expert support.	Project based on people-centered principles of residents' participation, role in housing and site decisions. Program flexibility allows wide support.
Nov. 2013 Typhoon Yolanda in Tacloban City. Including multiple actors and providers, housing recovery is coordinated by Tacloban City. Target beneficiaries are informal coastal residents. Most housing recovery support in the provision of new housing in resettlement sites to the north. NGOs are also involved with housing provision, including in coordination with the City's relocation plan, as well as independently.	The largest number of housing units will be provided by the National Housing Authority, more than 13,000 units planned.	NHA standard 22.5m^2 townhouse design, built by contractors. Residents not included in design or construction.	No participation in the settlement design, but some attempts at community-building.	The goal of the project is to provide stronger houses; housing provided as a relief object without including people.
	Two NGOs (UPA and CRS) supporting construction of new housing resettlement sites close to city center. Around 500 houses planned for each.	Residents included in planning for house and site design, and included in construction	Participation in decision making (home owners association) and construction	Housing program designed to support needs of residents-including location and livelihood support activities.
GEJE-Japan government. After the Great East Japan Earthquake and tsunami, Japanese government housing support include the provision of public housing and lots for private reconstruction in collective relocation programs.	Government' planned support for housing reconstruction includes 30,000 units of disaster recovery public housing and 19,500 residential lots for private reconstruction.	Although there have been attempts to design public housing units that are responsive to residents' needs, people are not directly involved in the design.	Various levels of community-building related to moving public housing; efforts are made, but are not part of housing provision.	Public housing is constructed by the government and provided to residents as a social welfare commodity; they are not involved in planning.

Based on previous experiences in post-disaster housing recovery in Indonesia, the REKOMPAK program is a rare case of community-based housing recovery support from the national government. Similar to successful examples of owner-driven housing reconstruction that have been proven successful in Gujarat and Pakistan, and also used in previous housing recovery after the 2006 Java Earthquake in the same region, the REKOMPAK program in Merapi provided building materials, along with technical experts and advisors, to residents who took a primary role in the rebuilding process. With large component of relocation, housing recovery in Merapi also including the provision of land in resettlement sites, or compensation for land that residents acquired themselves. As shown in Table 1, the REKOMPAK program demonstrates People-Centered Housing Recovery principles at all 3 scales: individual houses, community participation, and disaster-area-wide policies [21].

Other factors also contribute to the relative success of the REKOMPAK-Merapi as a housing resettlement program in comparison to the other cases. Distances that residents were relocated was minimal-usually less that 5 kilometers from their former land and resettlement site, which allowed them to return to their land and is many case continue to use them for farming. In contrast, the ongoing relocation in Tacloban is moving people with very little economic capital to an area with few options for livelihood development as of now. In Japan, the distance from former neighborhoods to relocation areas is not that large, but the larger challenge stems from economic decline, depopulation and aging population in the region. Another factor is resident involvement in reconstruction as part of local culture where this is a normalized activity; from communities discussing and deciding their housing needs as group, and then being directly involve in the reconstruction process could be implemented in Indonesia, whereas in Japan this kind of direct participation in housing provision is unfamiliar. In the case of Tacloban, the smaller scale NGO-driven projects of UPA and CRS demonstrate that while it is possible for the local residents to be involved from the planning to construction states, the main housing provision still relies on contractor-driven housing construction by NHA without residents' involvement.

8. Conclusion

Although *build* back better directly relates to the reconstruction of housing and settlements, the interpretation of BBB has not been clearly defined in relation to other key principles of housing recovery, although there are connections principles of human rights in housing recovery, housing and development, and disaster risk reduction. The term Build Back Better (BBB) has been widely used to describe the goals of recovery plans and projects, yet the specific meaning and application of BBB is often unclear. In terms of housing relocation, there is a potential for BBB to be interpreted narrowly, prioritizing avoidance of risk over other factors that support residents' life and livelihood recovery.

The three cases discussed in this paper all include the provision of houses to reduce future disaster risk; in Merapi and Tacloban, houses are designed with seismic reinforcement, and in all three cases, residential relocation is intended to move people out of areas at risk from future hazards. With the increased safety and high quality of housing construction, these houses can all be considered to represent Build Back Better (BBB). However, when considered in relationship to the ideas of People Centered Housing Recovery, the limited role and participation of affected residents becomes clear.

Along with 'participation' in the recovery process, most recent literature on housing recovery post-disaster clearly supports the benefit of 'putting people at the center of recovery.' In this paper, 'people-centered housing recovery' was defined as a framework with multiple aspects (policy, process, and housing form) applied to multiple scales (disaster area, community, and individual household). Although policies dealing with housing recovery in these three cases and countries vary, as do the respective cultural and economic contexts, using the framework of 'people-centered housing recovery,' allowed for a comparison of relevant aspects of each case study, and the identification of respective strengths and weaknesses of different aspects of each project in terms of the impacts on residents' lives. Compared to BBB, "people centered housing recovery" could represent a more detailed way to consider the varied aspects of housing recovery projects.

REFERENCES

[1] Clinton, W. "Key Propositions for Building Back Better." Office of the UN Secretary-General's Special Envoy for Tsunami Recovery (2006) 2.
[2] UNISDR. Sendai Framework for Disaster Risk Reduction 2015-2030 (2015).
[3] Fan, L. Disaster as opportunity? Building Back Better in Aceh, Myanmar and Haiti. ODI: London. (2013).
[4]Pakistan's Earthquake Reconstruction & Rehabilitation Authority (ERRA) website.,. http://www.erra.pk/. Accessed 3/23/2016.
[5]Government of Nepal, National Planning Commission *Post Disaster Needs Assessment.* Kathmandu: Government of Nepal (2015)
[6] Kennedy et. al. The Meaning of 'Build Back Better': Evidence from Post-Tsunami Aceh and Sri Lanka. Journal of Contingencies & Crisis Management, 16 (1) (2008) 24-36.
[7]UNDRO (United Nations Disaster Relief Organization), Shelter after Disaster, Guidelines for Assistance (1982).55.
[8] Turner, J.F.C. Housing by People: Towards Autonomy in Building Environments. Marion Boyers: London (1976).

[9] Jha, A et al. Safer Homes, Stronger Communities: A Handbook for Reconstructing after Natural Disasters. World Bank (2010).
[10] UNHabitat. *Shelter and Housing, UN HABITAT in Disaster and Conflict Contexts,* (2011).
[11] Lyons, M. et al. Building Back Better: Delivering People-centred Housing Reconstruction at Scale. Rugby, UK: Practical Action (2010).
[12] Schilderman, T. and Parker, E., Eds. Still Standing: Looking back at reconstruction and disaster risk reduction in housing. Rugby, UK: Practical Action (2014).
[13] IFRC and OCHA. Shelter after Disaster, 2nd Edition (2015).
[14] Davis, I. and Alexander, D. Recovery after Disaster. London: Routledge (2016).
[15] Davis, I. Ed. (1981) Disasters and the Small Dwelling, Pergamon: Oxford. Pp13-16.
[16]United Nations Development Program (UNDP) (2011). *People-centred Development: Empowered lives. Resilient nations.*
[17] Korten, D. (1987). "Third Generation NGO Strategies: A Key to People-centered Development," *World Development,* Vol. 15, 145-159.
[18] IFRC (2015). *Shelter after disaster,* 2nd Ed.
[19] Lankatilleke, L. (2010) "The people's process: The viability of an international approach," in Lyons, M. et al. Building Back Better: Delivering People-centred Housing Reconstruction at Scale. Rugby, UK: Practical Action.
[20] Schilderman, T. (2010), "Putting people at the centre of reconstruction," in Lyons, M. et al. Building Back Better: Delivering People-centred Housing Reconstruction at Scale. Rugby, UK: Practical Action.
[21] Maly, E., Iuchi, K. and Nareswari, A. (2015). Community-based Housing Reconstruction and Relocation: REKOMPAK program after the 2010 Eruption of Mt. Merapi, Indonesia. Institute of Social Safety Science Journal, Volume 27, 205-214.
[22] Iuchi, K. and Maly, E. (Forthcoming 2016). Residential Relocation Processes in Coastal Areas: Tacloban City after Typhoon Yolanda. Chapter 14 in: Sapat, A. and Esnard, A-M. (Eds.). *Coming Home after Disaster: Multiple Dimensions of Housing Recovery.* Routledge: Boca Raton.
[23] Maly, E and Iuchi, K. (2016), "The Role of NGOs in Post-Disaster Housing Provision for a People Centered Recovery," Conference Paper presented at the 11th International Symposium on Architectural Interchanges in Asia (ISAIA), Sept. 21, 2016.
[24] Iuchi, K., Maly, E., and Johnson, L. (2015). Three years after a mega-disaster: Recovery policies, programs and implementation after the Great East Japan Earthquake. In Post-tsunami hazard reconstruction and restoration, ed. by V. Santiago-Gandino, Y. A. Kontar, and Y. Kaneda. London: Springer.
[25] Reconstruction Agency (2016), "Current situation of provision of public housing and support for private housing reconstruction after the Great East Japan Earthquake," (in Japanese), Aug. 3. 2016. http://www.reconstruction.go.jp/topics/main-cat1/sub-cat1-15/20160831_jutakukyokyuu.pdf
[26] Tamiyo Kondo, 2012a, Annual Change of Housing Rebuilding in the Disaster Area-A study on housing rebuilding in City of New Orleans after Hurricane Katrina (2005), Journal of Architecture and Planning vol.679, pp.2283-2292.

IOP Conf. Series: Earth and Environmental Science **56** (2017) 012026 doi:10.1088/1755-1315/56/1/012026

A conceptual framework for evaluating tsunami resilience

Dinil Pushpalal

Graduate School of International Cultural Studies, Tohoku University, Japan

Abstract. As many coastal towns in the northeast coast of Japan were destroyed by tsunami accompanied with the Great East Japan Earthquake, a few of them were survived or little damaged with no or less casualties due to some reasons. Yoshihama in Iwate prefecture is one of such little damaged communities and is known as "Lucky Beach." There were such "lucky" and "unlucky" regions in Indonesia and Sri Lanka too, which were affected by Indian Ocean Tsunami. Identification of reasons for vulnerability or resilience is the primary consideration of this article. It presents pragmatic conceptual framework for evaluating resilience, based on author's first-hand experience on above both tsunamis. Integral resilience of a given area has been considered after dividing into three phases namely, onsite resilience, instantaneous survivability, and recovery potentiality of the area. The author assumes that capacity of each phase depends on socioeconomic, infrastructural and geographical factors of the area considered. The paper moves forward, arguing appropriateness of the framework by giving examples collected from Japan, Indonesia and Sri Lanka. The framework will be useful for evaluating resilience of coastal townships and also planning resilient townships, specifically focusing on tsunami.

1. Introduction

News about large disasters becomes frequent in recent years. In Asia the largest two such tsunamis happened within seven years, challenging the knowledge on disasters. The Indian Ocean Tsunami (IOT) caused by magnitude 9.1 earthquake occurred on 26th December 2004. It affected more than 15 countries; Indonesia and Sri Lanka were the most seriously affected countries. Around 127,720 people were killed and 93,285 were still missing, 139,195 houses were destroyed, 85,000 houses were damaged and 635,384 people were displaced in Indonesia. In Sri Lanka, around 35,399 people were killed, 114,069 housing were destroyed or damaged and 480,000 people were displaced [1,2]. Around six years later, on 11th March, 2011 the Great East Japan Earthquake and Tsunami (GEJET) caused by magnitude 9.0 earthquake hit Japan. Around 18,703 people were killed and 2,674 people were missing, 126,574 houses were fully destroyed, and more than 450,000 people had to be evacuated [3].

Among various stories and studies regarding damage and lost caused by both tsunami disasters, there are some stories about survived or little damaged coastal areas due to some reasons. These "lucky beaches" exist in Indonesia, Sri Lanka and Japan. In Indonesia several coastal areas such as Lhok Pawoh & Ladang Tuha Villages in South Aceh District and Moawo & Pasar Lahewa Villages in Nias Island, were survived or no casualties. Lhok Pawoh and Ladang Tuha Villages were protected by geographical factors, while social factors contributed for lifesaving of Moawo and Pasar Lahewa Villages' residents [1,4]. In Sri Lanka, some areas such as Galle Fort were survived due to infrastructural factors, while the surroundings were severely damaged. In Japan, Yoshihama in Iwate prefecture reported a little damage because of social and geographical factors [5]. These facts lead to the question, why some areas were survived or a little damaged while other areas were not?

The main purpose of this article is to propose a general conceptual framework, which can be used to assess and analyse the vulnerability or resilience of townships when facing tsunami. It will analyse some factors that contribute to the fate of "lucky beach," "unlucky beach" or other relevant coastal areas.

2. The Conceptual Framework

UNISDR defines resilience as the ability of a system, community or society exposed to hazards to resist, absorb, accommodate to and recover from the effects of a hazard in a timely and efficient manner, including through the preservation and restoration of its essential basic structures and functions [6]. Accordingly, the proposed framework assumes that an "ideal resilient region" should fulfil three requisites namely onsite resilience, instantaneous survivability, and recovery potentiality. The level of fulfilment in a given requisite can basically be evaluated by socioeconomic, infrastructural, and geographical capital of the region in question (see figure 1). Onsite resilience is the "hereditary" ability of a given place to withstand tsunami even before tsunami comes. Instantaneous survivability is the ability to survive during the climax of the disaster. Recovery potentiality is the ability to recover soon after the disaster, even though the region was destroyed by tsunami.

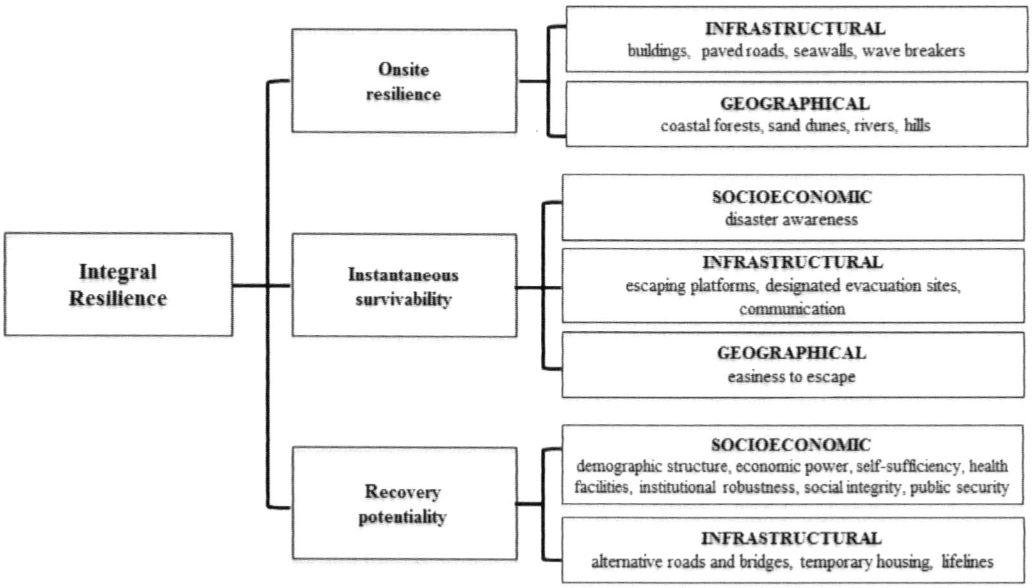

Figure 1. The conceptual framework.

2.1. Socioeconomic factor

The proposed framework hypothesizes that the resilience is partially a socially constructed ability. Priority 3 of Sendai Framework for Disaster Risk Reduction encourages enhancement of social resilience through investing in disaster risk prevention and reduction [7]. It encourages a broader and a more people-centred preventive approach to disaster risk.

Many authors have investigated on measuring socioeconomic factors of vulnerability or resilience. Some of the very relevant and comprehensive research works will be introduced here. Resilience index proposed by Kusumastuti et al. [8] considered that social resilience is a function of demographic characteristics and access to resources, while community resilience is related to the attributes of the area that promote population wellness, quality of life, and emotional health; and economic resilience is related to the economic vitality of the community. Furthermore, institutional resilience is related to the efforts of local government to raise the awareness and preparedness of the residents toward disaster. The socioeconomic vulnerability index (SeVI) proposed by Ahsan and Warner [9] has used a Composite Indicator Framework method, which consists of three main dimensions: adaptive capacity, sensitivity and exposure. These dimensions were further divided into five domains: demographic, social, economic, physical, and exposure to natural hazards. Socioeconomic factors in Ahsan and Warner [9] study have been evaluated through population density and growth, male/female ratio, percentage of illiterate households, participation for elections and volunteering, percentage of unemployed households and households below poverty line etc. Antwi et al. [10] measured the socioeconomic index through 6 indicators, namely population density, livelihood diversification, vulnerable groups, knowledge on climate, migration rate, and access to social service. In the same study political/governance index has been measured using 5 indicators namely, politically influential persons, stakeholder organizations, number of assembly members, and local participation in district activities.

Based on above previous literature the socioeconomic factor of our framework will be considered some quantitative dimensions such as demographic structure, economic power, self-sufficiency, and health facilities, which can be often gathered by census organizations, both government and private. Qualitative dimensions such as institutional robustness, social integrity, public security and disaster awareness can be collected through field surveys.

2.2. Infrastructural factor

Experience of past major tsunamis has proven that infrastructures such as seawalls and large buildings protected the coastal communities from tsunamis. At least those structures reduce the impact of tsunami. One of such examples is Dutch fort of Galle city in Sri Lanka. Built on a promontory, with walls to withstand cannon fire, the ramparts and bastions surrounding the 350-year old Dutch fort of Galle acted as a wave breaker against 2004 Indian Ocean tsunami, saving the old part from devastation. The coral, stone and stucco wall form its bastions and ramparts served to protect the approximately 400 buildings inside it on the 26th December 2004 when the tsunami broke through the lower gates of the Fort. The waves flowed out again very quickly and left no major structural damage, while surrounding of the fort was inundated above 3.5 m and destroyed the new part of the city.

The role of existing buildings on disaster mitigation is paid attention after the Great East Japan Earthquake (GEJE). Many authors have investigated damage conditions of buildings and justified the reasons for survival or failure [11,12,13]. GEJE fully destroyed around 126,000 buildings, and the number of half-destroyed buildings was counted as 272,000. However, some large buildings survived and had sometimes served to reduce the impact of tsunami while giving the temporary shelter to the affected inhabitants. One of the examples is a group of buildings in Sendai sewage purification centre in Miyagi Prefecture. All of those buildings were located at a distance of 325-637 m from the coastline and at altitudes between 2.2 and 3.1 m. On March 11, 2011, tsunami attacked the premises and ran up to 14.7 m [14,15]. None of the buildings were destroyed or submerged. Despite of close proximity to the sea, these reinforced concrete buildings survived and they were praised by the neighbours for protecting their lives. These structures could be used as escaping platforms and designated evacuation sites too.

Infrastructural factor of the proposed framework can be evaluated by field surveys connected with GIS on existing artificial structures such as reinforced concrete buildings, dams, seawalls etc.

2.3. Geographical factor

The geographical factor weighs the role of coastal forests, sand dunes, rivers and hills to reduce the impact caused by tsunami. It is well known that moving from low-lying valleys to higher lands saves the damages by tsunami. On the other hand if an area is located on higher lands, it is geographically resilience. One of the very clear examples is Yoshihama in Iwate Prefecture, Japan. Despite of 18-meter high wave rampaged from the Pacific Ocean into the valley, just four houses were lost and one person died on March 11, 2011. Yoshihama became a safer place because of mass relocation to high ground after having experience of tsunami in 1896 and 1933, in which Yoshihama experienced 204 and 17 fatalities respectively [5].

Researchers are split over the opinions about coastal forests. Some of them believe uprooted trees made more harms than benefits at the event of GEJE, because of huge power of waves [11]. However, others determined that homestead grooves (*igune*) in Sendai Plain protected nearby houses by reducing the impact of tsunami, as well as stopping the flowing debris, damaging the houses [16]. Ujiie et al. [17] have conducted interviews in 53 households with homestead grooves to investigate the function of these groves against the tsunami. Of the households surveyed, 64.2 per cent believed that the homestead grooves protected their homes from the tsunami; 18.9 per cent indicated that the homestead grooves did not protect their homes from the tsunami; and 7.5 per cent indicated that the homestead grooves unfortunately increased the damage done to their homes by tsunami. The 64.2 per cent households who had a positive image on homestead grooves insisted three merits of homestead grooves. Firstly, their opinion was that homestead grooves protected their homes from debris, uprooted trees and vehicles washed out by the tsunami. Secondly, homestead grooves, especially to the east of houses (seaward), decreased the impact of the tsunami so that their homes could not be washed out. Thirdly, they mentioned that homestead grooves, especially to the west of houses (landward), protected their belongings in their houses and soil erosion [14,18].

Sand dunes and mangroves have been studied by Tanaka et al. [19] on the role of coastal vegetation in tsunami protection based on field observations carried out after the Indian Ocean tsunami on December 26, 2004, in Sri Lanka. The study has concluded that combination of the sand dune followed by vegetation toward landside played an important role in retarding tsunami. In particular, if a mangrove forest is located behind a sand dune, it could trap most of the debris and prevent the buildings behind the forest from being damaged.

Based on these information we can anticipate that if an area has coastal forests, sand dunes, and hills, it is resilient. These features can be mapped using GIS and transformed to numbers.

3. Three Phases of Conceptual Framework

Numerous previous literature have proposed many conceptual models for evaluating resilience or vulnerability. Despite their differences, a number of common elements have been found: (a) the examination of vulnerability from a social-ecological perspective; (b) the importance of place-based studies; (c) the conceptualization of vulnerability as an equity or human rights issue and (d) the use of vulnerability assessments to identify hazard zones, thereby forming the basis for pre-impact and hazard mitigation planning [20]. The proposed framework in this paper, integral resilience of a given area has been considered, dividing into three phases (requisites) namely, onsite resilience, instantaneous survivability, and recovery potentiality of the area. It is a modification of previous models, which is built by introducing three requisites to be met for labelling a place as an ideal tsunami resilient area. Each phase (requisite) is dependent on two or three factors, which can be measured by different dimensions using obtainable independent variables. The framework assumes that an area uninhabited, but fully protected by infrastructural and geographical factors is the ideal, scoring the maximum in the integral resilience. On the other hand, existence of inhabitants gives a negative impact to the integral resilience. Therefore, socioeconomic factor which is totally dependent of human activity is not included in onsite resilience (see figure 1). The validity of this concept can be proved by social vulnerability score measured by Cutter et al. [21] for the Yellowstone National Park County in USA. The low social vulnerability score for Yellowstone National Park County is a result of its protected status with a very small population that has little ethnic, racial, or gender diversity.

As shown in figure 1, the dimensions to be considered under the socioeconomic, infrastructural, and geographical factors are varied according to the requisite. For example, while the geographical factor of onsite resilience will be evaluated by availability of coastal forests, sand dunes, rivers, and hills; geographical factor of instantaneous survivability is only a dependent of easiness to escape. This assumes that escaping is only the way remaining for saving life at the climax stage of a tsunami attack, which prevails in less than an hour. Recovery potentiality in figure 1 consists of two factors, socioeconomic and infrastructural. In recovery potentiality geographical factor is not important anymore, because coastal forests, sand dunes, hills etc. are only needed to resist tsunami when it approaches, or provide an easy way to escape during a tsunami attack. Socioeconomic factor in recovery potentiality is one of the most important that can determine the recovery level of post-disaster phase. It could be evaluated by demographic structure, economic power, self-sufficiency, health facilities, institutional robustness, social integrity, and public security, while it assumes disaster awareness does not matter in this stage.

4. Summary

This article forwarded a pragmatic conceptual framework for evaluating resilience, based on author's first-hand experience on tsunamis. The proposed model paid considerable attention on (a) dividing the disaster phenomena into three phases in relation to the temporality of tsunami, (b) defining which factors influence on a given phase and (c) selecting influential dimensions on each factor. One dimension has been carefully placed in only one phase to avoid double counting. The paper also provided some practical examples gathered through Indian Ocean tsunami and Great East Japan Earthquake in order to prove appropriateness of some dimensions. The next step is to apply the model into practical situations in order to verify the applicability of the model, and finally be instrumental in building resilient coastal townships.

5. References

[1] Danar O R and Pushpalal D 2014 Building community resilience: conceptual framework and its application in post tsunami resettlement *Elsevier Procedia Economic and Finance* **18** pp 489-96

[2] International Recovery Platform 2016 Brief overview of the disaster by countries http://www.recoveryplatform.org/countries_and_disasters/disaster/15/indian_ocean_tsunami_2004 (Accessed 31.10.2016)

[3] Pushpalal D, Rhyner J, Hossini V 2014 *The Great Eastern Japan Earthquake, 11 MARCH 2011: Lessons Learned and Research Questions* (Bonn: United Nations University) p 10

[4] Onrizal 2008 The stories of Tsunami Survivors (in Bahasa Indonesia). http://onrizal.wordpress.com/2008/10/05/kisah-kampung-yang-selamat-dari-tsunami/ (Accessed 13.5.2014)

[5] Corkill E 2011 Heights of survival. http://www.japantimes.co.jp/life/2011/06/12/general/heights- of-survival/#.WAg1-un_pol (Accessed 31.10.2016)

[6] UNISDR Terminology on Disaster Risk Reduction 2009 (Geneva: UNISDR) p 24

[7] Sendai Framework for Disaster Risk Reduction 2015 (Geneva: UNISDR) p 18

[8] Kusumastuti R D, Viverita, Husodo Z A, Suardi L, Danarsari D N 2014 Developing a resilience index towards natural disasters in Indonesia *International Journal of Disaster Risk Reduction* **10** pp 327-40

[9] Ahsan M N and Warner J 2014 The socioeconomic vulnerability index: A pragmatic approach for assessing climate change led risks–A case study in the south-western coastal Bangladesh *International Journal of Disaster Risk Reduction* **8** pp 32-49

[10] Antwi E K, Boakye-Danquah J, Owusu A B, Loh S K, Mensah R, Boafo Y A, Apronti P T 2015 Community vulnerability assessment index for flood prone savannah agro-ecological zone: A case study of Wa West District, Ghana *Weather and Climate Extremes* **10** pp 56-69

[11] Suppasri A, Shuto N, Imamura F, Koshimura S, Mas E, Yalciner A C 2013 Lessons Learned from the 2011 Great East Japan Tsunami: Performance of Tsunami Countermeasures, Coastal Buildings, and Tsunami Evacuation in Japan *Pure Appl. Geophys.* **170** pp 993-1018

[12] Matsutomi H, Yamaguchi E, Naoe K, Harada K 2012 Damage Conditions to Reinforced Concrete Buildings and Coastal Black Pine Trees in the 2011 Off the Pacific Coast of Tohoku Earthquake Tsunami (in Japanese). *Coastal Engineering Journal* **68(2)** pp. 351-55

[13] Gokon H and Koshimura S 2012 Mapping of building damage of the 2011 Tohoku earthquake tsunami in Miyagi Prefecture *Coastal Engineering Journal* **54(1)** pp 1250006-1-12
[14] Ogata K 2014 The Role of Obstructions in Disaster Risk Reduction: Focusing on the Great East Japan Earthquake, Master Thesis, Department of International Resources Policy, Graduate School of International Cultural Studies, Tohoku University pp 100-131
[15] Pushpalal D and Ogata K 2014 The Role of Buildings in Disaster Risk Reduction: Focusing on the Great East Japan Earthquake *Elsevier Procedia Economic and Finance* **18** pp 483-88
[16] Koganezawa T and Umikawa K 2012 The function of tsunami risk management on the coastal forest and wind break forest in Sendai coastal region (in Japanese) *Bulletin of Miyagi University of Education* **47** pp 1-9
[17] Ujiie S and others 2013 Study on the damage situation of Igune from the salt of the tsunami and the disaster prevention function of Igune in the Sendai Plain – A case study from the Sangenjaya District, Tamaura area, Iwanuma, Miyagi Prefecture (in Japanese) *Journal of the City Planning Institute of Japan* **48(1)** pp 100-109
[18] Ogata K and Pushpalal D 2014 Japanese Literature Survey of Ecosystem Services in Disaster Risk Reduction *The Great Eastern Japan Earthquake, 11 MARCH 2011: Lessons Learned and Research Questions* Eds. Pushpalal D, Rhyner J, Hossini V (Bonn: United Nations University) http://collections.unu.edu/view/UNU:1904 pp 68-75
[19] Tanaka N, Sasaki Y, Mowjood M I M 2006 Effects of Sand Dune and Vegetation in the Coastal Area of Sri Lanka at the Indian Ocean Tsunami *Advances in Geosciences Vol. 6: Hydrological Science* Eds. Namsik Park et al. World Scientific Publishing Company pp 149-59
[20] Cutter S L, Barnes L, Berry M, Burton C, Evans E, Tate E, Webb J 2008 A place-based model for understanding community resilience to natural disasters *Global Environmental Change* **18** pp 598–606
[21] Cutter S L, Boruff B J, Shirley W L 2003 Social Vulnerability to Environmental Hazards *Social Science Quarterly* **84(2)** pp 242-61

Acknowledgments

This work was supported by JSPS KAKENHI Grant Number JP 16H05648, and the content of article is basically based on above proposal for grant-in-aid.

Institute of Physics
Dirac House, Temple Back
Bristol BS1 6BE UK

ISSN: 1755-1307
ISBN 978-1-5108-3712-6

3rd International Summer School and Workshop on Complex and Magnetic Soft Matter Systems: Physico-Mechanical Properties and Structure (CMSMS'17)

Journal of Physics: Conference Series Volume 994

Dubna, Russia
28 - 30 June 2017

IOP
Institute of Physics